Environmental Science

Series Editors: R Allan • U. Förstner • W. Salomons

Jan Vermaat, Laurens Bouwer,
Kerry Turner and Wim Salomons
(Eds.)

Managing European Coasts

Past, Present and Future

With 62 Figures

 Springer

JAN VERMAAT
WIM SALOMONS
LAURENS BOUWER
INSTITUTE FOR ENVIRONMENTAL
STUDIES
VRIJE UNIVERSITEIT AMSTERDAM
DE BOELELAAN 1087
1081 HV AMSTERDAM
THE NETHERLANDS

E-mail: jan.vermaat@ivm.falw.vu.nl
    wim.salomons@home.nl
    laurens.bouwer@ivm.falw.vu.nl

KERRY TURNER
CSERGE
SCHOOL OF ENVIRONMENTAL SCIENCES
UNIVERSITY OF EAST ANGLIA
NORWICH UK
NR4 7TJ UNITED KINGDOM

E-mail: r.k.turner@uea.ac.uk

ISSN 1431-6250
ISBN 3-540-23454-3 **Springer Berlin Heidelberg New York**

Library of Congress Control Number: 2004114233

Springer is a part of Springer Science+Business Media
springeronline.com
© Springer-Verlag Berlin Heidelberg 2005
Printed in Germany

Cover design: E. Kirchner, Heidelberg
Production: A. Oelschläger
Typesetting: Camera-ready by PTP-Berlin
Printing: Mercedes-Druck, Berlin
Binding: Stein + Lehmann, Berlin
Printed on acid-free paper   30/2132/AO 5 4 3 2 1 0

Foreword

Coastal zones play a key role in Earth System functioning and form an "edge for society" providing a significant contribution to the life support systems. Goods and services derived from coastal systems depend strongly on multiple transboundary interactions with the land, atmosphere, open ocean and sea bottom. Increasing demands on coastal resources driven by human habitation, food security, recreation and transportation accelerate the exploitation of the coastal landscape and water bodies. Many coastal areas and human activities are subject to increasing risks from natural and man-induced hazards such as flooding resulting from major changes in hydrology of river systems that has reached a global scale. Changes in the hydrological cycle coupled with changes in land and water management alter fluxes of materials transmitted from river catchments to the coastal zone, which have a major effect on coastal ecosystems. The increasing complexity of underlying processes and forcing functions that drive changes on coastal systems are witnessed at a multiplicity of temporal and spatial scales.

Demographic pressure has resulted in an acceleration of human interventions that impact natural processes taking place in the coastal zone. The demands for coastal resources and human security are further exacerbated by broad scale changes of climate patterns and oceanic circulation. This combination of anthropogenic drivers/pressures combined with natural system oscillation and natural change keeps changing our environment to an extent that has culminated in what is now described as the "Anthropocene". However, even today our understanding of regional and global changes that impact coastal systems is still hampered by traditional disciplinary fragmentation. In order to maintain or restore a sustainable delivery of goods and services for humankind, science is challenged to better inform society, decision-makers and planners about:

- Global changes that are part of natural cycles of change, such as climate, and those due to changes in the global economy/trade and policy;
- Regional (trans-boundary and supra-national) changes as a result of multi national and regional drivers and pressures in the coastal zone; and
- Regional changes at often transboundary catchment level, which affect the downstream coastal zone and the near-shore marine environment.

Consequently the regional or local perspective of coastal change becomes increasingly important simply by recognition that coastal people are more and more seen as an integral part of the system. On European scales policy making has identified the relevance of the river basin or catchment scale for coastal change in one of the most recently launched regional legal "instruments", the European Water Framework Directive (Directive 2000/60/EC, L 327/1, Brussels, 23 Oct.

2000. Coastal management gaps and needs have been reviewed in form of an extended multiple site pilot project in Europe resulting in the EU recommendations concerning the implementation of Integrated Coastal Zone Management in Europe (2002/413/EC, L 148/24, Brussels 30 May 2002). For their Common Implementation Strategy, both initiatives require a profound scientific underpinning that is capable of conceptualizing the coastal zone against driver – change relations. Global and regional drivers and their interplay with social and societal choices need to be considered if scenarios shall be developed that can inform both the policy maker and manager.

The ELOISE thematic cluster is the European Union's contribution to the Land Ocean Interactions in the Coastal Zone, LOICZ, core project of the International Geosphere and Biosphere Program, IGBP. After 9 years of collaborative regional research, ELOISE has made an effort to synthesise its findings form 60 multi-national and often trans-disciplinary projects and to highlight major directions towards future sustainability in the coastal zone. ELOISE stems originally from the Environment & Climate and the MAST (Marine Science and Technology) Research Programmes under the 4th EU RTD Framework Programme. Acting in con-cert with the Programme for International Co-operation (INCO) and the research programmes of the Member States it continued under the 5th Framework Programme.

In 2003, in order to enhance the "Community Added Value" of the ELOISE cluster and to synthesise its science, the ELOISE consortium and secretariat has carried out three thematic workshops on:

1. Upscaling and demands at the European and global levels,
2. Integration into European Policy, and
3. Developing coastal futures for Europe.

These workshops feature a mix of fundamental and applied science encapsulated in a harmonized and effective synthesizing and communication mechanism based on a "Dahlem Conference Approach". The goal was, through a retrospective, current and future perspective, to identify information needs, instruments and frameworks that enable the science community to inform the coastal management in Europe on all relevant scales.

The book presented here reports on the latter two workshops, that were held back-to-back. It focuses on four major areas. In the first chapter, Laure Ledoux and her co-authors review the general relevance and applicability of ELOISE science for and in European coastal policymaking. Not surprisingly in recognition of the rather curiosity-driven origin of the ELOISE research, they identify a visible mismatch in the policy information needs and the products provided. Rarely has the mostly fundamental science been able to acknowledge the multiple and partly variable temporal and spatial scales of coastal change and environmental and human interaction. Finding ways to properly upscale the various "case studies" still remains a challenge that calls upon the science to develop and use typological approaches that allow an issue based categorisation of land ocean interactions. The involvement of the human dimension has been running behind and so has the recognition and reflection of the different views of coastal stakeholders on "their" coastal zone. This fragmentation in people's perception is basically symptomatic also of the traditional scientific work. The outcome of the first workshop is summarised in the second chapter of this book by Peter Herman and his co-authors.

A second block (Themes 2-3) reviews one of the most progressive and complex legal instruments, the European Water Framework Directive (WFD), its relevance for Coastal Zone Management, the data needs and methodological implications. It also considers the question of how to support the implementation of such a complex instrument and its multi-scale and transboundary effects by appropriate institutions and capacity building. Being highly innovative the Water Framework Directive still faces persistent technical problems in determining the environmental objectives and in interpreting key concepts such as "ecological status". Findings underline the need to apply typological and model-based approaches to derive the reference conditions and to classify aquatic systems. The necessary monitoring needs to be underpinned by appropriate indicators that can capture system functioning and state change across the relevant scales. The consequences of the WFD are also examined within the more specific context including marine protected areas (MPAs). A successful implementation of the WFD will rely strongly on promoting communication and closer collaboration between scientists, economists and other stakeholders including the public from the onset and on their involvement in the decision making process. The relevant scale here is the water continuum encompassing catchment managers as well as coastal managers.

Exogenous drivers such as climate change and globalization are reviewed in terms of their effects on European coastal zones and means to effectively manage the coast (Theme 4). Decisions made for management and those considered in a more proactive context need to be informed by scenarios that rely on appropriate valuation of both the environmental and human values. Cases for scenario use are presented. Strong sustainability options will be carefully weighed between the three provisions of human safety, economic development and ecological integrity. However, climate and sea level change as well as economic development pose considerable uncertainty on any prediction a fact that not only calls for sophisticated scientific response but again for a continuous involvement of the public and the media.

Integrated assessment, its capacity to provide the multidisciplinary information for scenario development and its shortcomings are reviewed and examples are provided featuring a variety of traditional (tourism) and rather recent coastal land and sea uses (windparks). The authors assess three different scenarios (1) a world market perspective, (2) global sustainability and (3) a regional, environmental stewardship for a variety of natural and anthropogenic driver/pressure settings to provide a forward look at European coastal areas. The most relevant current state changes in the coastal environment on a regional scale are habitat loss (including coastal squeeze); changes in biodiversity; and the loss of fisheries productivity. Others such as eutrophication, contamination and erosion are thought to be of lo-cal or moderate importance. It is expected that while the key characteristic driving forces will continue, climate change will have additional, often related impacts. Ultimately, under a globalization scenario impacts seem to be increasing while under the other two more appropriate response form society may help mitigate the impacts and lead to better sustainability. This thematic section provides a variety of cases including the Humber and Rhine River where integrated assessment is demonstrated and where the multiple scales relevant for scientific investigation and management become very obvious.

The book paves the way to an integrated view on the complex issues of coastal zone management. It showcases the shortcomings of existing scientific information mostly due to a miss-match in scales on which it is provided. The need for integrated approaches and participation from the onset is underlined and reviewed under various perspectives. By doing this the ELOISE book provides an experience and science-based rationale that provides a strong argument for a serious re-view of science, research design, science management and funding policy. It also underlines the need for improved networking and communication across the scientific disciplines as well as the funding agencies, the stakeholders and public. Mismatch of scales and lacking ownership are symptomatic for the perception that science so far rather rarely informed the policy and public awareness process appropriately.

The book strongly supports the fundamental change that the LOICZ project, the global interface of ELOISE, is undergoing in transition from its first decade of mostly curiosity-driven global change research towards an issue-driven scientific assessment, synthesis and communication platform. This is highlighted by the recently approved draft Science Plan and Implementation Strategy (http://www.loicz.org). It puts LOICZ much more than in the past into the position to deliver both up-scaled information needed to improve our earth system understanding on global scales as well as issue driven information that can be downscaled and used in management and awareness raising on local and regional scales. This long-lasting transition has been nourished substantially by the experiences made in the LOICZ core project ELOISE. The discussions and papers presented provide a good picture of how the bridges between traditional and future sciences need to be shaped.

Hartwig Kremer and Hartmut Barth
LOICZ IPO, Texel, and EC, DG-Research, Brussels

Preface and Acknowledgement

This monograph is the result of workshops which were organized as part of the "Eloise-Tender" contract from the DG-research (contract number EVK1-CT2002-70001). This monograph is primarily the outcome of two workshops organized in Egmond aan Zee, The Netherlands, June 2-5, 2003, but also incorporates the results of an earlier workshop held in Goes, The Netherlands, May 7-10, 2003. For these three workshops, overall themes (upscaling, policy relevance and "coastal futures") were identified to synthesize relevant results from the EU ELOISE research cluster.

The adopted format for the workshops was the "Dahlem-framework". This framework involves in-depth debate without formal presentations but informed by written material that is distributed among participants beforehand. Papers were contributed by invitation as well as through an open call. Each session was organised via a number of sequential meetings and led to a group report that was peer-reviewed.

We thank all workshop participants and fellow authors for their contribution during the workshops in June 2003, as well as for their patience and care during the subsequent editing process. Special thanks are due to Laure Ledoux, who had to step back from the editorial team for personal reasons but made important contributions at the inception of this work, to Ann Dixon and Corrie Zoll, who did much of the final style-and-detail checks, as well as to Agata Oelschlaeger of Springer, for managing our book at the publisher's end. Sadly, we want to commemorate our colleague and partner Giuseppe Bendoricchio who passed away.

We believe that the workshops and the input of the participants have generated a synthesis which appeals to both scientists and coastal managers.

Amsterdam, May 2004   Jan Vermaat, Kerry Turner,
                                  Wim Salomons, and Laurens Bouwer

Table of Contents

Contributing Authors and Workshop Participants

ANDREWS, Julian: School of Environmental Sciences, University of East Anglia, Norwich NR4 7TJ, United Kingdom

AVILA, Conxita: Centro de Estudios Avanzados de Blanes (CEAB), Carrer d'accés a la cala St. Francesc 14, 17300 Girona, Spain

BÄCK, Saara: Finnish Environment Institute, P.O. Box 14, 00251 Helsinki, Finland

BARRETT, Kevin: Norwegian Institute for Air Research (NILU), P.O. Box 100, 2027 Kjeller, Norway

BARTH, Hartmut: Commission of the European Communities, DG XII: Science, 200 Rue de la Loi, 1049 Brussels, Belgium

BEAUMONT, Nicola: Plymouth Marine Laboratory, Prospect Place, The Hoe, Plymouth PL1 3DH, United Kingdom

BENDORICCHIO, Giuseppe: Padua University, Via Marzolo 9, 35131 Padua, Italy

BEHRENDT, Horst: Institute of Freshwater Ecology and Inland Fisheries, Müggelseedamm 310, 12623 Berlin, Germany

BOUWER, Laurens M.: Institute for Environmental Studies (IVM), Vrije Universiteit Amsterdam, De Boelelaan 1087, 1081 HV Amsterdam, The Netherlands

BRESSER, Ton H.M.: National Institute of Public Health and the Environment (RIVM), P.O. Box 1, 3720 BA Bilthoven, The Netherlands

BROUWER, Roy: Institute for Inland Water Management and Waste Water Treatment (RIZA), P.O. Box 17, 8200 AA Lelystad, The Netherlands

BRUIN, Erwin F.L.M. de: Grontmij Advies en Techniek BV, P.O. Box 119, 3990 DC Houten, The Netherlands

CAVE, Rachel: School of Environmental Sciences, University of East Anglia, Norwich NR4 7TJ, United Kingdom

CIESLAK, Andrzej: The Maritime Office in Gdynia, Chrzanowskiego 10, 81-338 Gdynia, Poland

COLIJN, Franciscus: GKSS Research Centre, Institute for Coastal Research, Max Planckstrasse 1, 21502 Geesthacht, Germany

EDWARDS, Tony: Environment Agency, 1 Viking Close, Great Gutter Lane East, Willerby, Hull HU10 6DZ, United Kingdom

EISMA, Marc: Maritime Development Department, Rotterdam Municipal Port Management; P.O. Box 6622, 3002 AP Rotterdam, The Netherlands

GEORGIOU, Stavros: CSERGE, School of Environmental Sciences, University of East Anglia, Norwich NR4 7TJ, United Kingdom

GUPTA, Joyeeta: Institute for Environmental Studies (IVM), Vrije Universiteit Amsterdam, De Boelelaan 1087, 1081 HV Amsterdam, The Netherlands

HEIP, Carlo H.R.: Netherlands Institute of Ecology (NIOO), Centre for Estuarine and Marine Ecology, P.O. Box 140, 4400 AC Yerseke, The Netherlands

HERMAN, Peter M.J.: Netherlands Institute of Ecology (NIOO), Centre for Estuarine and Marine Ecology, P.O. Box 140, 4400 AC Yerseke, The Netherlands

JASPERS, Frank G.W.: UNESCO-IHE – Institute for Water Education, P.O. Box 3015, 2601 DA Delft, The Netherlands

JICKELLS, Tim: School of Environmental Sciences, University of East Anglia, Norwich NR4 7TJ, United Kingdom

KABUTA, Saa: National Institute for Coastal and Marine Management (RIKZ), P.O. Box 20907, 2500 EX The Hague, The Netherlands

KANNEN, Andreas: Research and Technology Centre (FTZ) Westcoast, Hafentörn, 25761 Büsum, Germany

KLEIN, Richard J.T.: Potsdam Institute for Climate Impact Research (PIK), P.O. Box 601203, 14412 Potsdam, Germany

KONTOGIANNI, Areti D.: University of Aegean, Alkaiou 1, 81100 Mytilene, Greece

KREMER, Hartwig: Royal Netherlands Institute for Sea Research (NIOZ), P.O. Box 167, 1790 AD Den Burg, Texel, The Netherlands

LAANE, Remi P.W.M.: National Institute for Coastal and Marine Management (RIKZ), P.O. Box 20907, 2500 EX The Hague, The Netherlands

LEDOUX, Laure: Macaulay Institute, Craigiebuckler, AB15 8QH Aberdeen, United Kingdom

LENHART, Hermann-Josef: Institut für Meereskunde, Hamburg University, Troplowitzstrasse 7, D-22529 Hamburg, Germany

LINDEBOOM, Han: Alterra and Royal Netherlands Institute for Sea Research (NIOZ), P.O. Box 167, 1790 AD Den Burg, Texel, The Netherlands

LISE, Wietze: Institute for Environmental Studies (IVM), Vrije Universiteit Amsterdam, De Boelelaan 1087, 1081 HV Amsterdam, The Netherlands

MARQUENIE, Joop M.: Nederlandse Aardolie Maatschappij (NAM), P.O. Box 28000, 9400 HH Assen, The Netherlands

MEE, Laurence D.: Institute of Marine Studies, University of Plymouth, Drake Circus, Plymouth PL4 8AA, United Kingdom

MONCHEVA, Snejana: Institute of Oceanology, P.O. Box 152, 9000 Varna, Bulgaria

MORA, Joan: Centro de Estudios Avanzados de Blanes (CEAB), Carrer d'accés a la cala St. Francesc 14, 17300 Girona, Spain

MOSCHELLA, Paula S.: The Marine Biological Association, Citadel hill, Plymouth PL1 2PB, United Kingdom

NICHOLLS, Robert J.: School of Civil Engineering and the Environment, University of Southampton, Highfield, Southampton SO17 1BJ, United Kingdom

NUNNERI, Corinna: GKSS Research Centre, Institute for Coastal Research, Max Planckstrasse 1, 21502 Geesthacht, Germany

O'RIORDAN, Tim: CSERGE, School of Environmental Sciences, University of East Anglia, Norwich NR4 7TJ, United Kingdom

ROCHELLE-NEWALL, Emma: Station Zoologique, BP 28, Observatoire Oceanologique, 06234 Villefranche-sur-Mer, France

SALOMONS, Wim: Institute for Environmental Studies (IVM), Vrije Universiteit Amsterdam, De Boelelaan 1087, 1081 HV Amsterdam, The Netherlands

SARDÁ, Rafael: Centro de Estudios Avanzados de Blanes (CEAB), Carrer d'accés a la cala St. Francesc 14, 17300 Girona, Spain

SKOURTOS, Mihalis S.: University of Aegean, Alkaiou 1, 81100 Mytilene, Greece

STIVE, Marcel J.F.: Department of Civil Engineering, Delft University of Technology, P.O. Box 5048, 2600 GA Delft, The Netherlands

TETT, Paul: School of Life Sciences, Napier University, 10 Colinton Road, Edinburgh EH10 5DT, United Kingdom

TIMMERMAN, Jos: Institute for Inland Water Management and Waste Water Treatment (RIZA), P.O. Box 17, 8200 AA Lelystad, The Netherlands

TURNER, R. Kerry: CSERGE, School of Environmental Sciences, University of East Anglia, Norwich NR4 7TJ, United Kingdom

VELLINGA, Tiedo: Maritime Development Department, Rotterdam Municipal Port Management; P.O. Box 6622, 3002 AP Rotterdam, The Netherlands

VERMAAT, Jan E.: Institute for Environmental Studies (IVM), Vrije Universiteit Amsterdam, De Boelelaan 1087, 1081 HV Amsterdam, The Netherlands

VLAS, Jaap de: Nederlandse Aardolie Maatschappij (NAM), P.O. Box 28000, 9400 HH Assen, The Netherlands

VOSS, Maren: Baltic Sea Research InstituteSeestrasse 5, 18119 Rostock, Germany

WINDHORST, Wilhelm: Ökologie-zentrum, Kiel University, Olshausenstrasse 40, 24098 Kiel, Germany

YSEBAERT, Tom: Netherlands Institute of Ecology (NIOO), Centre for Estuarine and Marine Ecology, P.O. Box 140, 4400 AC Yerseke, The Netherlands

1. ELOISE research and the implementation of EU policy in the coastal zone

Laure Ledoux[1], Jan E. Vermaat, Laurens M. Bouwer, Wim Salomons, and R. Kerry Turner

Abstract

This paper presents a timely review of European coastal research as brought together in the ELOISE programme, at the end of its third phase of funding. The programme is intended to be the response of the EC to the challenge highlighted by the Land-Ocean Interactions in the Coastal Zone research project (LOICZ). Following a review of policy issues in the European coastal zones, and EU initiatives to address them, we assess the actual and potential contributions of research project findings to ELOISE objectives, and to the implementation of EU policy legislation affecting the coast. We identify several discrepancies between the project outputs of the ELOISE programme and the information needs arising from the implementation of the relevant directives. We suggest underlying causes for these discrepancies, and propose new research priorities to mitigate the information gap problem.

Introduction

The ELOISE (European Land-Ocean Interaction and Shelf Exchange Studies) research programme has been formulated as the contribution of the European Community to the challenges described in the Coastal Zone core project of the International Geosphere-Biosphere Programme (Cadée et al. 1994). It also represents a research contribution to the EU intitiative on Integrated Coastal Zone Management. The ELOISE programme has been guided by a Science Plan, which was drafted by a discussion panel of experts in the Roosendaal workshop (Cadée et al. 1994).

---

[1] Correspondence to Laure Ledoux: l.ledoux@macaulay.ac.uk

J.E. Vermaat et al. (Eds.): Managing European Coasts: Past, Present, and Future, pp. 1–19, 2005.

Research funding in the EU is currently undergoing a major reorientation in both funding mechanisms and focus with the launching of the 6th Framework. A review of the productivity of the ELOISE programme in previous framework programmes appears timely, particularly where it concerns contributions to policy implementation in the coastal zone.

Two previous Implementation Reports (Nolan et al. 1998; Barthel et al. 1999) describe Phases I and II of ELOISE and the efforts to mould a coherent package of research projects. This paper aims to provide an assessment of the achievements of ELOISE with respect to (i) the key objectives of the cluster as described in the first two implementation reports, and (ii) the contribution of the ELOISE projects to the implementation of EU policies in the coastal zone; in particular with respect to Water Framework Directive, the Habitats Directive, and the Bathing Water Directive. The evaluation is based on an overview of the ELOISE projects and data derived from a brief questionnaire to project coordinators (see annexe 1).

The authors first summarise the vision and objectives of the ELOISE programme, and then review the current policy issues in the European coastal zone, before presenting the output of the ELOISE evaluation. The paper finally concludes with suggested new priorities for research.

The ELOISE programme

The ELOISE vision

The general aim of ELOISE, as described in the ELOISE Science Plan (Cadée et al. 1994), is "to develop a coherent European [coastal zone] research programme of high scientific value and relevance to human society'. As such, it constitutes the European contribution to the Land-Ocean Interactions in the Coastal Zone (LOICZ) project, a core project of the IGBP Global Change Programme established in 1993, designed to elucidate issues concerning the role of coastal areas in the global climate system, and the potential response of coastal systems to all sources of global change (Cadée et al. 1994). More specific objectives, agreed during the Rosendaal workshop, which brought together European scientists and representatives of the European Commission and LOICZ in 1994, are also described in the Science Plan: (i) to determine the role of coastal seas in land-ocean interactions (including shelf-sea interactions along the shelf edge) in the perspective of global change (*Global Cycles*); (ii) to determine the regional and global consequences of human impact through pollution, eutrophication, and physical disturbance on land-ocean interactions in the coastal zone (*Human Impacts*); (iii) to formulate a strategic approach to the management of sustainable coastal zone resource use and development, and to investigate information, policy and market failures that hamper sustainable coastal resources management (*Socio-economic Development*); (iv) to determine which methodology – including technologies, data management and modelling – and instrumentation is needed to implement ELOISE (*Infrastructure and Implementation*). These sub-objectives determine the four Research Foci of the ELOISE programme.

The programme is intended to contribute to other activities of the Commission in the fields of integrated coastal zone management and of spatial planning. The means to realise this contribution, however, remain unspecified, other than the topics of the four ELOISE foci, used to bring different research projects together.

Programme

The complete ELOISE programme consists of a considerable number of research projects in the 4th and 5th framework (29 in total 1999, about 53 by the end of FP5) plus a number of additional activities and accompanying measures. An important activity has been the annual ELOISE Scientific Conferences, of which 5 have been organised so far.

The ELOISE programme was jointly implemented in the fourth framework programme by the MAST and the ENVIRONMENT AND CLIMATE Programmes and continued under FP5 in Thematic Programme 4 (Energy, Environment and Sustainable Development) in the key actions "Sustainable marine ecosystems" and "Sustainable Management and Quality of Water".

In addition to the grouping of projects in four foci, ELOISE research is coordinated through cross-project working groups, which approximately match with the foci: (1) biogeochemical cycles and fluxes; (2) ecosystem structures and functioning, human impacts; (3) modelling and data management; and (4) coastal zone management and integration of natural and socio-economic science. The working groups identified the remaining gaps after phases I and II (Barthel et al. 1999). One of the most important aspects was the lack of socio-economic research. It was identified as a priority for FP5, along with the need to "identify and assess societal and policy responses for sustainable management of coastal zones and their resources.

Policy issues in the European coastal zone

The current situation

The main environmental concerns in the European coastal areas were identified in the European Commission Communication on Integrated Coastal Zone management strategy, and later described in more details in the DOBRIS assessment report (Stanners and Bourdeau 2001). The primary concerns can be categorised as: habitat and biodiversity loss, including fisheries; water quality; sea level rise and coastal erosion. Behind these environmental changes are socio-economic and physical drivers, investigated by Turner et al. (1998b) and also reviewed in the DOBRIS report. These include climate-related pressures, pressures resulting from anthropogenic actions, related to urbanisation and demographic changes, tourism, port and harbour development, agricultural intensification, industrial development, marine aggregates extraction, and fisheries and aquaculture.

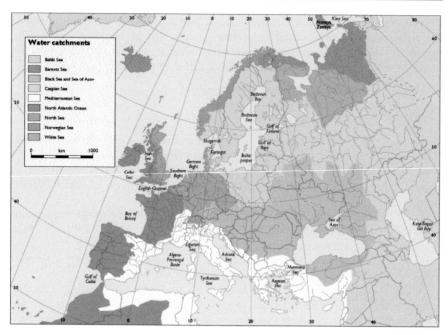

Fig. 1. Europe's seas with subsidiary seas and bays and catchments. (From Stanners and Bourdeau 1991)

Given geographical and cultural differences, the priorities clearly vary across European coastal regions (Fig. 1). The Dobris report provides an overview of the regional differences in the main environmental concerns in maritime and coastal zones. Table 1 extracts from the literature and summarises the main issues and their spatial relevance, as well as the drivers behind change, and policy responses at the European level.

Some of the environmental problems, such as toxic contamination, are widespread across Europe, others such as oil spill damages, and bacteriological quality issues are more localised. Eutrophication affects most seas, particularly the North, Irish, Baltic and Black Seas, whereas it is more localised in the Mediterranean (Adriatic Sea; Ærteberg et al. 2001) and the North Atlantic. The report concludes that the Mediterranean, Baltic, Black and North seas are the seas receiving consistently the highest loads of land-based or riverine contaminants. The northern seas (White, Barents and Norwegian seas) consistently receive small loads of contaminants. This was largely confirmed by the foresight exercise reported in Nunneri et al. (this volume).

Most environmental problems identified by leading experts have reached the attention of policy makers and have provoked a policy response, a few examples are included in Table 2. The European Union has produced a number of initiatives affecting the coastal zone, including specific directives. This policy regime is reviewed in more detail in the next section.

Table 1. Major environmental issues in European coastal waters and associated drivers and responses at the European level (adapted from Stanners and Bourdeau 1991)

Environmental Issues (Impacts)	Drivers	Pressures	Spatial Extent	Response at European level
Eutrophication	Agriculture, Urbanisation, Industry	Diffuse pollution (N,P), waste emissions	Most seas. Relatively less important in North Atlantic Ocean, Norwegian, Barents and White seas	Water Framework Directive, Nitrates Directive, Urban Waste Water Directive
Overfishing, loss of biodiversity	Fisheries, population growth	Fish catches, fishing gear	All seas. Especially North Sea, Wadden Sea, Black Sea, Barent, North sea	Common Fisheries Policy
Deterioration of bacteriological quality, health impacts	Agriculture, urbanisation, industry	Waste emissions, agricultural run off	Mediterranean, Black Sea, North sea	Bathing Directive
Habitat loss	Agriculture, Tourism, Climate Change (atmospheric emissions)	Habitat conversion (e.g. drainage), ports and touristic development, coastal erosion, sea level rise	European regions with high tourism and intensive agriculture, low lying coasts and deltas (sea level rise)	Birds and Habitats Directives
Toxic contamination (loss of biodiversity, health risk)	Industry, urbanisation, transport	Emissions of contaminants (heavy metals, synthetic organic compounds), contaminated sediments	All seas, especially around major European estuaries (less Barent and Norwegian sea)	Water Framework Directive, dangerous substances Directive, Seveso II Directive, IPCC Directive
Oil spill related ecological impacts	Maritime transport	Dumping, shipping accidents	Mediterranean, Black, Caspian, Norwegian, North sea	Regulation on prohibition of transport of heavy oils in single-hulled tankers; Erika I and II legislation packages.

EU policy in the coastal zone

In effect, most EU policies and instruments have some impact on the coastal zone. This section broadly describes these interactions before selecting the main areas of recent policy initiatives, which have most relevance to the evaluation of ELOISE projects.

A review of the influence of European policies on the evolution of coastal zones (IEEP 1999) concluded that EU policies have had far ranging consequences on European coasts. Policies encompassing significant drivers, such as the Common Agricultural Policy and Common fisheries policy indirectly influence coastal environments. The Structural and Cohesion Policy fund has also had a significant impact through the targeting of funds towards less developed coastal regions. This section describes EU policy initiatives in the Coastal Zone before focusing on specific legislation, which have had a particular influence on the coast.

EU initiatives in the coastal zone

EU activities concerning the coastal zone were initiated through international conventions covering its regional seas (Fig. 2). During the 1970s, the EU became for example a signatory of the Oslo Convention for the Prevention of Marine Pollution by Dumping from Ships and Aircraft (1972); the Paris Convention for Prevention of Marine Pollution from Land-Based Sources and the Helsinki Convention for the protection of the Marine Environment of the Baltic Sea (1974); and the Barcelona Convention for the Protection of the Mediterranean Sea against Pollution (1976). The Oslo and Paris conventions later merged into the Convention for the Protection of the Marine Environment of the northeast Atlantic (OSPAR) in 1992, while the Helsinki and Barcelona conventions were revised in 1992 and 1995 respectively. Integration of policies progressed in the 1980s, with the adoption of a European Coastal Charter in 1983.

It wasn't until 1992, however, with the new environmental remit brought by the Maastricht treaty, that a Council resolution calling for the development of a European strategy on coastal zones was adopted. A three-year demonstration programme on integrated coastal zone management lead to a European Commission Communication entitled "Towards a European Integrated Coastal Zone Management (IZM) Strategy. General principles and Policy Options. A reflection Paper" (EC 1999), and a proposal for a European Parliament and Council Recommendation concerning the implementation of Integrated Coastal Zone Management in Europe (COM/00/545 of 8 Sept. 2000). The European Parliament and Council adopted this recommendation in 2002 (2002/413/EC). The ICZM demonstration programme generated some agreed general principles for good management of coastal zones (Box 1).

The Strategy defines Integrated Coastal Zone Management (ICZM) as a "dynamic, continuous and iterative process designed to promote sustainable management of coastal zones" (EC 1999). Following on from the conclusions of the demonstration programme, the ICZM Strategy recommends to: (i) promote ICZM within the member States and at the "Regional Seas" level; (ii) make EU policies compatible with ICZM; (iii) promote dialogue between European Coastal Stakeholders; (iv) develop best ICZM practice; (v) generate information and knowledge about the coastal zone; (vi) disseminate information and raise public awareness.

Box 1. General principles for good management of coastal zones (EC 1999)

- Take a wide-ranging perspective
- Build on an understanding of specific conditions in the area of interest
- Work with natural processes
- Ensure that decisions taken today do not foreclose options for the future
- Use participatory planning to develop consensus
- Ensure the support and involvement of all relevant administrative bodies
- Use a combination of instruments

The Strategy also underlines that because of the diverse physical, economic, cultural and institutional characteristics of Member States, the response adopted should be flexible and problem-oriented. The philosophy underpinning the strategy is one of governance by partnership with civil society, with the EU providing leadership and guidance to support implementation at other levels. Where relevant, the Strategy builds on existing instruments and programmes, which often have not been necessarily designed with coastal zones in mind.

The Recommendation of the European Parliament and of the Council resulting from the European Commission's communication recommends that Member States take a strategic approach to the management of their coastal zones based on: (i) the protection of the coastal environment, following an ecosystem-based approach; (ii) the recognition of the threats of climate change and sea level rise to coastal zones; (iii) appropriate and ecologically responsible measures; (iv) sustainable economic opportunities and employment options; (v) a functional social and cultural system in local communities; (vi) adequate accessible land for the public; (vii) the maintenance or promotion of cohesion in the case of remote coastal communities; (viii) improved coordination of the actions of all relevant authorities, both at sea and on land. Member States should conduct or update an overall stocktaking to analyse which major actors, laws and institutions influence the management of their coastal zone. Based on the result of this stock-taking exercise, Member States should develop a national strategy, or where appropriate several strategies, following the principles of ICZM as described in the European Strategy. These strategies might be specific to the coastal zone, or be part of a geographically broader programme for promoting integrated management of a wider area, and should include a number of steps (Box 2).

The Commission is to review this Recommendation within 55 months following the date of its adoption and submit an evaluation report accompanied if appropriate by a proposal for further Community action.

Box 2. National Strategies for ICZM (OJEC L 14, pp 24-27)

National strategies should:

- Identify the roles of the different administrative actors whose competence includes activities or resources related to the coastal zone, as well as mechanisms for their coordination;
- Identify the appropriate mix of instruments for implementation of ICZM principles.

In particular Member States should consider:

- Develop national strategic plans for the coast;
- Include land purchase mechanisms and declarations of public domain;
- Develop contractual or voluntary agreements with coastal zone users;
- Harness economic and fiscal incentives;
- Work through regional development mechanisms;
- Develop or maintain national/regional/local legislation or policies and programmes addressing marine and terrestrial areas together;
- Identify measures to promote bottom-up initiatives where needed, and examine how to make best use of existing financing mechanisms both at European and national levels;
- Identify mechanisms to ensure full and coordinated implementation and application of Community legislation and policies that have an impact on coastal areas;
- Include adequate systems for monitoring and disseminating information to the public about their coastal zone;
- Determine how appropriate national training and education programmes can support implementation of ICZM principles in the coastal zone.

EU legislation in the coastal zone

Although there is no specific European legislation concerning the coastal zone, a number of directives have had an indirect impact (Fig. 2). For example, the Sewage Sludge and the Landfill Directives control activities that might lead to deterioration of coastal waters. The Environmental Impact Assessment Directive (EIA), and the Strategic Environmental Assessment (SEA) Directive require that significant environmental impacts of projects (EIA) and policies, plans and programmes (SEA) are identified and assessed and taken into account in the decision-making process to which the public can participate. This applies to projects and policies affecting the coastal zone and can therefore be expected to have a significant impact. In the most recent phase of EU legislation, two Directives have had or are expected to have very significant impacts on the coast, and they are described here in more detail.

The WFD and daughter directives

The Water Framework Directive is one of the few examples of policy response addressing water quality issues at the catchment scale. Adopted in June 2000, it integrates previously existing water legislation, updates existing directives according to new scientific knowledge, and strengthens existing legal obligations to ensure better compliance (Kaika and Page 2002). Earlier legislation on water (see

Fig. 2) had gone through two distinct phases (Kallis and Butler 2001, Kaika and Page 2002). The first one (1975-1987) was primarily concerned with public health, and setting standards for water quality for different uses (drinking, fishing, shellfish and bathing). In the second phase (1988-1996), priorities shifted towards pollution control, in particular for urban wastewater and agricultural run-off, with an effort to set emission limit values for different pollutants in water bodies. The third phase, which saw the birth of the Water Framework Directive, came after a state of the environment report showed that these policies had been effective in terms of reducing point source pollution, but that diffuse pollution remained a major problem (EEA 1998, Kaika et al. 2002). The new Directive is an attempt at more integrated and sustainable water management, expanding the scope of water protection for the first time to all waters, from surface water to ground water, and from freshwater ecosystems to estuaries and coastal waters. It encapsulates the new directions in European environmental policy institutionalised in the Maastricht treaty in 1992 and further reinforced by the Amsterdam treaty in 1997. The Member States agreed to sustainable development as a Community policy, to the Community being responsible for environmental policy within the limits of subsidiarity, and to the integration of environmental policy into other community policies. More specifically the precautionary principle, the principle of prevention of pollution at source, and the polluter-pays principle were all adopted (Barth and Fawell 2001).

Kallis and Butler (2001) point out that the directive introduces both new goals, and new means of achieving them (new organisational framework, and new measures). The overall goal is a "good" and non-deteriorating "status for all waters (surface, underground and coastal). This includes a "good" ecological and chemical quality status for surface water. Ecological status involves criteria for assessment divided into biological, hydromorphological and supporting physico-elements for rivers, lakes, transitional and "heavily modified" water bodies. For groundwater, the goal is a "good status" defined in terms of chemical and quantitative properties. A principle of "no direct discharges" to groundwater is also established, with some exemptions (e.g. mining). In addition, "protected zones", including areas currently protected by European legislation such as the Habitats Directive, should also be established, with higher quality objectives.

Organisation-wise, measures to achieve the new goals will be co-ordinated at the level of river basin districts, i.e. hydrological units and not political boundaries. Authorities should set up River Basin Management Plans, to be reviewed every 6 years, based on identifying river basin characteristics, assessing pressures and impacts on water bodies, and drawing on an economic analysis of water uses within the catchment. Monitoring is also an essential component, determining the necessity for additional measures. Finally, an important innovation introduced by the Directive is to widen participation in water policy-making: river basin management plans should involve extensive consultation and public access to information.

Following the Driver Pressure State Impact Response terminology (Turner et al. 1998a), the main "response" element of the directive is the programme of measures. "Basic" measures should be incorporated in every river basin management plan, at a minimum including those required to implement other EU legislation for the protection of water (see Fig. 2). If this doesn't suffice to achieve good water status, additional measures should be introduced, following a "combined approach", which brings together two existing strategies of Environmental Quality Standards (EQS – the legal upper limits of pollutant concentrations in water bodies) and Emission Limit Values (EVL – the upper limits of pollutant emissions into the environment). ELVs are first applied, through the introduction of best available technology for point source pollution, or best environmental practice for diffuse pollution. If this is not enough to reach EQSs, more stringent ELVs must then be applied in an iterative process. Furthermore, Member States should follow the principle of full cost recovery of water services, ensuring that water pricing policies are in place to "provide adequate incentives " for efficient use of water.

Although it does not target coastal zones specifically, the Directive does cover coastal water quality in its objective for good quality status, and provides a good example of integrated catchment management, addressing the issue of diffuse pollution of coastal waters.

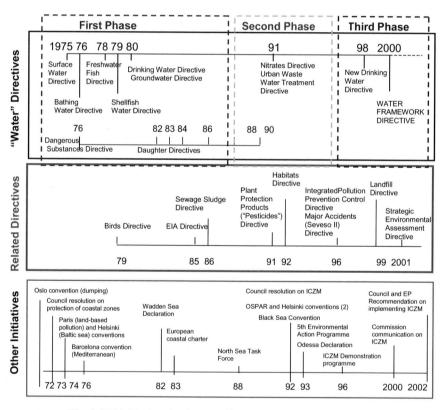

Fig. 2. EU initiatives having an effect on water and coastal zones

The Habitat and Birds Directives

As a signatory of the 1992 United Nations Convention on Biological Diversity, the EU is obliged under Article 6 to draw up a strategy to predict, prevent and tackle at source biodiversity loss in Western Europe. The two most important planks of EU biodiversity policy upon which the current Strategy builds are the 1979 Birds Directive and the 1992 Habitats Directive (Ledoux et al. 2000). Together, they aim to create a network of designated areas (Natura 2000) to protect habitats and species of community-wide importance, on a biogeographical basis[2]. It is, in effect, a "no-net-loss" policy, in so far as it requires all Natura 2000 areas to be protected from deterioration and damage.

The Wild Birds Directive, adopted in 1979, requires Member States to maintain populations of wild birds, to protect their habitats, to regulate hunting and trading, and to prohibit certain methods of killing. The establishment of special protected areas (SPAs) is a central component of the philosophy of the Directive. The Directive, as subsequently revised on a number of occasions since 1979, identifies a priority list of over 170 birds. Under Article 4, Member States are required to identify "the most suitable territories" (SPAs) under their jurisdiction in order to protect these species, and do all they can to ensure that the SPAs are not degraded, polluted or otherwise disturbed. Implementation of the Directive has, however, been extremely poor (Wils 1994).

The Habitats Directive was intended to remedy some of the deficiencies of the Birds Directive and extend the level of protection to a wider range of species and habitat types. The Directive aims to achieve a "favourable conservation status" for a long list of habitat types and species included in two extensive lists of habitat types (Annex I) and species (Annex II) of Community importance. The ecological term 'favourable conservation status' is defined with reference to such factors as the amount of habitat remaining, population dynamics and trends in the natural range of species. To these ends, Member States are required to identify and protect Special Areas of Conservation (SACs) in which the necessary steps are taken to ensure that the priority habitats and species therein are maintained at, or restored to, a favourable conservation status.

The Member States are required to take all appropriate steps to avoid the deterioration of those habitats and species for which protection is required. Under articles 6(3), a plan or project likely to have a significant effect on a Natura 2000 site must undergo assessment to determine whether it would damage the nature conservation interest of the site. If the plan or project is thought to impose a significant threat, it can only go ahead if (1) there is *no alternative solution;* (2) its implementation is of *overriding public interest*; (3) member states must provide compensatory measures which may include habitat restoration or recreation of the same type of habitat on the same site or elsewhere. Where the site hosts species and/or habitats listed as a priority by the Directive, under Article 6(4), development is permitted only on the grounds of: (1) human health and public safety; (2) "beneficial consequences of primary importance for the environment"; (3) (subject

---

[2] The selection of designated areas is not done on a country-by-country basis, but takes into account their biogeographic specificities. Six biogeographic regions were identified within EU countries.

to an opinion by the Commission), "other imperative reasons of overriding public interest."

A significant number of habitat types listed in Annex II of the Directive are located in the coastal fringe (dunes, mud flats, coastal lagoons, coastal freshwater wetlands, etc.). In addition, the Habitats Directive specifically establishes Marine Special Areas of Conservation. The Habitats Directive can therefore be expected to have a major impact on the coast. In its strict interpretation, the compensation requirement for displaced habitats also applies to habitats lost through natural, or semi-natural causes, such as sea level rise and coastal erosion, which is likely to have far reaching consequences given the current climate change predictions. In the UK, for example, relevant authorities are anticipating this need for compensation and are planning ahead by recreating coastal habitats through managed realignment – realigning existing hard defences further inland thereby recreating intertidal habitats (Ledoux et al. 2003).

Research support for policy implementation: The ELOISE contribution

In this section, we present an evaluation of the ELOISE cluster contribution to EU policy implementation. All coordinators of past and ongoing ELOISE projects were contacted to assess the direct and indirect relevance of current and recent coastal research for European policy and management (57 projects in total). 7 additional projects outside the ELOISE cluster were also selected for inclusion in the analysis to avoid identifying gaps that were covered outside this programme. The research objectives of the projects published on the CORDIS database were compared with the foci identified within ELOISE as well as with policy objectives in the EU directives relevant to European coastal waters identified above and summarised in Table 2.

The results were compiled in a spreadsheet that was sent to all coordinators. Coordinators were asked to check whether they agreed with the way the objectives of their project were assessed, and update them if necessary. They were also asked to provide in their own words 3 key points where they thought their research was contributing to future coastal zone management and policy. A reminder was sent to coordinators before the deadline. Overall, 18 replies were received out of the sixty-two projects identified, which represents a response rate of 29%, which is close to the average response rate in postal surveys. The analysis of the spreadsheet relies on the data updated by coordinators for the 18 replies received, and on our own assessment of the research objectives for the remainder of the projects. For the sake of transparency, we list the names of projects that provided a direct input in the survey (Appendix 1).

The results of the survey are presented in Fig. 3. The figure shows quite clearly that the majority of projects address the global cycles and human impacts ELOISE foci. Although one can expect some progress since the last evaluation, there are still a minority of projects looking at practical approaches for sustainable coastal zone resource use and development (socio-economic development) and the methodology and instrumentation need to implement ELOISE (infrastructure).

Table 2. The four ELOISE foci and major policy components of the three relevant EU-directives

ELOISE foci/topics:	Global Cycles
	Human Impacts
	Socio-Economic Development
	Infrastructure
Water framework directive objectives:	Drivers, Pressures and Impacts
	Economic Analysis
	GQS: transitional and coastal waters
	GQS: surface waters
	GQS: groundwater
	Heavily modified water bodies
	Geographical information systems
	Participatory approaches
	Integrated River Basin Management
	Monitoring and assessment tools
	Intercalibration
Bathing water directive:	Bacteriological quality assessment
	Economic analysis of policy measures
Habitat directive objectives:	Biodiversity assessment
	Management plans/policy measures
	Impact of activities on biodiversity

ELOISE related projects are quite narrowly focused in terms of their contribution to the implementation of European policy. The majority of projects contribute to identifying drivers and pressures of environmental change, and to developing monitoring and assessment tools. This is a positive point as identifying the sources of change is key to developing policy instruments for environmental protection. Monitoring is also a core element of the Water Framework Directive. It is not surprising that a very large majority of projects contribute to identifying good quality status in transitional and coastal waters, since the main focus of ELOISE on coastal issues. Surface water, groundwater, and heavily modified water body issues are probably covered in other clusters or research programmes. However, it is quite clear from the results that not enough research is devoted to economic analysis, participatory approaches and integrated management. Other key tools like GIS and intercalibration methods are also lacking. Not much research seems to address bacteriological water quality issues, and given the forthcoming revision of the bathing water directive, this is likely to need further attention. Finally, not enough projects were identified as contributing directly to the implementation of the Habitats Directive, especially regarding management issues. It is probable that a number of biodiversity projects were funded under other programmes, but given the likely impact of the Habitats Directive in the coastal zone; ELOISE should perhaps play a greater role in this area.

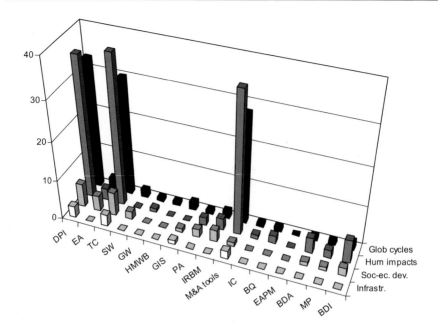

Fig. 3. Allocation of the number of projects per ELOISE focus and policy objectives of EU directives. The four ELOISE foci are global cycles, human impacts, socio-economic development and infrastructure. Further legend: Water Framework Directive objectives: DPI: drivers, pressures and impacts; EA: economic analysis; TC: good quality status of transitional and coastal waters; SW: good quality status of surface waters; GW: good quality status of groundwater; HMWB: heavily modified water bodies; GIS: geographical information systems; PA: participatory approaches; IRBM: integrated river basin management; M&A tools: monitoring and assessment tools; IC: Intercalibration. Bathing Water Directive objectives: BQ: Bacteriological quality assessment; EAPM: economic analysis of policy measures. Habitat Directive objectives: BDA: biodiversity assessment; MP: management plans/policy measures; BDI: impact of environmental change and human activities on biodiversity.

In interpreting these results, it is important to keep in mind that this evaluation inevitably contained some element of subjective interpretation – either from the project coordinators, or from the authors of this report. A good understanding of the meaning and scope of the ELOISE foci and EU policy objectives is also assumed (e.g. the contents of the infrastructure focus might not have been clear to all). Nevertheless, the sharp contrast and clarity of the results mean that while recognising that there is some degree of subjectivity, the overall result is probably robust.

The results of this survey also need to be viewed alongside a review of the published papers produced by ELOISE scientists, laid down in Chapter 2 (Herman et al). This review highlights the fact that significant advances in individual scientific topics have been made but that a common scientific infrastructure (including databases) has yet to be established. Such an infrastructure would form the necessary foundation for future applications of applied research in the context of EU policy and legislation.

New priorities

Research into coastal zone issues is vital to implementing EU policy. The EU ICZM strategy includes a requirement to generate information and knowledge about the coastal zone. While, along with the authors of the previous ELOISE evaluations, we recognise that research funding has been largely based on expert-based judgement of project quality and only to a limited extent on the existing science plan, future research should to have a stronger focus on policy implementation needs.

In terms of areas of policy, we have identified that the bulk of the research contributes to specific areas of implementation of the Water Framework Directive, e.g. the understanding of drivers, pressures and impacts (see also Herman et al. Chapter 2). The Directive is an ambitious piece of legislation, and the research needs are indeed huge (e.g. Ledoux and Burgess 2002). There is some basis for recommending however, that some research funds are also targeted towards implementation of the Bathing Water Directive, especially in the light of the ongoing revision, and of the Habitats Directive with a specific focus on the coastal zone, where ecosystems are very dynamic and management issues likely to be significant.

As far as specific research tools and methodologies are concerned, more attention needs to be placed on translating and integrating natural sciences into decision-making processes. Intercalibration, Geographical Information Systems (GIS), economic analysis, participatory approaches and integrated assessment methodologies all need to be developed further to achieve this integration, and equip managers with the right decision tools to face future coastal zone management challenges. The papers selected for this workshop provide examples of application and an opportunity to assess and discuss opportunities for further development.

For the shorter-term needs of European coastal research, we conclude from the above that:

1. A better orchestration of the investment of resources is required to meet implementation research needs. A change in evaluation in funding and evaluation procedures might be necessary. The Framework Programme 6 is an opportunity to bring these changes about;
2. For a successful implementation of the Water Framework Directive and other European legislation, integration of natural sciences, economic analysis and participatory approaches, for example within the framework of integrated assessment requires further attention (Turner, 2000). This needs willingness and ability to operate across monodisciplinary boundaries at individual and institutional level, but also time and resources. European centres and networks where longer-term interdisciplinary research effort into coastal science and management is institutionalised can play a key role. Again, FP6 instruments and the

new European Research Area can be useful in supporting and encouraging the appropriate research structures;

3. The catchment component of 'catchment-coast interactions', as identified in the LOICZ science plan and adopted by ELOISE, has lagged behind and should receive a stronger focus, within the perspective of regional seas;

4. Global change is the backdrop of the whole LOICZ programme and consequently of ELOISE. Its implications for the understanding and management of European coastal seas, however, remain largely unaddressed.

We argue that a new vision for the longer-term development of the European coasts is needed. Reaching the goals of the WFD, namely the achievement of a good ecological station for all European waters, will require mutual interplay of policy makers, coastal management and the coastal science community, building on cooperation, multidisciplinarity and a better understanding of regional seas and societal needs.

References

Ærteberg G (ed) (2001) Eutrophication in Europe's coastal waters. European Environment Agency Topic report no 7, Copenhagen

Barth F, Fawell J (2001) The Water Framework Directive and European Water Policy. Ecotoxicol Environ Saf 50:103-105

Barthel KG, Barth H, Murray CN, Suranyi M, Dubois K (1999) ELOISE implementation report – Phase II, Ecosystems research report no 33. European commission, DGXII/D1, Environment Directorate

Cadee N, Dronkers J, Heip C, Martin JM, Nolan C (1994) ELOISE (European Land-Ocean Interaction Studies) Science Plan. Ecosystems research report no. 11, European Commission EUR 15608,Luxemburg

European Environment Agency (1998) Europe's environment: the second assessment. EEA, Copenhagen

Kaika M, Page B (2002) The making of the EU Water Framework Directive: shifting choreographies of governance and the effectiveness of environmental lobbying. University of Oxford, School of Geography and the Environment. Economic Geography Research Group. WPG 02-12

Kallis G, Butler D (2001) The EU Water Framework Directive: measures and implications. Water Policy 3:125-142

European Commission (1999) Towards a European Integrated Coastal Zone Management (IZM) Strategy. General principles and Policy Options. A reflection Paper

European Commission (2000) Communication from the Commission to the Council and the European Parliament on "Integrated Coastal Zone Management: A Strategy for Europe" (COM/00/547 of 17 Sept. 2000)

Institute for European Environmental Policy (1999) The Influence of EU Policies on the Evolution of Coastal Zones. Thematic Study E, Study Contract ERDF No 98.00.27.049, ICZM Demonstration Programme

Ledoux L, Burgess D (2002) Science for Water Policy (SWAP) – implications of the Water Framework Directive, Proceedings of a Euroconference organised at UEA, Norwich, UK, 1-4 September 2002

Ledoux L, Crooks S, Jordan A, Turner RK (2000) Implementing EU biodiversity policy: UK experiences. Land Use Policy 17:257-268

Ledoux L, Cornell S, O'Riordan T, Harvey R, Banyard L (2003) Managed realignment: towards sustainable flood and coastal management. CSERGE Working Paper, forth-coming

Murray CN, Bidoglio G, Zaldivar J, Bouraoui F (2002) The Water Framework Directive: the challenges of implementation for River Basin–Coastal Zone research. Fresenius Environ Bull 11(9):530-541

Nolan C, Barth H., Barthel KG, Cameron F (1998) ELOISE Implementation Report Phase I. Ecosystems Research Report N. 22. European Commission.

Stanners D, Bourdeau P (2001) Europe's Environment - The Dobris Assessment. European Environment Agency

Turner, RK (2000) Integrating natural and socio-economic science in coastal management, J Mar Sci 25:447-460

Turner RK, Adger WN, Lorenzoni I (1998a) Towards integrated modelling and analysis in coastal zones: principles and practices. LOICZ reports and studies N. 11. LOICZ, Texel, Netherlands

Turner RK, Lorenzoni I, Beaumont N, Bateman IJ, Langford IH, McDonald AL (1998b) Coastal management for sustainable development: analysing environmental and socio-economic changes on the UK coast. Geogr J 164:269-281

Wils W (1994) The Birds Directive 15 years later. J Environ Law 6:219-242

Appendix 1: List of projects that provided input in the survey

Project Acronym	Full title of the project	Project Coordinator	Project duration
ELOISE Projects			
COSA	Costal Sands as Biocatalytical Filters	Dr. Markus Huettel	2002-05
DANUBS	Nutrient Management in the Danube Basin and its impact on the Black Sea	Prof. D. Helmut Kroiss	2001-05
DOMAINE	Dissolved organic matter (DOM) in coastal ecosystems: transport, dynamics and environmental impacts	Prof. Morten Søndergaard	2001-03
EROS-21	Biogeochemical Interactions between the Danube River and the North-Western Black Sea.	Dr. Jean-Marie Martin	1996-98
EUROCAT	European Catchments - Catchments changes and their impact on the coast	Prof. Willem Salomons	2001-04
M&MS	Monitoring & Managing of European seagrass beds	Ass. Prof. Jens Borum	2001-04
MEAD	Marine Effects of Atmospheric Deposition	Prof. Tim Jickells	2000-03
MERCYMS	An integrated approach to assess the mercury cycling in the Mediterranean basin	Prof. Nicola Pirrone	2003-05
MOLTEN	Monitoring long-term trends in eutrophication and nutrients in the coastal zone: Creation of guidelines for the evaluation of background conditions, anthropogenic influence and recovery	Dr. Daniel Conley	2001-04
NTAP	Nutrient dynamics mediated through turbulence and plankton interactions	Dr. Celia Marrase	2001-04
PROTECT	PRediction Of The Erosion of Cliffed Terrains	Dr. Jonathan Busby	2001-04
SIGNAL	Significance of External / Anthropogenic Nitrogen for Central Baltic Sea N-Cycling	Dr. Maren Voss	2000-03
STREAMES	Human effects on nutrient cycling in fluvial ecosystems: Development of an Expert System to assess stream water quality management at reach scale.	Dr. Franesc Sabater	2001-04
TIDE	Tidal Inlets Dynamics and Environment	Dr. Marco Marani	2002-05

Non-ELOISE projects			
BIOBS	Evaluation of coastal pollution status and bioindicators for the Black Sea	James Wilson	2002-05
DINAS-COAST	Dynamic and interactive assessment of national, regional and global vulnerability of coastal zones to climate change and sea-level rise	Richard Klein	2001-04
EUROSION	A European initiative for sustainable coastal erosion management	Stephane Lombardo	2002-03
EVALUWET	European valuation and assessment tools supporting wetland ecosystem legislation	Ed Maltby	2001-04

2. Land-ocean fluxes and coastal ecosystems – a guided tour of ELOISE results

Peter M.J. Herman[1], Tom Ysebaert, and Carlo H.R. Heip

Abstract

This chapter provides an overview of ELOISE projects that have concentrated on biogeochemical cycles. We will address the question what new insight we have gained from the ELOISE research, and how this fits in with an evolving scientific view on the role of coastal systems in land-ocean interaction. Much of that discussion will be on biogeochemical cycles of carbon and nutrients, and will keep the problem of eutrophication and its consequences for the coastal systems in the focus of attention. We will end our contribution with a discussion on how these findings could be used for improved management of the coastal systems, and what are the focal points for future research.

Introduction

After the definition of its science plan in 1994, the ELOISE cluster of projects was set up as the European contribution to the IGBP program LOICZ. The aim was to bundle efforts of the European scientific community in elucidating some of the outstanding scientific problems in the study of the role of coastal systems in the interaction between the terrestrial and the oceanic realms.

Coastal systems, defined as estuaries and coastal seas, occupy a minor proportion of the surface of the world's seas, but contribute disproportionally to the cycles of carbon and nutrients, since on average the intensity of processes is much higher than in the deep oceans. They are also an important ecotone between the terrestrial and the oceanic systems, and much of their characteristics and ecosystem processes can only be understood in the context of this gradient.

In 1994, the ELOISE science plan (Cadée et al. 1994) was published as a result of a workshop of European coastal scientists. The explicit aim of this science plan

---

[1] Correspondence to Peter Herman: p.herman@nioo.knaw.nl

J.E. Vermaat et al. (Eds.): Managing European Coasts: Past, Present, and Future, pp. 21–58, 2005.

was to set a roadmap for the European contribution to the international IGBP project LOICZ. The science plan highlighted the following objectives:

- Significance of coastal seas in global change. Emphasis in this section was on origin and fate of organic matter, nutrients, trace elements, sediments and biogases. The approach stressed the biogeochemical functioning of coastal ecosystems, and its interaction with the structure (biological structure, e.g. species composition, but also geomorphological structure) of coastal ecosystems;
- Human impact on coastal seas. The approach emphasised the regional and global consequences of human impact through pollution, eutrophication and physical disturbance;
- Socio-economic development and coastal seas. Here the focus was on a strategic approach to the management of sustainable coastal zone resource use and development. Much emphasis was placed on multidisciplinary approaches of natural and socio-economic sciences, and on the analysis of management failures as a basis for better management in the future;
- Methodology and Implementation of ELOISE. The science plan pleaded for the development of a European scientific infrastructure for coastal zone research tools and data management.

The implementation of the ELOISE science plan had to follow standard procedures for European R&D projects in the fourth and fifth frameworks. Project proposals were invited for (amongst others) the topics proposed in the ELOISE science plan. These proposals were selected on the basis of their scientific excellence and their contribution to European scale economic and social development. Successful proposals were accepted as projects and evaluated on the basis of their ability to achieve the proposed objectives. Coordination of the ELOISE project cluster did not take place at the level of proposal selection. It was only *after* the selected projects were known, that the projects fitting within the ELOISE science plan were clustered within ELOISE. Coordination between the projects in the cluster was achieved primarily through the ELOISE Open Science Meetings organised by the European Commission. These meetings were also a means to stimulate discussions and generate joint activities between projects.

After ten years of implementation, around fifty projects have been clustered as ELOISE projects. About half of them are still running, and some of the running projects have not yet published all their results. For the completed projects, most of the scientific results have been published. These publications have received an ELOISE publication number. The complete list, with abstracts, is made available at the ELOISE web site. This overview will largely be based on these published results. It may therefore miss information from on-going projects. However, we judged it was better to base conclusions on peer-reviewed published results than to use preliminary reports.

The main part of this paper will focus on the first of the four scientific objectives from the ELOISE science plan. We will address the question what new insight we have gained from the ELOISE research, and how this fits in with an evolving scientific view on the role of coastal systems in land-ocean interaction. Much of that discussion will be on biogeochemical cycles of carbon and nutrients, and will keep the problem of eutrophication and its consequences for the coastal systems in the focus of attention. We will end our contribution with a discussion on how these findings could be used for improved management of the coastal systems, and what are the focal points for future research coming out of the results.

A few of the completed ELOISE projects were so unrelated to the rest that we considered them as 'misclassified outliers' (to use a term from ecological clustering) and have not further discussed their results. This applied to the project CHABADA, which studied bacterial biodiversity, and the project CLICOFI on metabolic and physiological adaptations to changed temperature in fish. The DUNES project, which was directly aiming at improving dune management, also fell out of the scope of this review, but has been reviewed in Williams (2001). The BASIS project, studying the effect of climate change on terrestrial Arctic systems, also had little affinity with the other projects. The concerted action BBCS on the status of the Baltic mainly resulted in a review and discussion of the state of the environment in this region. The resulting papers of Jansson and Stålvant (2001) or Jansson (1987) could be consulted as a regional predecessor of the present attempt to summarise results. Finally, we omitted the projects POPCYCLE, MAMCS and MOE on atmospheric pollution by organic pollutants and mercury compounds. These topics will be summarised elsewhere.

Input of nutrients into the coastal zone

Atmospheric nutrient inputs

The ELOISE projects ANICE (de Leeuw et al. 2001) and MEAD have focused on nutrient inputs into coastal areas via deposition from the atmosphere. Atmospheric inputs are particularly important for nitrogen. Atmospheric transport and deposition of other essential macronutrients (phosphorus, silicium) is not quantitatively important compared to other inputs into the coastal sea.

Nitrogen is deposited from the atmosphere in different chemical forms. Many nitrogen species in the atmosphere and in aerosols are chemically very reactive, and therefore chemical dynamics within the atmosphere have to be taken into account in the models for nitrogen deposition.

The reactions in the atmosphere, moreover, have to take into account interactions with variable aerosol concentrations and composition. The further development of mathematical models for these interactions, combined with proper transport modelling, has been the core objectives of ANICE (de Leeuw et al. 2001). The project has not focused on a single model, but rather has developed several modelling approaches, each with their own scope at different scales. For the North Sea basin, Hertel et al. 2002 (ANICE project) estimated N deposition (Fig. 1), distinguishing different chemical forms and ways of deposition, as summarised in Table 1.

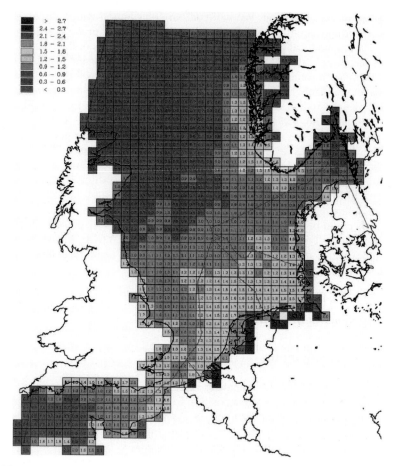

Fig. 1. Total atmospheric deposition of nitrogen (ton N km^{-2}) over the North Sea in 1999. The results reflect the distribution of sources on the one hand, and precipitation on the other hand. From Hertel et al. (2002 – ANICE project)

As can be seen from Table 1, wet deposition dominates N input into the North Sea. This is true for the adjacent watersheds too, and it complicates the calculation of the spatio-temporal distribution of atmospheric inputs at small spatial and temporal scales. Wet deposition is tightly linked to rain events, the interaction of fronts with polluted air masses and other short-term relatively local processes (Jickells, 1998; Spokes et al. 2000). It is estimated that yearly average figures are more reliable than local, short-term model estimates. However, with respect to interactions between atmospheric input and the biological system in the water column, timing effects can be very important. It can be expected that the influence of a nutrient input pulse in a nutrient-starved system has a much greater importance than in a nutrient replete situation. This non-linearity calls for a closer examination of the interaction between atmospheric inputs and the biological system. This is the subject of current research within the MEAD project.

Table 1. Estimation of the different chemical forms and deposition modes of atmospheric nitrogen into the North Sea. From Hertel et al. (2002)

	Wet deposition		Dry deposition		Total deposition	
	kt	%	kt	%	kt	%
NH_3	24	3	19	3	43	6
NH_4^+	191	27	36	5	227	32
NO_y gas	151	21	43	6	194	27
NO_3^-	210	30	35	5	245	35
N total	577	81	132	19	709	100

As for the origin of the nitrogen deposited via the atmosphere, two important sources are recognized (Hertel et al. 2002). For the North Sea it is estimated that 38% of the total deposition originates from emissions related to agricultural activity (NH_3 and NH_4^+) and 72% due to emissions from combustion sources (mainly traffic, industry and power production) (aerosol phase NO_3^- and gas phase NOy compounds). A remarkable finding of this study is that ship traffic is a very significant source of marine-deposited N.

Spatially, two pathways for atmospheric inputs are important: direct deposition onto the water surface of estuaries and coastal seas, and deposition in the watersheds. The latter source, as it operates over large surfaces with atmospheric concentrations of nitrogen, which are often elevated because they are close to emission sources, is quantitatively important, even though a relatively small fraction of the deposited N eventually reaches the rivers. In North-American estuaries, Castro and Driscoll (2002) calculated that between 15 and 42 % of the N input into the estuary was derived from atmospheric deposition in the watershed and river. It can be assumed that this fraction will be similar in European estuaries. Adding this fraction to the direct atmospheric input into the North Sea suggests that (direct and indirect) atmospheric input of N is responsible for a significant proportion (one-third to one-half) of the total (non oceanic) N input into the system.

Results from ELOISE and closely related research activities in Europe, demonstrate that state-of-the-art analytical tools and models are available, and that these are able to supply good estimates of atmospheric N input, as well as of the respective sources, at relatively coarse temporal and spatial scales. Application to other areas than the North Sea might be wanting, but the models should be generic enough to extrapolate easily. More research is needed at smaller scales, with particular emphasis on the non-linear interaction with biology. Moreover, the atmospheric chemistry needs further process study and development of models. As atmospheric input of nitrogen is dependent on human-influenced sources, continuous monitoring seems to be essential at the European scale.

Watershed processes

The ELOISE projects INCA and RANR have focused on the modelling of nutrient transformation processes in the watershed and the river network. The main aim of these projects was to relate statistics of land use and human (agricultural, domestic) practices of nutrient input into the system, to the load of rivers.

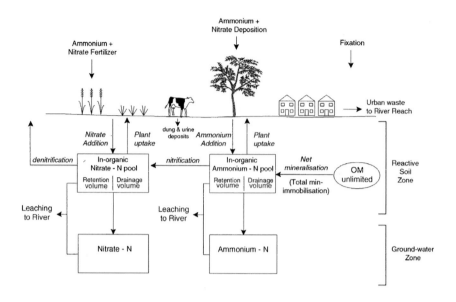

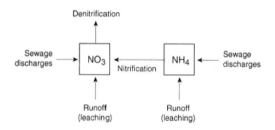

Fig. 2. Overview of land (upper panel) and instream (lower panel) processes in the INCA model. From Wade et al. (2002b)

The INCA project (see overview in Neal et al. 2002), has developed a generic deterministic model for application across Europe. The model includes land and river processes, and is driven by spatially explicit input data. Figure 2 illustrates the land and river processes in the INCA model.

Wade et al. (2002a) have developed a version of the model for phosphorus. The cycles of both nutrients have not yet been fully coupled, and this remains a topic for further research.

The INCA model has been applied across Europe to different catchments, in order to cover variability in climatic factors (rainfall patterns, snow, temperature), edaphic factors (peatland, sandy and loamy soils) and human-induced factors (sewage outfalls, land use, atmospheric inputs). Neal et al. (2002) conclude that overall a satisfactory result has been obtained, and that a good dynamic tool is available for simulation of N output to the coastal zone.

The model has allowed the calculation of a number of scenarios. Climatic change scenarios show a significant effect of the yearly rainfall pattern on N output from rivers (Limbrick et al. 2000). Jarvie et al. (2002) and Whitehead et al. (2002) show that changes in land use are important. Up to a factor four differences in instream N concentrations were found upon a change of land-use from arable land to wasteland. Flynn et al. (2002) describe similar effects of changes in land use, and also demonstrate that the effect of riparian buffer zones is limited.

Reductions of N input into rivers at sewage outlets have a direct and big effect on the N concentrations and loads of rivers, and appear to be the most effective measures to reduce riverine N load.

Although dynamic models such as INCA suffer from parameter and process uncertainty (Wade et al. 2002b), they offer the advantage over regression-based models that they capture within-year dynamics, and that they may also produce more meaningful extrapolations when calculating scenarios of land use change and climatic changes (Neal et al. 2002).

From the INCA project, a good and validated tool is available for dynamically calculating N loads from European catchments. What is lacking is a systematic application to all (or to the most important) European catchments. This is closely related to data scarcity, since the model requires extensive input and validation datasets. Further research is also needed on the coupling of organic loads and different nutrients, as well as on the representation of within-river dynamics of nutrients and organic matter.

The RANR project has concentrated on the role of groundwater in the coupling between watersheds and rivers. One of the basic problems tackled was the delay in response of major German rivers (case study was the Elbe) after the considerable reductions of N input in arable land in former East Germany from 1990 onwards. In contrast to the input curve of N to cropland, which shows a sudden drop in 1990 and a slow increase afterwards, the N load of the river Elbe shows a very gradual decrease over the decade. The delay in response is explained by the long residence times of groundwater reservoirs. In the project these residence times were calculated for some East-German watersheds, using spatially explicit groundwater models (Kunkel and Wedland, 1997). Grimvall et al. (2000) present a conceptual model with two reservoirs, one fast-responding and one with a long response time, to explain the apparent paradox that nitrogen export from rivers (1) reacts rapidly to increased input from point and diffuse sources and (2) has a very long lag time after a reduction of the input. For nitrogen, the time scale of response to reductions of input can be decades. For phosphorus, there is a rapid reduction in river concentrations from high to moderate upon reduction of the input, but further reduction to low concentrations may take many decades, due to continued leaching of phosphorus from sediments in the river system. A typical output from this conceptual model is illustrated in Fig. 3.

In contrast to the INCA project, the RANR project has not embarked on the construction of fully deterministic models for nutrient dynamics in watersheds. Instead it applied a combination of deterministic models, statistical models and a meta modelling based upscaling from one-dimensional vertical process models (Forsman and Grimvall, 2003). The latter approach was applied in scenario studies, and formalized in a decision support system (Forsman et al. 2003a, 2003b).

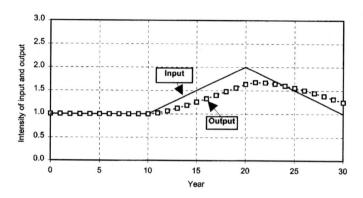

Fig. 3. Impact of changed input on the output of a model comprising two parallel compartments with different residence times (1 and 10 years, respectively). From Grimvall et al. (2000)

The mixed statistical-deterministic modelling approach in RANR is a promising avenue in data-dense situations. It offers a number of appealing applications where large sets of monitoring data, both in rivers and in groundwater, are available. This aspect makes it a robust tool to evaluate and validate scenarios for these watersheds. It may, at the same time, limit its value as an upscaling tool to the whole of Europe.

Groundwater dynamics are not only important within watersheds discharging in a river. The project SUBGATE studied the direct submarine groundwater discharge to the sea in a Baltic area (Kaleris et al. 2002 – SUBGATE). Direct submarine groundwater discharge can be a very important route of water transport to the sea, about 5 to 10 % globally but values up to 40 % of the river flow have been reported (Moore 1996). At the SUBGATE study site, discharge rates of approximately 0.05 $m^3.s^{-1}$ per km of land-sea interface were reported. Modelling showed that the pattern of discharge is spatially very variable, and that the process is very difficult to measure from field data alone. The importance of groundwater discharge for nitrogen fluxes to the sea is currently being investigated in the project NAME.

River processes

Within the EROS2000/EROS21 projects, a full coupling of models was established from the watershed, over the river, to the coastal area and the deep sea. The study case was the Danube as it influences the north-western shelf of the Black Sea and the whole of the Black Sea proper. The RIVERSTRAHLER model (Billen and Garnier, 1999, Garnier et al. 2002), which was first applied to the Seine, was used to describe nutrient and ecological dynamics in the Danube watershed and river. This model synthesises the hydrological network of a river basin by stream order, which reduces the computational load to a reasonable level. For different sub-basins, nutrient and organic inputs are derived from gross statistics (population density, type and intensity of industrial activity, fertiliser application, land use). The river model for the different stream orders of several sub-basins represents full ecological dynamics, including transformations of nutrients in the ecosystem.

The model has been used to calculate nutrient and organic loadings from the Danube before and after the big economic changes in the East. It has also been used to calculate silicate deliveries to the Black Sea. Garnier et al. (2002); to refute the hypothesis that the Iron Gate dams are responsible for the decrease of silicate loads (Humborg et al. 1997) and to propose the alternative explanation that increases of productivity as a consequence of increased nitrogen and phosphorus loading are responsible for most of the silicate retention in the basin.

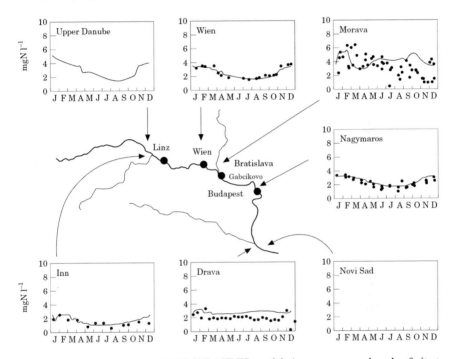

Fig. 4. Example output of the RIVERSTRAHLER model. Average seasonal cycle of nitrate concentrations in the upper course of the Danube for the period 1988-1990. Observational data are given for comparison. From Garnier et al. (2002)

The applications of the RIVERSTRAHLER model, with minor modifications, to several rivers (Seine: Billen et al. 2001; Danube: Garnier et al. 2002; Schelde: Billen et al. in preparation) demonstrate the portability of the model across different systems. It remains to be tested in more extreme European environments. The dynamic coupling between biological and nutrient dynamics makes the model a most promising tool, also for scenario studies. It can take into account changes in nutrient ratios of anthropogenic inputs, and deliver outputs that are useful for estuarine and coastal sea ecological models.

Estuarine transformations of nutrients and organic carbon

Nitrogen cycle

Estuaries and coastal areas play an important role in the nitrogen cycle, as transformers of nitrogen (through oxidation of reduced inorganic nitrogen forms, mineralisation of organic nitrogen input from terrestrial and riverine sources) and as nitrogen sinks. The latter role depends critically on the process of denitrification. Since this is an anaerobic process, it is mostly concentrated in deeper layers of the sediment. The estuarine nitrogen cycle has been the subject of many studies (see e.g. reviews by Nedwell et al. 1999 and Herbert, 1999) for a number of reasons. Nitrification, the oxidation of ammonium to nitrate, can be a dominant process in oxygen consumption in heavily loaded estuaries (e.g. Soetaert et al. 1995). Estuarine denitrification contributes to the removal of nitrogen, often the limiting nutrient, from the biological cycle. In a first order, it has been described as a function of estuarine residence time and nitrate loading (Dettman, 2001). Nitrogen transformations also contribute to the formation of the greenhouse gas N_2O.

Several ELOISE projects have concentrated on the nitrogen cycle in estuarine and coastal habitats. Results of these studies have altered the view on nitrogen cycling, stressing the link between nitrogen and carbon cycles and the importance of incorporation (and subsequent immobilisation) of nitrogen into biomass. In essence, these studies have modified the view on nitrogen cycling to make them more dependent on the physical and biological setting in different systems, and to stress the mutual dependence of light, physical conditions, nitrogen loading and the effectiveness of estuaries as a 'filter' for nitrogen inputs.

Nitrogen fixation by pelagic cyanobacteria

The BASIC project concentrated on nitrogen fixation, and on the food web and nutrient relations between nitrogen fixing cyanobacteria and other elements of the microbial food web. Stal and Walsby (2000) monitored a N fixing cyanobacterial bloom in the Baltic during a period of calm and stratified conditions, interrupted by a deep mixing event as a consequence of a short-lasting storm. From the population shifts they evaluated the relative importance of light, mixing and temperature in regulating N-fixing cyanobacteria and non-N-fixing picocyanobacterial populations. They show that the seasonal cycle in water column stratification, as well as the seasonal cycle in light availability, are the structuring forces for N fixing blooms. The effect of temperature is secondary. These authors also demonstrate the importance of picocyanobacteria in primary production, and show that upon mixing these populations are able to sequester very rapidly the available nitrogen. Part of that nitrogen originates from excess fixation by the N fixing cyanobacterial populations.

An important problem that remains to be explained is why N fixing blooms only occur in brackish temperate waters, and are confined to tropical and subtropical areas at full marine salinity.

Gallon et al. (2002) show that for a bloom of cyanobacteria in the Baltic, rates of primary production and N_2 fixation are out of phase, especially at longer time scales. During the development of a bloom, the relative needs of the cells for fixed carbon and nitrogen may change.

Repka et al. (2001) showed that nutrient concentrations (in particular phosphate) affect the growth rate of the N-fixing cyanobacteria *Nodularia*, but not its toxin concentration. Engström et al. (2000) studied grazing by copepods on *Nodularia*. They showed that the ability of the copepod *Acartia* to selectively avoid grazing on *Nodularia*, provides a considerable advantage over other species.

Nitrogen cycle in vegetated sediments

The effect of plant (microalgae or macrophytes) growth on the N cycles in benthic systems was the subject of many studies in the NICE and ROBUST projects. Within NICE both microalgal-dominated sediments and sediments inhabited by macroalgae and macrophytes (seagrass) were studied. The studies on microalgae-dominated sediments (e.g. Sündback et al. 2000; Thornton et al. 1999; Ottosen et al. 2001) were summarised by Risgaard-Petersen (2003) in an extensive meta-analysis, complemented by laboratory experiments to detect causal factors. Until recently, and mostly based on theoretical considerations, it was assumed that the oxygen release into the sediments by benthic plants (directly by benthic algae, *via* aeration of the rhizosphere by macrophytes) created favourable conditions for coupled nitrification-denitrification. Ammonia produced by ammonification in the sediment can be oxidised to nitrate in oxic sediment layers and can diffuse into the anoxic zone and be denitrified. In principle, the extension of oxic-anoxic interfaces in sediment therefore increases the probability of coupled nitrification-denitrification. Risgaard-Petersen (2003) convincingly showed for studies of sediments inhabited by microalgae that this anticipated effect is overruled by ammonium (and nitrate) uptake by the algae. Competition for the nitrogen substrate decreases the amount of inorganic nitrogen available for coupled nitrification-denitrification. The consumption of DIN is direct by the algae, but also indirect by heterotrophic bacteria that quickly incorporate extracellular polysaccharides produced by the algae (Middelburg et al. 2000).

At a community level, the autotrophy-heterotrophy status of sediment communities will determine whether the sediment as a whole acts as a sink or source of nutrients. Only when the sediment is net heterotrophic, can it be anticipated that sufficient DIN will be set free and be available for denitrification. Several studies demonstrated that, at least during productive seasons and especially in the light, sediments dominated by microphytobenthos are net autotrophic. Serôdio and Catarino (2000) developed a mathematical model for microphytobenthic production, and showed that most variation is on relatively short time scales (hours to fortnights). Miles and Sundbäck (2000, NICE) measured primary production by microphytobenthos at three sites across Europe. Their study shows that tidal regime is the most important factor explaining between-site differences. Yearly integrated productivity was generally comparable across sites. Barranguet et al. (1997 – ECOFLAT) used pigment analysis to derive the fate of microphytobentic production. They show that bacterial degradation in the sediment is the dominant fate in spring, while grazing becomes more important in summer. Only in exposed sediments, was resuspension a dominant loss factor. Benthic-pelagic exchange of microphytobenthos was further discussed by Lucas and Holligan (1999 – ECOFLAT), Lucas et al. (2000 – ECOFLAT) and Middelburg et al. (2000 – ECOFLAT). In general, export of microphytobenthic biomass was very limited at muddy sites, but even in sandy sites the major part of the production is consumed *in situ*.

Studies on sediments dominated by seagrass, both within the NICE and ROBUST projects, confirm the trend that 'vegetated' sites are autotrophic or only slightly heterotrophic, and that N uptake and storage in biomass is a dominant factor in the sediment N cycle, decreasing the potential denitrification (Welsh et al. 2000, 2001 – NICE; Ottosen et al. 1999 – NICE; Hansen et al. 2000 – ROBUST; Risgaard-Petersen et al. 1998; Risgaard-Petersen and Ottosen, 2000 – NICE). Nielsen et al. (2001 – ROBUST) showed elevated nitrogen fixation in the rhizosphere of *Spartina* and *Zostera* compared to the bulk sediment. The study suggests that sulphate reducing bacteria in the rhizosphere are a shunt between the carbon and nitrogen cycles, as they contribute to nitrogen fixation based on carbon flows derived from the plants.

Boschker et al. (2000 – ROBUST) show that the available organic carbon for bacteria in *Zostera* sediments is mainly of algal origin. At least for this study site, *Zostera* detritus can therefore be assumed to be either buried or exported from the beds, which again stresses the importance of biomass as a key factor in the N cycle. Similar findings have been reported by Gacia et al. (2002 – PHASE) in *Posidonia* seagrass meadows.

The comparison between different macrophytes highlights important differences in rhizosphere oxygenation (Heijs et al. 2000 – ROBUST; Azzoni et al. 2001- ROBUST). The dynamics of nitrogen uptake and release in macroalgae, e.g. Naldi and Wheeler (2002, NICE) and Naldi and Viarola (2002, NICE) generally showed much faster rates of uptake, storage and release of nitrogen by *Ulva* and other macroalgae than by seagrass. This fast turnover and benthic-pelagic exchange of nitrogen may provide a habitat for much 'faster', opportunistic species and occasionally for intense phytoplankton blooms.

Nitrogen sequestration in bacterial biomass in pelagic estuarine systems

There is a striking similarity between the results discussed earlier for vegetated sediments and results on N cycles in estuarine waters. Middelburg and Nieuwenhuize (2000 – BIOGEST) showed very high nitrogen uptake rates by heterotrophic bacteria in the Thames. The process was responsible for 50-90 % of the uptake of different inorganic and organic nitrogen forms, and decreased the estimated turnover time of nitrogen in the water column by an order of magnitude compared to estimates based on phytoplankton uptake only. The study also demonstrates the role of organic forms of nitrogen. Amino acids were preferred over ammonium, urea and nitrate as nitrogen sources to heterotrophic bacteria, although due to concentration differences the uptake of nitrate was highest in absolute value. In different European estuaries, Middelburg and Nieuwenhuize (2001 – BIOGEST) used stable isotope ratios to distinguish 'true' from 'apparent' conservative behaviour of nitrate in estuaries. They showed that significant turnover can occur at short time scales, even if concentrations may appear 'conservative'. This restricts the usefulness of property-salinity plots to derive estimates of estuarine processes (see also Regnier et al. 1998 – BIOGEST on that problem, more from a physical point of view). In contrast to studies on vegetated sediments, however, storage of nitrogen in biomass of pelagic heterotrophic bacteria may have short time scales and not constitute a removal from the biological cycling for long time periods.

The results suggest a closer link between the cycles of nitrogen and those of dissolved and particulate organic matter. This was also confirmed in mesocosm experiments by Havskum et al. (2003 – NTAP – see below).

Coupling between pelagic nitrogen cycle and other biogeochemical processes

Studies of pelagic biogeochemistry during the BIOGEST project revealed some unexpectedly strong couplings between the nitrogen cycle and other biogeochemical processes. Abril and Frankignoulle (2001 – BIOGEST) showed that nitrification had a significant influence on alkalinity of the water in the Schelde estuary. Therefore the process influences pH of the water, and the inorganic carbon buffer. It therefore indirectly influences the output of CO_2 to the atmosphere in the estuary.

A remarkable link between suspended sediment dynamics and denitrification was described for the Gironde by Abril et al. (2000a – BIOGEST). Fluid mud layers that form at every slack tide around neap tides, entrap every time a quantity of nitrate that is very effectively denitrified as the fluid mud rapidly becomes anoxic. The process is probably not extremely important for the estuary-wide nitrogen budget, but demonstrates very nicely how denitrification depends on spatially and temporally variable oxic-anoxic interfaces. Moreover, it has a significant effect on the N_2O production rate in estuaries.

The rapid uptake of dissolved inorganic nitrogen by pelagic heterotrophic bacteria, as well as by benthic primary producers (see above), provides a closer link than previously thought between particulate organic nitrogen and dissolved inorganic nitrogen in estuaries. Also, due to repeated sinking and re-suspension of particle-associated bacteria, it may intensify the coupling between pelagic and benthic N cycles.

New and poorly known processes in the nitrogen cycle

Several published ELOISE studies document the occurrence and rate of Dissimilatory Nitrate reduction, a process consuming nitrate and reducing it to ammonium. Welsh et al. (2001 – ROBUST) document it from a seagrass meadow, and stress the importance of the process as a possible source of N_2O to the atmosphere. The process was also measured by Christensen et al. (2000 – NICE) under trout cages in a Danish fjord. The authors show that significant DNRA only occurs under the heavy organic loading of the sediment, occurring right underneath the cages. Little is know about the factors determining the occurrence and rate of the process. This point needs further study.

Of similar or even larger importance may be the recently discovered ANAMMOX process (Dalgaard and Thamdrup, 2002; Thamdrup and Dalsgaard, 2002), which is a reaction involving nitrate and ammonium, and leading to the production of N_2 and thus removal of reactive nitrogen from the system. The process was described to be important at least at a few study sites. Depending on its (unknown) importance for estuarine systems, it may require a thorough revision of our views on coastal N cycles. This discovery illustrates the importance of continued fundamental research, even on relatively well studied problems as the environmental nitrogen cycle. Current concepts and models may have to be adjusted to incorporate new discoveries.

A coupled view of sediment biogeochemistry

Sediment biogeochemistry is determined by a complex of (chemical, microbiological and macrobiological) processes that closely interact. Although many processes have been studied in great detail, the outcome of the whole, and especially the changes in fluxes and rates as a consequence of changed forcing, remain difficult to predict. The effect of 'vegetation' on nitrogen fluxes, as discussed above, is just one example. Within ELOISE projects, other biogeochemical reactions have been extensively studied. The ROBUST and ISLED projects devoted special attention to sulphate reduction, and the occurrence of free sulphide as a toxic agent to higher plants and animals. Heijs et al. (2000 – ROBUST) show the importance of radial oxygen loss in preventing the occurrence of free sulphide in the sediment. The potential capacity for microbiological sulphide oxidation is high, but the realised rate is mainly limited by oxygen. When, in *Ruppia* meadows, the radial oxygen loss becomes too small after spring, the Fe buffer for sulphide shows a quick overflow, and the free sulphide causes damage to the plants, leading to a further reduction of sulphide oxidation capacity. This leads to a strong release of (originally Fe-bound) phosphorus from the sediment, and the system further collapses. For the saltmarsh plant *Spartina anglica*, Holmer et al. (2002 – ISLED) show that it is able to oxidise sediments and reduce sulphate reduction rate in the sediments significantly, even when the sediment is permanently waterlogged. These authors therefore predict little impact of sea level rise on this species, except perhaps in the most seaward stands where sulphate reduction rates in the sediment are highest – and where the presence of the plant is most important to prevent erosion. Gribsholt and Kristensen (2002 – ISLED) demonstrate the large influence of both the plant *Spartina anglica* and the worm *Nereis diversicolor* on the oxygen distribution in sediments. Root oxygenation and bioturbation by the worm both reduce the relative importance of sulphate reduction as a mineralisation pathway, although they hardly influence the total sediment metabolism. The plants appear to be superior competitors to the worms in mesocosms where both were incubated.

De Wit et al. (2001), in an overview of the ROBUST project, discuss the link between Fe, Ca, sulphide and phosphorus dynamics in sediments. They propose the existence of different 'buffers' in the sediment system, which can gradually be filled. As an example, Fe can react with free sulphide and thereby detoxify the sulphide. Fe and Ca bound complexes of phosphorus sequester phosphorus in the sediment, preventing its release and re-use for further primary production. Higher plants buffer against sulphide and P release by oxygenating the sediment. Sulphide-oxidising bacteria (in the presence of oxygen) also remove sulphide. When these different buffer systems overflow, mechanisms slowing down the rates of nutrient turnover are shunted, and the entire ecosystem will change structure and come into a mode of faster production, mineralisation and nutrient turnover. Several pieces of evidence for this hypothesis have been brought forward, but a further elaboration of this model of alternative stable states, as well as a direct experimental test, would be very worthwhile.

Within the EROS-2000 and EROS-21 projects, sediment biogeochemistry was studied as an important part of the whole ecosystem response of the north-western Black Sea shelf to variations in nutrient and organic input from the Danube and other major rivers. Friedl et al. (1998 – EROS) present results of benthic lander incubations. Sediment-water fluxes of oxygen, ammonium, silicate, orthophosphate, iron, manganese and sulphide were simultaneously measured at sites along an onshore-offshore transect. In general, a decrease of all benthic fluxes with distance from the coast was observed. Benthic regeneration of phosphate and silicate was very important. It contributed fluxes to the shelf system of the same order of magnitude as the Danube river fluxes. However, the N:P ratio of benthic fluxes was drastically different from that in the Danube outflow. Strong benthic denitrification led to an N-deficient outflux. Friedrich et al. (2002 – EROS) present similar results, but add seasonal dynamics. They show that oxygen depletion in summer leads to enhanced iron and manganese outfluxes from the sediments. Wijsman et al. (2001 – EROS) show that these iron effluxes from the shelf are sufficient to explain the trapping and deposition of iron in the anoxic basin of the Black Sea. Wijsman et al. (1999) discuss the relation between sediment biogeochemical processes and the structure and function of the macrobenthic animal community on the Black Sea shelf. For the shelf sediments, Wijsman et al. (2002 – EROS) provide a coupled diagenetic model. They predict from model runs that there are critical organic loading levels of sediments, where the sediment chemistry suddenly switches from oxic mineralistion to iron/manganese dominated mineralisation, and from these to sulphate reduction dominated mineralisation. The responses are highly non-linear due to the dynamics of re-oxidation of the reduced reaction products, which decrease redox potential and push the system further into the more reduced state.

A very generic non-linear coupled diagenetic model was developed within the ECOFLAT project (Meysman et al. 2003a,b). Application of this modelling tool to some of the datasets produced within this and other ELOISE projects would be a most meaningful exercise. Soetaert et al. (2000, 2001 – METROMED) discuss the general problem of coupling sediment diagenetic models to ecological models for the water column. As the full dynamic calculation of diagenesis for every cell of the water model is too costly, they propose a number of alternative efficient schemes that allow for a very reasonable representation of benthic processes in coupled ecosystem models at a relatively low computational cost.

Metal biogeochemistry – the extreme case of the Tinto/Odiel rivers

The Tinto/Odiel rivers and estuary in southern Spain, the case study of the TOROS project, has very extreme biogeochemical characteristics. As such it provided an excellent study area to investigate metal biogeochemistry (e.g. Cossa et al. 2001). These rivers, with a pH < 3, drain the largest sulphide mineralisation of the world, and have been subject to mining for over 4500 years (Elbaz-Polichet et al. 2001a). In addition, the estuary receives drainage from a phosphogypsum deposit, and during the study period there was an accidental mine tailings spill (Achterberg et al. 1999 – TOROS). The follow-up of this accident showed that a combination of meteorological conditions, human removal measures and estuarine processes reduced the short-term impact of the disaster considerably. On the long term, however, the estuary is a constant sources of metals to the Mediterranean Sea (Elbaz-Polichet et al. 2001b). The transport and fate of metals flowing out of the estuary has been successfully modelled using a coupled 3D model.

Release of biogases

Carbon dioxide

The project BIOGEST concentrated on the role of coastal and estuarine ecosystems in the global cycles of carbon dioxide (CO_2) and other greenhouse gases. In an extensive review, Gattuso et al. (1998) discussed the autotrophic / heterotrophic status of whole coastal ecosystems. Although considerable uncertainty remains due to lack of data and difficulties in upscaling from individual measurements to the scale of the ecosystem, they tentatively concluded that all coastal systems, with the notable exception of estuaries, are (slightly) autotrophic. Estuaries in general are sites of concentrated heterotrophic activity, fuelled by external organic inputs of terrestrial and riverine origin. Besides organic loading, also nutrient loading and light conditions (influenced by suspended sediment loading and dynamics) influence the outcome of the trophic balance. Cabeçadas et al. (1999 – BIOGEST) summarise the role of nutrients, light and biogeochemical transformations in the European estuaries Scheldt, Sado and Gironde. These relations were formalised in a coupled biogeochemical model for the Scheldt (Vanderborght et al. 2002 – BIOGEST). Frankignoulle et al. (1998 – BIOGEST) use their estimates of the CO_2 balance in the major European estuaries, to extrapolate to the whole estuarine surface in Europe. They conclude that the CO_2 emission from estuarine waters corresponds to 12 % of the anthropogenic CO_2 emission in Europe, and is therefore highly significant for the regional CO_2 budget. Part of the CO_2 emitted from estuarine waters is advected from the river, and released in the estuary as a consequence of pH changes in the water. However, this percentage was relatively low (around 10 % of total emission) for the Scheldt estuary (Abril et al. 2000b – BIOGEST). The majority of the enhanced CO_2 emission is locally produced from heterotrophic transformations of advected organic matter. Stable isotope ratios of suspended and dissolved organic matter in estuaries show clear indications of the important transformations taking place within the estuary (Middelburg and Nieuwenhuize, 1998).

Data for the Southern Bight of the North Sea (Borges and Frankignoulle, 1999, 2002 – BIOGEST) show a gradual transition from a heterotrophic plume close to the estuarine mouth, to an autotrophic zone more offshore. This is demonstrated in Fig. 5, which shows the transition from oversaturation to undersaturation in CO_2 around the mouth of the Scheldt estuary. Offshore, it can be assumed that nutrients brought into the coastal waters enhance productivity, whereas water transparency increases beyond the zone of settlement of fine particles. A similar pattern was described in relation to functioning of benthic communities off the Danube mouth in the Black Sea by Wijsman et al. (1999 – EROS). It is a matter of intense research at this moment if, and how, these patterns extrapolate to the whole of the North Sea and other coastal seas.

Benthic-pelagic exchange, as well as functioning of the pelagic ecosystem, are also strongly influenced by sedimentation and resuspension of particulate matter. Lemaire et al. (2002 – BIOGEST), in developing a typology of phytoplankton communities in European estuaries, revealed that suspended particulate matter content is one of the prime factors determining community composition. When SPM is sufficiently low to allow for primary production, estuarine residence time of the water determines whether genuine estuarine communities can develop. At sufficiently low SPM and sufficiently high residence times, nutrients come into play. In the other cases, the estuaries will be highly heterotrophic and export their nutrients to the coastal sea.

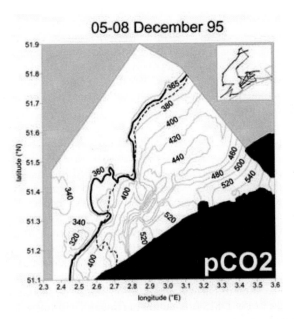

Fig. 5. Isolines of pCO_2 in the plume of the Scheldt estuary (Southern Bight of the North Sea), demonstrating the transition from over- to undersaturation as one progresses from the estuary to the coastal sea. From Borges and Frankignoulle (2002).

Apart from extending to more offshore waters, a major challenge in the quantification of the autotrophy/heterotrophy balance of estuaries and coastal systems is in downscaling to smaller spatial scales than the entire estuary. In particular, the relative importance of different subsystems, such as the sediment communities of tidal flats, shallow subtidal vegetated areas and deep gullies, and the pelagic system in different salinity zones or vertical strata of estuaries remains an open question, which is being studied in the ongoing project EUROTROPH. Relatively many data have been collected for different intertidal or shallow subtidal areas in European estuaries, but extensive GIS databases are needed in order to extrapolate measurements to the whole ecosystem. Middelburg et al. (2003) summarize the available data on benthic respiration.

Methane

Methane (CH_4) emissions from estuaries to the atmosphere have also been estimated as part of the BIOGEST project. The order of magnitude of the emissions showed that estuarine emissions are relevant with respect to total marine emissions, but not with respect to global emissions that are largely dominated by terrestrial and freshwater wetland systems.

Apart from their importance in global cycles, the relevance of measurements of estuarine methane concentration is that they point to poorly described processes that are potentially important for estuarine biogeochemical functioning. Middelburg et al. (2002 – BIOGEST) report possible influences of processes on intertidal flats, of groundwater release into rivers and estuaries, and of tidal pressure differences influencing ebullition of methane. More direct investigations are clearly needed to quantify fluxes associated with these phenomena.

Riparian vegetation has a considerable influence on methane oxidation in the oxidised rhizosphere. The effect differs between seasons and also between species (Van der Nat and Middelburg, 1998a, 1998b). The net overall effect of marsh vegetation on methane emission is positive, i.e. fluxes are enhanced (van der Nat and Middelburg, 2000).

Dimethylsulfide

The physiological and biochemical processes involved in dimethylsulfide (DMS) and DMSP production, as well as the food web interactions leading to DMS release in seawater, were the subject of the ESCAPE project. Stefels (2000 – ESCAPE) reviews the biochemical regulation of DMSP formation, as well as the influence of environmental factors on the process. She proposes as a basic model that DMSP production forms part of an overflow mechanism, where phytoplankton have to channel excess carbon fixed when growth is difficult due to nutrient limitation. Archer et al. (2000, 2001, 2003 – ESCAPE), showed directly in natural waters that the ingestion of particulate DMSP by microzooplankton could account for the measured rates of DMS production in seawater. However, also non-grazing mortality processes, such as viral lysis, enhance DMS concentrations in the water (Malin et al. 1998). Simo et al. (1998a, 1998b, 2000) discuss the biological production and consumption of DMSO.

Jonkers et al. (2000 – ROBUST) measured DMS production in sediments with and without *Zostera*. They showed that light and oxygenation reduced production rates, whereas increased organic loading, darkness and anoxia increased the rate. *Zostera* had a net decreasing influence on DMS production. Welsh (2000 – ROBUST) reviewed the role of DMSP as one of the organic metabolites that can be used to osmoregulate, and serve at the same time other roles (e.g. anti-predation).

Nitrous oxide

Nitrous oxide (N_2O) is released as a by-product of several transformations in the nitrogen cycle, including nitrification, denitrification and dissimilatory nitrate reduction. Its release is thus closely linked to the intensity of these processes. It is, however, also under environmental control. In the Scheldt, de Bie et al. (2002 – BIOGEST) demonstrate that in particular nitrification under suboxic conditions leads to enhanced N_2O production. Such conditions can easily be met around the Maximum Turbidity Zone in estuaries.

Marty et al. (2001 – METROMED) discuss N_2O and CH_4 production in two oceanic shelves, the Gulf of Lions and Thermaikos Gulf and Abril et al. (2000 – BIOGEST) report N_2O production in fluidised muds of the Gironde estuary.

Structure and function of ecosystems under anthropogenic pressure

Nutrient loading and the response in coastal pelagic communities

Over the past decades, the scientific view on the pelagic food web has changed from a linear food chain model to a food web model in which the microbial 'loop' plays an essential role. This paradigm change has important consequences for modelling the effects of eutrophication on coastal ecosystems. In the COMWEB project pelagic food web changes upon nutrient enrichment were analysed along two dimensions. Spatially, communities from the Mediterranean Sea, the North Sea, the NE Atlantic and the Baltic were studied. Temporally, short-term responses in mesocosm experiments, seasonal-scale responses in an experimentally enriched lagoon, and long-term responses in eutrophicated coastal communities in the North Sea were compared. Formal food web analysis was applied to all study systems, using inverse modelling. Short-term responses in the microbial food web in all communities were small. It was concluded that the microbial food web is more or less in steady state between production and consumption in these communities (a reasonable assumption since the incubations were started with summer communities), and that it adjusts its internal equilibrium very fast. Nearly linear responses were detected in the larger components, both in biomass and primary production (Fig. 6).

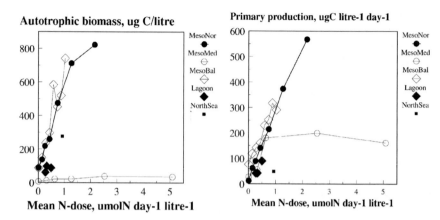

Fig. 6. Responses of coastal ecosystems on nutrient additions. Biomass of autotrophs (left) and primary production (right) expressed as a function of N dose in mesocosm experiments, an experimentally enriched lagoon, and the Southern North Sea. Results of the COMWEB project. From Olsen et al. 2001

The experimental time scale was too short for the grazers to adapt to the changed nutrient inputs. One notable exception was in the Mediterranean Sea mesocosm experiments, where production initially rose with nutrient additions, but then levelled off quickly, while biomass did not respond at all. Increased cell lysis and production of DOC (Agusti et al. 1998) explains this non-response to nutrients. It was suggested that nutrient ratios may alter the response, with N enriched inputs (relative to other nutrients) capable of provoking a biomass increase (Olsen et al. 2001). This result is unexpected in P-limited systems and not well explained. Medium-term responses in the lagoon were qualitatively similar, but lower in magnitude, since larger grazers had been able to adapt to the new situation. It is suggested that adjustment of the benthic compartment could further reduce the response at longer scales.

Long-term responses in the North Sea showed consistent changes in biomass and production with the medium-term responses of the lagoon. However, closer examination of the North Sea coastal system (Gasparini et al. 2000 – COMWEB; Rousseau et al. 2000 – COMWEB) shows that the major response to nutrient additions in disequilibrium (much higher N additions than P and Si) is translated into a bloom of *Phaeocystis globosa*, a species that is not grazed by copepods and actually inhibits copepod grazing on diatoms. *Phaeocystis* production is mainly processed by the microbial foodweb, and transfer of this production (via microzooplankton) to mesozooplankton is particularly poor: only 1.6 %, compared to 34 % transfer efficiency from diatom production to mesozooplankton grazing. A scheme for the foodweb structure and flows in spring is given in Fig. 7 (from Rousseau et al. 2000).

Mesocosm experiments by Havskum et al. (2003 – NTAP) demonstrate that increased primary production by diatoms follows upon nutrient addition in the presence of silicate. This increase in primary production occurs independent of the addition of glucose to the mesocosms. In silicate-deplete mesocosms, however, glucose addition results in bacterioplankton that can successfully compete with (non-diatom) phytoplankton for mineral nutrients. The importance of bacteria-phytoplankton competition for nutrients, and the success of large, relatively un-grazable phytoplankton in this competition, provide a partial mechanistic under-standing of whole-community enrichment experiments.

The EULIT project investigated the response of hard substrate littoral commu-nities to nutrient enrichment. Between the treatments, little or no change in bio-mass and primary production were found (Bokn et al. 2001; Kersting and Lind-blad, 2001). Barrón et al. (2003) showed that the community is highly autotrophic, due to a high DOC and POC export. They suggest that this high export prevents the community from showing eutrophication symptoms. Alternatively, the harsh physical environment and space competition could limit primary production, with nutrients being only of secondary importance. In this respect, these communities could be comparable to light-limited communities in turbid estuaries.

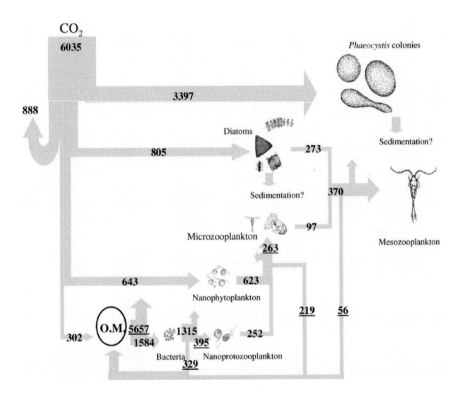

Fig. 7. Carbon budget for a station in the Southern Bight of the North Sea, for the period 26 February – 6 June 1998. Flows expressed as $mgC.m^{-3}.period^{-1}$. Underlined figures are not directly measured but estimated. O.M. represents pool of organic matter. From Rousseau et al. (2000).

An integrated study of biogeochemical processes in the Black Sea, concentrating on the gradient from the Danube mouth to the outer shelf (EROS project) revealed the complex ecosystem reactions to the riverine loading. Ragueneau et al. (2002 – EROS) documented an efficient biotic and abiotic removal of phosphorus from the inflowing river water, strong benthic denitrification, high benthic dissolved silicate regeneration, mortality of freshwater phytoplankton, intense nitrification as the main processes resulting in strong changes of nutrient ratios along a spatially limited gradient of salinity in the river plume.

Phosphorus deficiency has unexpected results, in that it seems to be limiting heterotrophic bacterial activity, leading to a seasonal accumulation of dissolved organic carbon (Becquevort et al. 2002 – EROS – see also Saliot et al. 2002 – EROS). Results from these measurements were used in a coupled ecosystem model by Lancelot et al. (2002 – EROS). The model was used to hindcast the changes in the Black Sea shelf ecosystem over the past decades. It demonstrates in a dramatic way the importance of nutrient ratios in the input waters for the functioning and structure of the ecosystem. A peak in the eutrophication of the Black Sea shelf was reached in 1991, when nutrient inputs were high and well equilibrated. In accordance with the COMWEB experimental results (see above) this well-equilibrated eutrophication resulted in an enhancement of the linear food chain. Most of this production in the Black Sea at the time went to the gelatinous zooplankton, where it resulted in the strong bloom of the introduced species *Mnemiopsis*. It is probable that strong overfishing has prevented this production from going to fish (Gucu, 2002 – EROS). Less balanced nutrient ratios in the eutrophication prevailed in the 1980's and the late 1990's. N or P deficiency in the nutrient input favoured the microbial food web, with primary production going to the microzooplankton and only a small fraction transferred to higher trophic levels.

The model has high biological resolution (about 30 state variables) at the expense of spatial resolution. However, a clever coupling with a detailed 3-D hydrodynamic model (Beckers et al. 2002 – EROS) provided a sound physical basis for its application. The high-resolution hydrodynamical model, coupled to a simple biological model, was able to describe the importance of mesoscale phenomena (frontal structures, coastal exchange) in the distribution of chlorophyll over the entire shelf. This proved very useful input in explaining patterns in macrobenthos (Wijsman et al. 1999).

The benthic food web

In contrast to the pelagic microbial food web that has been intensively studied for the last decades relatively little is known about the microbial food web in sediments. Numerous studies, also in the ELOISE context, have addressed the rates and regulations of biogeochemical processes, such as denitrification, sulphate reduction etc. (see above). The fate of the (bacterial) biomass produced during these processes, as well as its relative importance as food for benthic heterotrophic eukaryotes, in comparison with detrital organic matter deposited onto the sediment, remains largely unresolved. Also for macrobenthos, for which extensive autecology studies are available, it remains unclear which fraction of the total organic matter in sediments can be considered as food resources. At the scale of entire estuaries, a literature review has revealed a tight correlation between macrobenthic biomass and primary production in the system (Herman et al. 1999 – PHASE/ECOFLAT). This relation suggests a bottom-up control on macrobenthos and has consequences for possible effect of eutrophication – in particular it also predicts a decrease of (harvestable) benthic populations upon eutrophication abatement. However, not more than a fourth to a third of the organic matter sedimenting seems to be of any use to macrobenthos. It is unsure what determines the magnitude of this fraction, although one of the influences may be the macrobenthos' own bioturbation activity. Kristensen and Holmer (2001 – ISLED) described markedly higher decomposition rates of organic matter as a consequence of sediment oxygenation due to bioturbation. However, conflicting results were obtained by Dauwe et al. (2001 – ECOFLAT).

The ECOFLAT project has devoted considerable attention to the structure and functioning of the (microbial) food web in the sediments of an intertidal flat. Feeding relations between bacteria, microphytobenthos, heterotrophic nanoflagellates, ciliates, nematodes and macrofauna have been elucidated, using a combination of field observations, lab experiments and field experiments (Moens et al. 1999a, 1999b, 1999c, 2000, 2002; Hamels et al. 1998, 2001a, 2001b; Middelburg et al. 2000; Herman et al. 2000, 2001). Turnover of microphytobenthos was dependent on sediment granulometry, being much faster at a sandy than at a muddy site (Middelburg et al. 2000). This was related to more intense grazing by microfauna and meiofauna at the sandy site (Hamels et al. 1998). Nematodes grazed directly on microphytobenthos (Moens et al. 2002), but are also important predators on ciliates that graze on the benthic algae. Both meiofauna (Moens et al. 2002) and macrofauna (Herman et al. 2000) very selectively ingest POC derived from microphytobenthos. Their natural stable isotope ratio is very near to that of the algae, and very different to the ratio of the bulk POC in the sediment (Fig. 8).

The crucial role of microphytobenthos for the benthic food web on intertidal flats and shallow (euphotic) subtidal sediments complements the results on their important role in the N budget of sediments. Microphytobenthos appears to be one of the key elements in the material and energy flow in these systems, and an important determinant of macrobenthic life. Besides this role, it is also a strong structuring factor, by its effect on stabilisation of the sediment (see below).

The benthic food web studies jointly indicate that organic matter quality is very important, and that there is a large difference, for benthic animals, between 'organic matter' and 'food'. Much work remains to be done to further characterise organic matter quality and estimate the feeding conditions for benthic animals better (e.g. Dauwe et al. 1998 – ECOFLAT). This fundamental work is a prerequisite for a better understanding of organic matter cycling in sediments upon organic or nutrient enrichment. The general problem is how to relate spatial distributions of benthic populations that can be described well with statistical models (e.g. Ysebaert et al. 2002; Ysebaert and Herman, 2002 – ECOFLAT; Thorin et al. 2001 – EUROSAM) to causal mechanism relating benthos to the general ecosystem functioning.

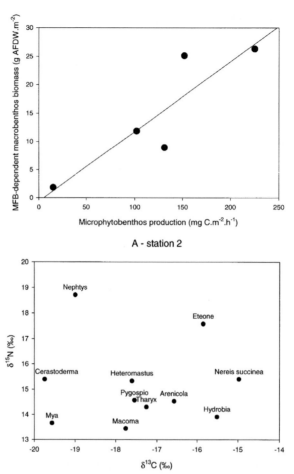

Fig. 8. Illustration of the role of microphytobenthos as food for intertidal macrobenthos. (A) stable isotope signatures of most macrobenthic species are in between those of phytoplankton ($\delta^{13}C \sim -20\ ‰$) and microphytobenthos ($\delta^{13}C \sim -15\ ‰$) but much higher than those of bulk POC in the sediment ($\delta^{13}C \sim -23\ ‰$). (B) The relation between the biomass of microphytobenthos-dependent macrobenthos and production by the benthic alga. Both figures from Herman et al. (2000)

Josefson et al. (2002 – KEYCOP) and Josefson and Hansen (2003 – KEYCOP) described vertical deposition of phytoplankton from a benthic point of view. They found that diatoms (a high proportion of which were viable cells) were a major component of the flux. They described a remarkably low use of this fresh material by the benthos in mesocosms. Reigstad et al. (2000 – ESCAPE) studied input of POC to the benthos in Norwegian fjords as a function of copepod grazing in the pelagic. Their study shows that a match (and occasional mismatch) between the timing of copepod advection into the fjord and blooming of the phytoplankton determines whether the POC will sink mainly as phytoplankton cells or as faecal pellets.

The role of physical forcing

Physical processes and ecosystem functioning interact at a multitude of scales, ranging from turbulence effects below the Kolmogorov length scale, to transport at the scale of ocean basins.

The NTAP project studies the effect of turbulence on feeding interactions in the microbial food web. Turbulence levels affect many vital rates of microorganisms (review by Peters and Marrassé, 2000). It also affects interspecific interactions, e.g. by increasing the particle size range grazed by microheterotrophs, thus reducing grazing on bacteria (Peters et al. 2002 – NTAP). In nutrient-enriched mesocosms, turbulence increased the relative importance of phytoplankton to bacteria, the phytoplankton species composition, and the stoichiometry of the particulate organic matter (Arin et al. 2002; Maar et al. 2002 – NTAP). Thus it can be expected that outcomes from nutrient enrichment will be different in coastal systems differing in their turbulence intensity, a feature that is related to tidal regime, wind stress and general hydrography.

Turbulence of the water column has a profound influence on vertical mixing, and thus on the benthic-pelagic coupling. The relative role of benthic suspension feeders as a grazing control of phytoplankton development critically depends on vertical mixing rates (review in Herman et al. 1999 – PHASE / ECOFLAT). Grazing by benthic suspension feeders can be very important, as these animals have the capacity to filter large volumes of water per unit of surface and time. Experiments within PHASE also demonstrated that benthic beds of filter feeding mussels can themselves enhance turbulent mixing of the water column, and therefore increase the fluxes of food towards the bed (Herman et al. 1999). The conditions under which benthic filter feeders can effectively act as eutrophication controls remains an important topic of study in predicting the response of diverse coastal systems to nutrient inputs. The effect of vertical mixing on the benthic-pelagic exchange cannot be uncoupled from its effects on the dynamics of phytoplankton. In deeper systems, phytoplankton blooms depend on stratification and reduced mixing length for the onset of the bloom, which effectively cuts them off from benthic grazing. In shallower systems, this coupling/uncoupling will be very different between tidally well-mixed systems and (partially) stratified, occasionally wind-mixed systems. Modelling these processes is the subject of the ongoing MABENE project.

Estuarine suspended particulate matter is, at least in part, under biological control (Herman et al. 2001 – ECOFLAT). Measurements of sediment erodability, among others in the ECOFLAT project, have shown that the development of benthic algal mats can greatly enhance sediment stability (Widdows et al. 2000). Grazing by macrobenthos, on the other hand, reduces sediment stability. Lucas and Holligan (1999) and Lucas et al. (2000) showed that these effects could be found back in the exchange of algal material between the bed and the water column. Van de Koppel et al. (2001) investigated the relation between sediment silt content, algal development and bottom shear stress. Their model suggests that alternative stable states occur in sediments, which can either be in the 'algae-silt-stable' state, or in the 'sand-no algae-dynamic' state, with strong positive feedbacks maintaining systems in one or the other state at similar external forcing.

Within the METROMED project, Redondo et al. (2001), and Thill et al. (2001) have also contributed to a better understanding of the dynamics of particles in estuaries, and the interactions with the bed. Karageorgis and Anagnostou (2001 – METROMED) investigated the effects of larger-scale horizontal advection and vertical sinking on the dynamics of particulate matter in shelf areas and the coastal ocean.

Basin-scale modelling

At a basin scale, coupling of 3-d hydrodynamic models with (simple) biogeochemical models has successfully been applied to the Black Sea within EROS (Stanev and Beckers, 1999; Stanev et al. 1999; Beckers et al. 2002). They demonstrated that a high spatial resolution in the model was needed to reproduce observed distributions of chlorophyll (Barale et al. 2002). A similar model type was used to estimate the transport of metals from the Tinto/Odiel system to the Mediterranean sea (Elbaz-Polichet et al. 2001a). Coupled modelling is also an important aspect in the ongoing OAERRE project.

Tusseau-Vuillemin et al. (1998 – METROMED) used a coupled 3D model for the Gulf of Lions. The model was calibrated using a 1-D vertical version. Model runs demonstrated that the shelf acted mostly as a sink for nitrate, except in winter when nitrate was exported to the open sea. This conclusion is in accordance with the hypothesis that shelf seas are usually autotrophic (see above).

Conclusions

In this section we follow the major objectives of the ELOISE programme, and discuss in how far the published ELOISE results fulfil the expectations.

A shifting view on coastal ecosystem processes

Human impact on coastal ecosystems through eutrophication and physical impacts is complex because any impact is translated into many non-linear ecological interactions. ELOISE research has contributed significantly to a better understanding of these relations, as highlighted above. It has become increasingly clear that a proper conceptual model for eutrophication should consider such aspects as nutrient sequestration in biomass and the turnover time of this biomass, competition for inorganic nutrients between bacteria and phytoplankton, the role of the microbial foodweb and the factors favouring the channelling of nutrients into this very inefficient food web, the high importance of nutrient ratios for ecosystem response and the complex riverine, estuarine and coastal processes affecting these ratios, the importance of physical processes in shaping the response of the pelagic system and in determining benthic-pelagic coupling, the complex and highly selective utilisation of organic input into benthic systems and the large role played by microphytobenthos in shallow benthic systems. Moreover, strong indications have been collected that coastal systems can switch states, e.g. from eelgrass-dominated to macroalgae-dominated, when critical thresholds ('buffer capacities') are exceeded, but that return to the original state can show strong hysteresis. These shifts in our views on ecosystem functioning and reaction to eutrophication stress or physical modifications, demonstrate that continued fundamental ecological studies are needed, because puzzling observations requiring paradigm shifts are still being collected. The many high-quality contributions from ELOISE projects to this research also show the success of the ELOISE approach in achieving the first of the ELOISE scientific objectives. Through ELOISE projects, European scientists have taken a leading role in studies of N cycling, microphytobenthos and benthic food web studies, micro- and mesocosm studies and certain areas of modelling.

The human impact

Over the past 10-15 years, the human impact on estuarine and coastal ecosystems has changed. We have seen a major change in nutrient and pollutant input from former Eastern European countries after 1990. The ecological responses to this decrease were often unexpected (e.g. very slow reduction in nitrogen input into rivers, despite sudden reduction of agricultural input; changes in nutrient ratios, rather than nitrogen or phosphorus levels, affecting the Black Sea most). During the 1990's, phosphorus reduction programmes in many European countries have significantly changed the N/P ratio of anthropogenic nutrient input. We are only beginning to realise the importance of these changes, and are not yet able to fully appraise or model their effects. Unexpected consequences of these changes may be anticipated. Yet a number of conclusions from existing studies can be drawn:

The effect of (reduction of) eutrophication will be different in physically different coastal water bodies. In this physical typology, suspended sediment load (related to tidal currents and sediment input from rivers), vertical mixing intensity and frequency, level of turbulence, history of eutrophication and sediment load, and existing structure of the ecosystem will all be important. Sufficient data should be available to achieve such a typology for all European coastal systems, but this will require a major database building efforts.

The effect of eutrophication will be highly dependent on the ratio between phosphorus, nitrogen, silicate and organic loading of the systems. Load reduction measures that respect nutrient ratios in such a way that the 'linear' food chain and long-living plants and animals are favoured, should receive strong emphasis. Models need to be further developed for this type of prediction.

Along the 'water continuum' from headlands to the coastal sea, important biogeochemical modifications in the nature and magnitude of nutrient and organic load take place. Integrated management should take all these modifications into account. Models for the whole continuum should be conceptually consistent and have sound interfaces. Good progress has been made in this development, but further development is called for.

Integration of socio-economic and natural sciences

In the finished ELOISE projects that have published their results, very little integration between socio-economic and natural sciences was to be found. This was not included in the projects' objectives. However, at least some ongoing projects pursue this subject in greater depth.

There are, nevertheless, great opportunities for analysing some of the existing projects *a posteriori* from a socio-economic science point of view. As nutrient reduction scenarios have a high societal cost and a variable expected outcome from an ecological point of view, several scenarios could be constructed using recent evidence from ELOISE projects. Such a study that builds on existing natural science results would have great potential, especially if it can dynamically incorporate new knowledge.

The importance of physical-ecological interactions highlighted in ELOISE research also provides excellent opportunities for coupled research. Many human operations (e.g. dredging, trawling, dam construction, land reclamation) directly affect the physical boundary conditions of coastal systems. Other human effects, including eutrophication, are modulated through the physical-biological interactions, and could open possibilities for effect reduction when combined with physical measures (e.g. enhancing benthic-pelagic exchange).

At a European to global scale, the evaluation of the contribution from European coastal systems to greenhouse gas production, as well as the thorough study of the factors determining this contribution, could form an essential element in a socio-economic / natural science evaluation of reduction scenarios.

European scientific infrastructure

It is very apparent from the published results of ELOISE projects that they have significantly contributed to scientific methodology. Both in field methods, laboratory analyses and modelling new high-level developments have been fostered by ELOISE research, and some of these developments have been highlighted in this review.

These developments have not lead, however, to a common scientific infrastructure as put forward in the science plan. Models for different parts of the 'water continuum', are usually 'tailor-made' for specific purposes, and cannot easily communicate with one another. This is in itself not abnormal or undesirable for research models, but there is now a need for translation of these research models into operational, management-oriented models.

Also, upscaling from individual study systems to the European scale calls for more and better mechanistic models. Although empirical models are often better at reproducing individual data sets, the need for knowledge at a regional scale calls for models that can be applied to a variety of systems without need for recalibration.

ELOISE, with its structure of isolated projects, has not lead to a comprehensive database describing the essential features of (most) European coastal systems. From ELOISE research it can be deduced what the minimal set of variables in such a description should be. The effort of collecting these data would seriously enhance the possibilities of upscaling of scientific results to the European level. It would also foster development and application of models that could then easily be tested on a multitude of systems.

As a general conclusion therefore, ELOISE projects until now have published a wealth of high-quality science, have contributed to significant shifts in view on coastal ecosystem processes and have made large progress in formalising this knowledge into formal models. After all this effort, plus the effort that will be forthcoming from running projects, it is time to seriously invest in the exploitation and use of this knowledge.

Acknowledgements

We thank the colleagues who discussed with us the relevance of ELOISE scientific results during two dedicated workshops in May-June 2003. Special thanks are due to Jack J. Middelburg for many constructive comments. This work and the workshops were financially supported by the European Commission. A full list of ELOISE-related publications can be found at the ELOISE website http://www.nilu.no/projects/eloise/.

References

Abril G, Riou SA, Etcheber H, Frankignoulle M, de Wit R, Middelburg JJ (2000a) Transient, tidal time-scale, nitrogen transformations in an estuarine turbidity maximum-fluid mud system (The Gironde, south-west France). Estuar Coast Shelf Sci 50:703-715

Abril G, Etcheber H, Borges AV, Frankignoulle M (2000b) Excess atmospheric carbon dioxide transported by rivers into the Scheldt estuary. Comptes rendus de l' Académie des sciences série ii fascicule a- sciences de la terre et des planètes 330:761-768

Abril G, Frankignoulle M (2001) Nitrogen-alkalinity interactions in the highly polluted Scheldt basin (Belgium). Water Res 35:844-850

Achterberg EP, Braungardt C, Morley NH, Elbaz-Poulichet F, Leblanc M (1999) Impact of Los Frailes mine spill on riverine, estuarine and coastal waters in southern Spain. Water Res 33:3387-3394

Agusti S, Satta MP, Mura MP, Benavent E (1998) Dissolved esterase activity as a tracer of phytoplankton lysis: Evidence of high phytoplankton lysis rates in the northwestern Mediterranean. Limnol Oceanogr 43:1836-1849

Archer SD, Stelfox-Widdicombe CE, Burkill PH, Malin G (2001) A dilution approach to quantify the production of dissolved dimethylsulphoniopropionate and dimethyl sulphide due to microzooplankton herbivory. Aquat Microb Ecol 23:131-145

Archer SD, Verity PG, Stefels J (2000) Impact of microzooplankton on the progression and fate of the spring bloom in fjords of northern Norway. Aquat Microb Ecol 22:27-41

Archer SD, Stelfox-Widdicombe CE, Malin G, and Burkhill PH (2003) Is dimethyl sulphide production related to microzooplankton herbivory in the southern North Sea? J Plankton Res 25:235-242

Arin L, Marrase C, Maar M, Peters F, Sala MM, Alcaraz M (2002) Combined effects of nutrients and small-scale turbulence in a microcosm experiment. I. Dynamics and size distribution of osmotrophic plankton. Aqua Microb Ecol 29:51-61

Azzoni R, Giordani C, Bartoli M, Welsh DT, Viaroli P (2001) Iron, sulphur and phosphorus cycling in the rhizosphere sediments of a eutrophic *Ruppia cirrhosa* meadow (Valle Smarlacca, Italy). J Sea Res 45:15-26

Barale V, Cipollini P, Davidov A, Melin F (2002) Water constituents in the north-western Black Sea from optical remote sensing and in situ data. Estuar Coast Shelf Sci 54:309-320

Barranguet C, Herman PMJ, Sinke JJ (1997) Microphytobenthos biomass and community composition studied by pigment biomarkers:importance and fate in the carbon cycle of a tidal flat. J Sea Res 38:59-70

Barron C, Marba N, Duarte CM, Pedersen MF, Lindblad C, Kersting K, Moy F, Bokn T (2003) High organic carbon export precludes eutrophication responses in experimental rocky shore communities. Ecosystems 6:144-153

Beckers JM, Gregoire M, Nihoul JCJ, Stanev E, Staneva J, Lancelot C (2002) Modelling the Danube-influenced North-western continental shelf of the Black Sea. I:Hydrodynamical processes simulated by 3-D and box models. Estuar Coast Shelf Sci 54:453-472

Becquevort S, Bouvier T, Lancelot C, Cauwet G, Deliat G, Egorov VN, Popovichev VN (2002) The seasonal modulation of organic matter utilization by bacteria in the Danube-Black Sea mixing zone. Estuar Coast Shelf Sci 54:337-354

Billen G, Garnier J (1999) Nitrogen transfer through the Seine drainage network: a budget based on the application of the 'RIVERSTRAHLER' model. Hydrobiologia 410:139–150

Billen G, Garnier J, Ficht A, Cun C (2001) Modelling the response of water quality in the Seine river estuary to human activity in its watershed over the last 50 years. Estuaries 24:977-993

Blaabjerg V, Mouritsen KN, Finster K (1998) Diel cycles of sulphate reduction rates in sediments of a Zostera marina bed (Denmark). Aquat Microb Ecol 15:97-102

Bokn TL, Hoell EE, Kersting K, Moy FE, Sorensen K (2001) Methods applied in the large littoral mesocosms study of nutrient enrichment in rocky shore ecosystems – EULIT. Cont Shelf Res 21:1925-1936

Borges AV, Frankignoulle M (1999) Daily and seasonal variations of the partial pressure of CO2 in surface seawater along Belgian and southern Dutch coastal areas. J Mar Syst 19:251-266

Borges AV, Frankignoulle M (2002) Distribution of surface carbon dioxide and air-sea exchange in the upwelling system off the Galician coast. Global Biogeochem Cycles 16:1020

Boschker HTS, Wielemaker A, Schaub BEM, Holmer M (2000) Limited coupling of macrophyte production and bacterial carbon cycling in the sediments of *Zostera* spp meadows. Mar Ecol Prog Ser 203:181-189

Cabecadas G, Nogueira M, Brogueira MJ (1999) Nutrient dynamics and productivity in three European estuaries. Mar Pollut Bull 38:1092-1096

Cadée N, Dronkers J, Heip C, Martin J-M, Nolan C (eds.) (1994) Eloise. European Land-Ocean Interaction Studies. Science Plan. Report of an international workshop organized jointly by the Environment and Mast Programmes of DG XII of the European Commission and the Netherlands Institute of Ecology, Centre for Estuarine and Marine Ecology, Roosendaal, January 12 to 14, 1994.

Castro MS, Driscoll CT (2002) Atmospheric nitrogen deposition to estuaries in the Mid-Atlantic and Northeastern United States. Environ Sci Technol 36:3242-3249

Christensen PB, Rysgaard S, Sloth NP, Dalsgaard T, Schwaerter S (2000) Sediment mineralization, nutrient fluxes, denitrification and dissimilatory nitrate reduction to ammonium in an estuarine fjord with sea cage trout farms. Aquat Microb Ecol 21:73-84

Cossa D, Elbaz-Poulichet F, Nieto JM (2001) Mercury in the Tinto-odiel estuarine system (Gulf of Cadiz, Spain): sources and dispersion. Aquat Geochem 7:1-12

Dalsgaard T and Thamdrup B (2002) Factors controlling anaerobic ammonium oxidation with nitrite in marine sediments. Appl Environ Microbiol 68:3802-3808

Dauwe B, Herman PMJ, Heip CHR (1998) Community structure and bioturbation potential of macrofauna at four North Sea stations with contrasting food supply. Mar Ecol Prog Ser 173:67-83

Dauwe B, Middelburg JJ, Herman PMJ (2001) The effect of oxygen on the degradability of organic matter in subtidal and intertidal sediments of the North Sea area. Mar Ecol Prog Ser 215:13-22

de Bie MJM, Middelburg JJ, Starink M, Laanbroek HJ (2002) Factors controlling nitrous oxide at the microbial community and estuarine scale. Mar Ecol Prog Ser 240:1-9

de Leeuw G, Cohen L, Frohn LM, Geernaert G, Hertel O, Jensen B, Jickells T, Klein L, Kunz GJ, Lund S, Moerman M, Muller F, Pedersen B, von Salzen K, Schlunzen KH, Schulz M, Skjoth CA, Sorensen LL, Spokes L, Tamm S, Vignati E (2001) Atmospheric input of nitrogen into the North Sea: ANICE project overview. Cont Shelf Res 21:2073-2094

Dettmann EH (2001) Effect of water residence time on annual export and denitrification of nitrogen in estuaries:A model analysis. Estuaries 24:481-490

de Wit R, Stal LJ, Lomstein BA, Herbert RA, van Gemerden H, Viaroli P, Cecherelli VU, Rodriguez-Valera F, Bartoli M, Giordani G, Azzoni R, Schaub B, Welsh DT, Donnelly A, Cifuentes A, Anton J, Finster K, Nielsen LB, Pedersen AGU, Neubauer AT, Colangelo MA, Heijs SK (2001) ROBUST:The ROle of BUffering capacities in STabilising coastal lagoon ecosystems. Cont Shelf Res 21:2021-2041

Elbaz-Poulichet F, Braungardt C, Achterberg E, Morley N, Cossa D, Beckers JM, Nomerange P, Cruzado A, Leblanc M. (2001a) Metal biogeochemistry in the Tinto-Odiel rivers (Southern Spain) and in the Gulf of Cadiz:a synthesis of the results of TOROS project. Cont Shelf Res 21:1961-1973

Elbaz-Poulichet F, Morely NH, Beckers J-M., Nomerange P (2001b) Metal fluxes through the Straite of Gibralter:the influence of the Tinto and Odiel rivers (SW Spain). Mar Chem 73:193-213

Engstrom J, Koski M, Viitasalo M, Reinikainen M, Repka S, Sivonen K (2000) Feeding interactions of the copepods *Eurytemora affinis* and *Acartia bifilosa* with the cyanobacteria *Nodularia* sp. J Plankton Res 22:1403-1409

Flynn NJ, Paddison T, Whitehead PG (2002) INCA modelling of the Lee system:strategies for the reduction of nitrogen loads. Hydrol Earth Syst Sci 6:467-483

Forsman Å, Andersson C, Grimvall A, Hoffmann M (2003) Estimation of the impact of short-term fluctuations in inputs on temporally aggregated outputs of process-oriented models. J Hydroinf (accepted)

Forsman Å, Grimvall A (2003) Reduced models for efficient simulation of spatially integrated outputs of one-dimensional substance transport models. Environ Model Software 18:319-327

Forsman Å, Grimvall A, Scholtes J, Wittgren HB (2003) Generic structures of decision support systems for evaluation of policy measures to reduce catchment-scale nitrogen fluxes. Phys Chem Earth 28:589-598

Frankignoulle M, Abril G, Borges A, Bourge I, Canon C, DeLille B, Libert E, Theate JM (1998) Carbon dioxide emission from European estuaries. Science 282:434-436

Friedl G, Dinkel C, Wehrli B (1998) Benthic fluxes of nutrients in the northwestern Black Sea. Mar Chem 62:77-88

Friedrich J, Dinkel C, Friedl G, Pimenov N, Wijsman J, Gomoiu MT, Cociasu A, Popa L, Wehrli B (2002) Benthic nutrient cycling and diagenetic pathways in the north-western Black Sea. Estuar Coast Shelf Sci 54:369-383

Gacia E, Duarte CM, Middelburg JJ (2002) Carbon and nutrient deposition in a Mediterranean seagrass (Posidonia oceanica) meadow. Limnol Oceanogr 47:23-32

Gallon JR, Evans AM, Jones DA, Albertano P, Congestri R, Bergman B, Gundersen K, Orcutt KM (2002) Maximum rates of N-2 fixation and primary production are out of phase in a developing cyanobacterial bloom in the Baltic Sea. Limnol Oceanogr 47:1514-1521

Garnier J, Billen G, Hannon E, Fonbonne S, Videnina Y, Soulie M (2002) Modelling the transfer and retention of nutrients in the drainage network of the Danube River. Estuar Coast Shelf Sci 54:285-308

Gasparini S, Daro MH, Antajan E, Tackx M, Rousseau V, Parent JY, Lancelot C (2000) Mesozooplankton grazing during the *Phaeocystis globosa* bloom in the southern bight of the North Sea. J Sea Res 43:345-356

Gattuso JP, Frankignoulle M, Wollast R (1998) Carbon and carbonate metabolism in coastal aquatic ecosystems. Annu Rev Ecol Syst 29:405-434

Giordani G, Azzoni R, Bartoli M, Viaroli P (1997) Seasonal variations of sulphate reduction rates, sulphur pools and iron availability in the sediment of a dystrophic lagoon (Sacco di Goro, Italy). Water Air Soil Pollut 99:363-371

Gribsholt B, Kristensen E (2002) Effects of bioturbation and plant roots on salt marsh biogeochemistry: a mesocosm study. Mar Ecol Prog Ser 241:71-87

Grimvall A, Stalnacke P, Tonderski A (2000) Time scales of nutrient losses from land to sea - a European perspective. Ecol Eng 14:363-371

Gucu AC (2002) Can overfishing be responsible for the successful establishment of *Mnemiopsis leidyi* in the Black Sea? Estuar Coast Shelf Sci 54:439-451

Hamels I, Sabbe K, Muylaert K, Barranguet C, Lucas C, Herman P, Vyverman W (1998) Organisation of microbenthic communities in intertidal estuarine flats, a case study from the Molenplaat Westerschelde Estuary), the Netherlands. Eur J Protistol 34:308-320

Hamels I, Muylaert K, Casteleyn G,Vyverman W (2001) Uncoupling of bacterial production and flagellate grazing in aquatic sediments:a case study from an intertidal flat. Aquat Microb Ecol 25:31-42

Hamels I, Moens T, Muylaert K, Vyverman W (2001) Trophic interactions between ciliates and nematodes from an intertidal flat. Aquat Microb Ecol 26:61-72

Hansen JW, Pedersen AGU, Berntsen J, Ronbog IS, Hansen LS, Lomstein BA (2000) Photosynthesis, respiration, and nitrogen uptake by different compartments of a *Zostera marina* community. Aquat Bot 66:281-295

Hansen JW, Udy JW, Perry CJ, Dennison WC, Lomstein BA (2000) Effect of the seagrass Zostera capricorni on sediment microbial processes. Mar Ecol Prog Ser 199:83-96

Havskum H, Thingstad TF, Scharek R, Peters F, Berdalet E, Montserrat Sala M, Alcaraz M, Bangsholt JC, Li Zweifel U, Hagström Â, Perez M, Dolan JR (2003) Silicate and labile DOC interfere in structuring the microbial food web via algal–bacterial competition for mineral nutrients: results of a mesocosm experiment Limnol Oceanogr 48:129–140

Heijs SK, Azzoni R, Giordani G, Jonkers HM, Nizzoli D, Viaroli P, van Gemerden H (2000) Sulfide-induced release of phosphate from sediments of coastal lagoons and the possible relation to the disappearance of *Ruppia* sp. Aquat Microb Ecol 23:85-95

Herbert RA (1999) Nitrogen cycling in coastal marine ecosystems. FEMS Microbiol Rev 23:563-590

Herman PMJ, Middelburg JJ, Widdows J, Lucas CH, Heip CHR (2000) Stable isotopes as trophic tracers:combining field sampling and manipulative labelling of food resources for macrobenthos. Mar Ecol Prog Ser 204:79-92

Herman PMJ, Middelburg JJ, Heip CHR (2001) Benthic community structure and sediment processes on an intertidal flat:results from the ECOFLAT project. Cont Shelf Res 21:2055-2071

Herman PMJ, Middelburg JJ, Van de Koppel J, Heip C (1999) The ecology of estuarine macrobenthos. Adv Ecol Res 29:195-240

Hertel O, Skjoth CA, Frohn LM, Vignati E, Frydendall J, de Leeuw G, Schwarz U, Reis S (2002) Assessment of the atmospheric nitrogen and sulphur inputs into the North Sea using a Lagrangian model. Phys Chem Earth 27:1507-1515

Holmer M, Gribsholt B, Kristensen E (2002) Effects of sea level rise on growth of Spartina anglica and oxygen dynamics in rhizosphere and salt marsh sediments. Mar Ecol Prog Ser 225:197-204

Holmer M, Gribsholt B, Kristensen E (2002) Effects of sea level rise on growth of Spartina anglica and oxygen dynamics in rhizosphere and salt marsh sediments. Mar Ecol Prog Ser. 225:197-204

Humborg C, Ittekkot V, Cociasu A, Von Bodungen B (1997) Effect of Danube River dam on Black Sea biogeochemistry and ecosystem structure. Nature 386:385-388

Jansson BO, Stalvant CE (2001) The Baltic Basin Case Study - towards a sustainable Baltic Europe. Cont Shelf Res 21:1999-2019

Jansson BO (1997) The Baltic Sea:current and future status and impact of agriculture. Ambio 26:424-431

Jarvie H P, Wade AJ, Butterfield D, Whitehead PG, Tindall CI, Virtue WA, Dryburgh W, McGraw A (2002) Modelling nitrogen dynamics and distributions in the River Tweed, Scotland:an application of the INCA model. Hydrol Earth Syst Sci 6:433-453

Jickells TD (1998) Nutrient biogeochemistry of the coastal zone. Science 281:217-222

Jonkers HM, van Bergeijk SA, van Gemerden H (2000) Microbial production and consumption of dimethyl sulfide (DMS) in a seagrass (*Zostera noltii*)-dominated marine intertidal sediment ecosystem (Bassin d'Arcachon, France). FEMS Microb Ecol 31:163-172

Josefson AB, Forbes TL, Rosenberg R (2002) Fate of phytodetritus in marine sediments:functional importance of macrofaunal community. Mar Ecol Prog Ser 230:71-85

Josefson AB, Hansen JLS (2003) Quantifying plant pignmets and live diatoms in aphotic sediments of Scandinavian coastal waters confirms a major route in the pelagic-benthic coupling. Mar Biol 142:649-658

Kaleris V, Lagas G, Marczinek S, Piotrowski JA (2002) Modelling submarine groundwater discharge: an example from the western Baltic Sea. J Hydrol 265:76-99

Karageorgis AP, Anagnostou CL (2001) Particulate matter spatial-temporal distribution and associated surface sediment properties:Thermaikos Gulf and Sporades Basin, NW Aegean Sea. Cont Shelf Res 21:2141-2153

Kersting K, Lindblad C (2001) Nutrient loading and metabolism in hard-bottom littoral mesocosms. Cont Shelf Res 21:2117-2125

Kristensen E, Holmer M (2001) Decomposition of plant materials in marine sediment exposed to different electron accepters (O_2^-, NO_3^-, and SO_4^{2-}), with emphasis on substrate origin, degradation kinetics, and the role of bioturbation. Geochim Cosmochim Acta 65:419-433

Kunkel R, Wendland F (1997) WEKU - a GIS-supported stochastic model of groundwater residence times in upper aquifers for the supraregional groundwater management, Environ. Geol. 30:1-9

Lancelot C, Staneva J, Van Eeckhout D, Beckers JM, Stanev E (2002) Modelling the Danube-influenced north-western continental shelf of the Black Sea. II:Ecosystem response to changes in nutrient delivery by the Danube River after its damming in 1972. Estuar Coast Shelf Sci. 54:473-499

Lemaire E, Abril G, De Wit R, Etcheber H (2002) Distribution of phytoplankton pigments in nine European estuaries and implications for an estuarine typology. Biogeochemistry 59:5-23

Limbrick KJ, Whitehead PG, Butterfoeld D, Reynard N (2000) Assessing the potential impacts of various climate change scenarios on the hydrological regime of the River Kennet at Theale, Berkshire, south-centra England, UK:an application and evaluation of the new semi-distributed model, INCA. Sci Total Environ 251/252:539–556

Lucas CH, Holligan PM (1999) Nature and ecological implications of algal pigment diversity on the Molenplaat tidal flat (Westerschelde estuary, SW Netherlands). Mar Ecol Prog Ser 180:51-64

Lucas CH, Widdows J, Brinsley MD, Salkeld PN, Herman PMJ (2000) Benthic-pelagic exchange of microalgae at a tidal flat. 1. Pigment analysis. Mar Ecol Prog Ser 196:59-73

Maar M, Arin L, Simo R, Sala MM, Peters F, Marrase C (2002) Combined effects of nutrients and small-scale turbulence in a microcosm experiment. II. Dynamics of organic matter and phosphorus. Aquat Microb Ecol 29:63-72

Malin G, Wilson WH, Bratbak G, Liss PS, Mann NH (1998) Elevated production of dimethylsulfide resulting from viral infection of cultures of *Phaeocystis pouchetii*. Limnol Oceanogr 43:1389-1393

Marty D, Bonin P, Michotey V, Bianchi M (2001) Bacterial biogas production in coastal systems affected by freshwater inputs. Cont Shelf Res 21:2105-2115

Meysman FJR, Middelburg JJ, Herman PMJ, Heip CHR (2003a) Reactive transport in surface sediments. I. Model complexity and software quality. Comput Geosci 29:291-300

Meysman FJR, Middelburg JJ, Herman PMJ, Heip CHR (2003b) Reactive transport in surface sediments. II. Media:an object-oriented problem-solving environment for early diagenesis. Comput Geosci 29:301-318

Middelburg JJ, Nieuwenhuize J, Iversen N, Hogh N, De Wilde H, Helder W, Seifert R, Christof O (2002) Methane distribution in European tidal estuaries Biogeochemistry 59:95-119

Middelburg JJ, Nieuwenhuize J (2001) Nitrogen isotope tracing of dissolved inorganic nitrogen behaviour in tidal estuaries. Estuar Coast Shelf Sci 53:385-391

Middelburg JJ, Nieuwenhuize J (1998) Carbon and nitrogen stable isotopes in suspended matter and sediments from the Schelde Estuary. Mar Chem 60:217-225

Middelburg JJ, Nieuwenhuize J (2000) Nitrogen uptake by heterotrophic bacteria and phytoplankton in the nitrate rich Thames estuary. Mar Ecol Prog Ser 203:13-21

Middelburg JJ, Barranguet C, Boschker HTS, Herman PMJ, Moens T, Heip CHR (2000) The fate of intertidal microphytobenthos carbon: an in situ 13C labelling study. Limnol Oceanogr 45:1224-1234

Miles A, Sundbäck K (2000) Diel variation of microphytobenthic productivity in areas with different tidal amplitude. Mar Ecol Prog Ser 205:11-22

Moens T, Herman P, Verbeeck L, Steyaert M, Vincx M (2000) Predation rates and prey selectivity in two predacious estuarine nematode species. Mar Ecol Prog Ser 205:185-193

Moens T, Luyten C, Middelburg JJ, Herman PMJ, Vincx M (2002) Tracing organic matter sources of estuarine tidal flat nematodes with stable carbon isotopes. Mar Ecol Prog Ser 234:127-137

Moens T, Van Gansbeke D, Vincx M (1999a) Linking estuarine nematodes to their suspected food. A case study from the Westerschelde Estuary (south-west Netherlands). J Mar Biol Assoc UK 79:1017-1027

Moens T, Verbeeck L, Vincx M (1999b) Preservation and incubation time-induced bias in tracer-aided grazing studies on meiofauna. Mar Biol 133:69-77

Moens T, Verbeeck L, Vincx M (1999c) Feeding biology of a predatory and a facultatively predatory nematode (*Enoploides longispiculosus* and *Adoncholaimus fuscus*). Mar Biol 134:585-593

Moore WS (1996) Large groundwater inputs to coastal waters revealed by ^{226}Ra enrichment. Nature 380:612–614

Naldi M, Viaroli P (2002) Nitrate uptake and storage in the seaweed *Ulva rigida* C. Agardh in relation to nitrate availability and thallus nitrate content in a eutrophic coastal lagoon (Sacca di Goro, Po River Delta, Italy). J Exp Mar Biol Ecol 269:65-83

Naldi M, Wheeler PA (2002) N-15 measurements of ammonium and nitrate uptake by *Ulva fenestrata* (chlorophyta) and *Gracilaria pacifica* (rhodophyta): Comparison of net nutrient disappearance, release of ammonium and nitrate, and N-15 accumulation in algal tissue. J Phycol 38:135-144

Neal C, Whitehead PG, Flynn NJ (2002) INCA:summary and conclusions. Hydrol Earth Syst Sci 6:607-615

Nedwell DB, Jickells TD, Trimmer M, Sanders R (1999) Nutrients in estuaries. Adv Ecol Res 29:43-92

Nielsen LB, Finster K, Welsh DT, Donelly A, Herbert RA, de Wit R, Lomstein BA (2001) Sulphate reduction and nitrogen fixation rates associated with roots, rhizomes and sediments from *Zostera noltii* and *Spartina maritima* meadows. Environ Microbiol 3:63-71

Olsen Y, Reinertsen H, Vadstein O, Andersen T, Gismervik I, Duarte C, Agusti S, Stibor H, Sommer U, Lignell R, Tamminen T, Lancelot C, Rousseau V, Hoell E, Sanderud KA (2001) Comparative analysis of food webs based on flow networks:effects of nutrient supply on structure and function of coastal plankton communities. Cont Shelf Res 21:2043-2053

Ottosen LDM, Risgaard-Petersen N, Nielsen LP, Dalsgaard T (2001) Denitrification in exposed intertidal mud-flats, measured with a new N-15-ammonium spray technique. Mar Ecol Prog Ser 209:35-42

Ottosen LDM, Risgaard-Petersen N, Nielsen LP (1999) Direct and indirect measurements of nitrification and denitrification in the rhizosphere of aquatic macrophytes. Aquat Microb Ecol 19:81-91

Pedersen AGU, Berntsen J, Lomstein BA (1999) The effect of eelgrass decomposition on sediment carbon and nitrogen cycling:A controlled laboratory experiment. Limnol Oceanogr 44:1978-1992

Peters F, Marrase C, Havskum H, Rassoulzadegan F, Dolan J, Alcaraz M, Gasol JM (2002) Turbulence and the microbial food web:effects on bacterial losses to predation and on community structure. J Plankton Res 24:321-331

Peters F, Marrasé C (2000) Effects of turbulence on plankton:an overview of experimental evidence and some theoretical considerations. Mar Ecol Prog Ser 205:291–306

Ragueneau O, Lancelot C, Egorov V, Vervlimmeren J, Cociasu A, Deliat G, Krastev A, Daoud N, Rousseau V, Popovitchev V, Brion N, Popa L, Cauwet G (2002) Biogeo-chemical transformations of inorganic nutrients in the mixing zone between the Da-nube River and the north-western Black Sea. Estuar Coast Shelf Sci 54:321-336

Redondo JM, de Madron XD, Medina P, Sanchez MA, Schaaff E (2001) Comparison of sediment resuspension measurements in sheared and zero-mean turbulent flows. Cont Shelf Res 21:2095-2103

Regnier P, Mouchet A, Wollast R, Ronday F (1998) A discussion of methods for estimating residual fluxes in strong tidal estuaries. Cont Shelf Res 18:1543-1571

Reigstad M, Wassmann P, Ratkova T, Arashkevich E, Pasternak A, Oygarden S (2000) Comparison of the springtime vertical export of biogenic matter in three northern Norwegian fjords. Mar Ecol Prog Ser 201:73-89

Repka S, Mehtonen J, Vaitomaa J, Saari L, Sivonen K (2001) Effects of nutrients on growth and nodularin production of *Nodularia* strain GR8b. Microb Ecol 42:606-613

Risgaard-Petersen N, Ottosen LDM (2000) Nitrogen cycling in two temperate *Zostera marina* beds:seasonal variation. Mar Ecol Prog Ser 198:93-107

Risgaard-Petersen N (2003) Coupled nitrification-denitrification in autotrophic and hetero-trophic estuarine sediments:on the influence of benthic microalgae. Limnol Oceanogr 48:93-105

Risgaard-Petersen N, Nielsen LP, Blackburn TH (1998) Simultaneous measurement of ben-thic denitrification with the isotope pairing technique and the N2 flux method in a con-tinuos flow-through system. Water Res 32:3371-3377

Rousseau V, Becquevort S, Parent JY, Gasparini S, Daro MH, Tackx M, Lancelot C (2000) Trophic efficiency of the planktonic food web in a coastal ecosystem dominated by *Phaeocystis* colonies. J Sea Res 43:357-372

Saliot A, Derieux S, Sadouni N, Bouloubassi I, Fillaux J, Dagaut J, Momzikoff A, Gondry G, Guillou C, Breas O, Cauwet G, Deliat G (2002) Winter and spring characterization of particulate and dissolved organic matter in the Danube-Black Sea mixing zone. Es-tuar Coast Shelf Sci 54:355-367

Serodio J, Catarino F (2000) Modelling the primary productivity of intertidal microphyto-benthos :time scales of variability and effects of migratory rhythms. Mar Ecol Prog Ser 192:13-30

Serodio J, da Silva JM, Catarino F (2001) Use of in vivo chlorophyll a fluorescence to quantify short-term variations in the productive biomass of intertidal microphytoben-thos. Mar Ecol Prog Ser 218:45-61

Simo R, Hatton AD, Malin G, Liss PS (1998a) Particulate dimethyl sulphoxide in sea-water:production by microplankton. Mar Ecol Prog Ser 167:291-296

Simo R, Malin G, Liss PS (1998b) Refinement of the borohydride reduction method for trace analysis of dissolved and particulate dimethyl sulfoxide in marine water samples. Anal Chem 70:4864-4867

Simo R, Pedros-Alio C, Malin G, Grimalt JO (2000) Biological turnover of DMS, DMSP and DMSO in contrasting open-sea waters. Mar Ecol Prog Ser 203:1-11

Soetaert K, Middelburg JJ, Wijsman J, Herman PMJ, Heip CHR (2001) Ocean margin early diagenetic processes and models. In: Wefer G, Billet D, Hebbeln D, Jorgensen, BB, Schluter M, van Weering T. (eds), Ocean Margin Systems, Springer-Verlag, Heidelberg

Soetaert K, Middelburg JJ, Herman PMJ, Buis K (2000) On the coupling of benthic and pelagic biogeochemical models. Earth-Sci Rev 51:173-201

Spokes LJ, Yeatman SG, Cornell SE, Jickells TD (2000). Nitrogen deposition to the eastern Atlantic Ocean:The importance of southeasterly flow. Tellus 52B:37-49

Stal LJ, Walsby AE (2000) Photosynthesis and nitrogen fixation in a cyanobacterial bloom in the Baltic Sea. Eur J Phycol 35:97-108

Stanev EV, Beckers JM (1999) Barotropic and baroclinic oscillations in strongly stratified ocean basins - Numerical study of the Black Sea. J Mar Syst 19:65-112

Stanev EV, Buesseler KO, Staneva JV, Livingston HD (1999) A comparison of modelled and measured Chernobyl Sr-90 distributions in the Black Sea. J Environ Radioactiv 43:187-203

Stefels J (2000) Physiological aspects of the production and conversion of DMSP in marine algae and higher plants. J Sea Res 43:183-197

Sundback K, Miles A, Goransson E (2000) Nitrogen fluxes, denitrification and the role of microphytobenthos in microtidal shallow-water sediments:an annual study. Mar Ecol Prog Ser 200:59-76

Thamdrup B, Dalsgaard T (2002) Production of N_2 through anaerobic ammonium oxidation coupled to nitrate reduction in marine sediments. Appl Environ Microbiol 68:1312-1318

Thill A, Moustier S, Garnier JM, Estournel C, Naudin JJ, Bottero JY (2001) Evolution of particle size and concentration in the Rhone river mixing zone:influence of salt flocculation. Cont Shelf Res 21:2127-2140

Thorin S, Radureau A, Feunteun E, Lefeuvre JC (2001) Preliminary results on a high east-west gradient in the macrozoobenthic community structure of the macrotidal Mont Saint-Michel bay. Cont Shelf Res 21:2167-2183

Thornton DCO, Underwood GJC, Nedwell DB (1999) Effect of illumination and emersion period on the exchange of ammonium across the estuarine sediment-water interface. Mar Ecol Prog Ser 184:11-20

Tusseau-Vuillemin MH, Mortier L, Herbaut C (1998) Modeling nitrate fluxes in an open coastal environment (Gulf of Lions):Transport versus biogeochemical processes. J Geophys Res (C Oceans) 103:7693-7708

Van de Koppel J, Herman PMJ, Thoolen P, Heip CHR (2001) Do alternate stable states occur in naural ecosystems? Evidence from a tidal flat. Ecology 82:3449-3461

Van der Nat FJ, Middelburg JJ (2000) Methane emission from tidal freshwater marshes. Biogeochemistry 49:103-121

Van der Nat FJWA, Middelburg JJ (1998a) Seasonal variation in methane oxidation by the rhizosphere of *Phragmites australis* and *Scirpus lacustris*. Aquat Bot 61:95-110

Van der Nat FJWA, Middelburg JJ (1998b) Effects of two common macrophytes on methane dynamics in freshwater sediments. Biogeochemistry 43:79-104

Vanderborght JP, Wollast R, Loijens M, Regnier P (2002) Application of a transport-reaction model to the estimation of biogas fluxes in the Scheldt estuary. Biogeochemistry 59:207-237

Viaroli P, Bartoli M, Fumagalli I, Giordani G (1997). Relationship between benthic fluxes and macrophyte cover in a shallow brackish lagoon. Water Air Soil Pollut 99:533-540

Wade AJ, Whitehead PG, Butterfield D (2002a). The Integrated catchments Model of Phosphorus dynamics (INCA-P), a new approach for multiple source assessment in heterogenous river systems:model structure and equations. Hydrol Earth Syst Sci 6:583-606.

Wade AJ, Durand P, Beaujouan V, Wessel WW, Raat KJ, Whitehead PG, Butterfield D, Rankinen K, Lepisto A (2002b) A nitrogen model for European catchments:INCA, new model structure and equations. Hydrol Earth Syst Sci 6:559-582

Welsh DT, Bartoli M, Nizzoli D, Castaldelli G, Riou SA, Viaroli P (2000) Denitrification, nitrogen fixation, community primary productivity and inorganic-N and oxygen fluxes in an intertidal *Zostera noltii* meadow. Mar Ecol Prog Ser 208:65-77

Welsh DT, Castadelli G, Bartoli M, Poli D, Careri M, de Wit R, Viaroli P (2001) Denitrification in an intertidal seagrass meadow, a comparison of N-15-isotope and acetylene-block techniques:dissimilatory nitrate reduction to ammonia as a source of N_2O? Mar Biol 139:1029-1036

Welsh DT (2000) Ecological significance of compatible solute accumulation by microorganisms:from single cells to global climate. FEMS Microbiol Rev 24:263-290

Whitehead PG, Lapworth DJ, Skeffington RA, Wade AJ (2002) Excess nitrogen leaching and C/N decline in the Tillingbourne catchment, southern England. Hydrol Earth Syst Sci 6:455-466

Widdows J, Brinsley MD, Salkeld PN, Lucas CH (2000) Influence of biota on spatial and temporal variation in sediment erodability and material flux on a tidal flat (Westerschelde, The Netherlands). Mar Ecol Prog Ser 194:23-37

Wijsman JWM, Herman PMJ, Gomoiu MT (1999) Spatial distribution in sediment characteristics and benthic activity on the northwestern Black Sea shelf. Mar Ecol Prog Ser 181:25-39

Wijsman JWM, Herman PMJ, Middelburg JJ, Soetaert K (2002) A model for early diagenetic processes in sediments of the continental shelf of the Black Sea. Estuar Coast Shelf Sci 54:403-421

Wijsman JWM, Middelburg JJ, Heip CHR (2001) Reactive iron in Black Sea sediments: implications for iron cycling. Mar Geol 172:167-180

Williams AT, Alveirinho-Dias J, Novo FG, Garcia-Mora MR, Curr R, Pereira A (2001) Integrated coastal dune management:checklists. Cont Shelf Sci 21:1937-1960

Ysebaert T, Meire P, Herman PMJ, Verbeek H (2002) Macrobenthic species response surfaces along estuarine gradients:prediction by logistic regression. Mar Ecol Prog Ser 225:79-95

Ysebaert T, Herman PMJ (2002). Spatial and temporal variation in benthic macrofauna and relationships with environmental variables in an estuarine, intertidal soft-sediment environment. Mar Ecol Progr Ser 244:105-124

3. Defining a good ecological status of coastal waters – a case study for the Elbe plume

Wilhelm Windhorst[1], Franciscus Colijn, Saa Kabuta, Remi P.W.M. Laane, and Hermann-Josef Lenhart

Abstract

The definition of a good ecological status of coastal waters requires a close co-operation between sciences (natural and socio-economic) and decision makers. An argument is presented for the use of ecosystem integrity assessment based on indicators of function and state. Ecosystem integrity is understood to be reflected in exergy capture (here expressed as net primary production), storage capacity (as nutrient input/outut balances for coastal sediments), cycling (turn-over of winter nutrient stocks), matter losses (into adjacent water), and heterogeneity (here the diatom/non-diatom ratio of planktonic algae is used). Its feasibility is assessed using ERSEM, an ecosystem model of the North Sea, for the Elbe plume, after prior satisfactory calibration. Three scenarios were applied corresponding to 80, 70 and 60% reduction of the riverine nutrient load into the German Bight, compared to a reference situation of 1995. The modelling effort suggested that drastic nutrient load reduction from the Elbe alone would have a limited effect on the larger German Bight: even a 60% reduction scenario would only lead to moderate changes in all five indicators. In conclusion, application of functional integrity indicators appears feasible for coastal seas at larger spatial scales (i.e. the German Bight), and, for the coast, would form a useful addition to the indicators presently proposed in the Water Framework Directive (WFD).

[1] Correspondence to Wilhelm Windhorst: wilhelm@ecology.uni-kiel.de

J.E. Vermaat et al. (Eds.): Managing European Coasts: Past, Present, and Future, pp. 59–74, 2005.
© Springer-Verlag Berlin Heidelberg 2005.

Introduction

The overall target of the Water Framework Directive (European Union 2000) is to achieve a good ecological status for coastal waters as well as for freshwater systems and aims thereby to reduce disturbing human impact as far as possible. According to this Directive, a "good ecological and chemical status" of waters is expected to be achieved after 15 years from the date (December 2000) of launching the Directive. The ecological status is defined by biological, physical and chemical characteristics of the ecosystem (see Moschella et al. this volume and Ledoux et al. this volume). While the chemical status is defined by the use of quality standards in relation to priority substances found in the system, reference biological conditions are those that prevailed under pristine conditions so that human impacts are excluded.

Even though the challenge is to define the level of human impact on aquatic ecological systems (the coastal zone) to achieve a good ecological status, two questions have to be answered: first: which amount of human impact on the ecosystem can be tolerated? In other words, how much of the ecological services[2] could be exploited whilst maintaining a good ecological status of the ecosystem? Secondly, within which range of quality can ecosystems be classified to be good? In other words within what tolerable margin can the ecosystem structure and dynamics be allowed to deviate from the pristine conditions while at the same time considered as being "good "? Both questions have to be answered in order to define suitable management plans for the use of for instance coastal zones. The target of the presented paper is to present an indicator and model based approach to combine information required to answer both questions.

The Water Framework Directive is geared towards addressing the responsibility of the human society. It therefore attempts to justify the amount of ecosystem services that are now in use whilst taking account of future management as well. The presented approach is an anthropocentric one. This is in accordance with the present philosophical discussion and also within the definitions for Ecological Quality and Ecological Quality Objectives given by the North Sea Task Force (NSTF), which consists of experts from both OSPAR[3] and the International Council for the Exploration of the Sea (ICES). Ecological Quality Objective (EQO) is "an overall expression of the structure and function of the marine ecosystem taking into account the biological community and natural physiographic, geographic and climatic factors as well as physical and chemical conditions including those resulting from human activities." The discussions about the health of the North Sea continued through 1990 after which the first ideas about EcoQOs were elaborated by OSPAR in the Quality Status Report in 1993.

---

[2] Ecosystem Services: the full range of benefits provided to society by ecosystems and their constituent biodiversity, encompassing more than just capital value of its constituent parts (5th Int. Conf. on the Protection of the North Sea, Bergen 2002).

[3] The OSPAR Commission has been established on basis of the „Oslo and Paris convention" and is an international body responsible for the protection of the marine environment of the North-East Atlantic.

But the societies and their decision makers have to be aware that ecosystem services encompass a broad range of issues that are partly contradicting with respect to the use of natural resources and are justified by a mix of ethical value settings (Barkmann 2000). Thus, the acceptable level of usage of ecosystem services is set by the power of different stakeholders to impose their will and societal regulations such as environmental laws. This means, that definitions and regulations of a "good ecological status" may vary with space and time, and even by cultures. For example, in the Adriatic Sea, fishermen would argue, that a higher level of eutrophication is beneficial for their haulage, while managers of tourism would prefer lower levels of eutrophication in order to minimise the effects on tourists. However, even if those stakeholders agree upon a common level of eutrophication, their expectations of a good ecological status can only be achieved if the ecological structure and processes of the coastal ecosystem are taken into account.

Ecosystem services and ecological impact: A theoretical background

In this section we will discuss interactions between the use of ecosystem services and its impact on ecological systems and present an approach to select suitable indicators to mirror the ecological impact. Thus, approaching the question ("what amount of human impact on the ecosystem can be tolerated?") means to study the functions of nature utilised by man and determines to which extend the activities of man can impact the ecological system.

Following the DPSIR[4] approach (Nunneri et al. this volume), and the analysis carried out as part of the EUROCAT-Project (Colijn et al. 2002) this level of analysis can be focused on several essential fields of the socio-economic system (human needs and activities), which in turn depends on different societal value settings. The EUROCAT[5] –project is an EU funded project that is commissioned between 2000 and 2004. It aims at achieving an effective and integrated management of river catchments through the integration of natural and social sciences. Its overall goal is to develop an integrated management approach for stakeholders (policy makers, regulatory agencies, environmental planners) acting at local, national and European levels.

Instead of the common procedure used by the Organisation for Economic Co-operation and Development (OECD) for Driver-analysis within the DPSIR framework, the EUROCAT consortium adopted a slightly different nomenclature for the DPSIR framework to suit the aim of the project (Colijn et al. 2002). Drivers, Pressures and Responses have been formulated for the river catchments as well as for the coastal areas in order to serve the needs of the EUROCAT project. As the focus of EUROCAT is to view the coastal zone as receptor area of catchment activities, State and Impact indicators have been developed only for the

---

[4] DPSIR = Driver-Pressure-State-Impact-Response.
[5] EUROCAT, European Catchments – catchment changes and their impact on the coast. European Commission, DG-Research, Contract No.: EVK1-CT-2000-00044.

coastal area and were subdivided into ecological State/Impact parameters and socio-economic State/Impact parameters (Colijn et al. 2002).

To identify the societal forces which drive the amount of ecosystem services which are used by man, the EUROCAT consortium selected six issues, namely Food Demand, Urbanisation, Energy Demand, Mobility and Transport, Industry and Housing, Nature conservation, causing pressures on ecosystems. These fields are consistent with the issues discussed in the Progress Report of the 5th Int. Conference on the Protection of the North Sea in Bergen 2002. The issues include the protection of ecosystems, biological diversity, hazardous substances, eutrophication, radioactive substances and offshore oil and gas activities. According to the EUROCAT approach the riverine nutrient loads (nitrogen and phosphorus) are selected as forcing function for the ecological change in the coastal zone.

Dealing with the second introductory question ("which level of divergences of ecosystem structure and function from the pristine conditions can be considered as 'good'") requires the analysis of ecological functions and structures as a means of maintaining the ecosystem services.

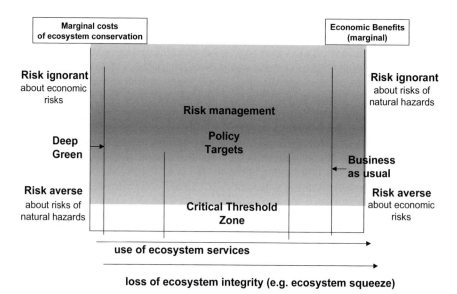

Fig. 1. Principal interactions between marginal costs of ecosystem conservation and ecological risks; based on Nunneri et al. (2002)

Barkmann and Windhorst (2000) introduced a specific interpretation of *ecological integrity* that aims at describing the relationship between the use of ecosystem services and unspecific ecological risks endangering the capacity of ecological systems to provide ecosystem services. The indication of the state of ecosystems has to provide strategies that give reliable information not only for local and short time developments, but also for the long-term integrity of the ecological life support system. According to Barkmann and Windhorst (2000) the latter is strongly connected with the self-organising capacity of ecosystems, which can be indicated with thermodynamic approaches (Baumann 2001). As shown in Fig. 1, a high level of self organising capacity, e.g. ecosystem integrity, is thereby thought to be beneficial as it maximises the possibilities of the ecosystem to provide ecosystem services and in parallel minimises the risk that the ecological system fails to provide the minimum level of natural resources needed by human societies. It is additionally assumed, that with an increasing use of ecosystem services socioeconomic risks decrease as the resource availability increases, which is in accordance with an attitude averting economic risks. In parallel, however, the ecosystem integrity is decreasing as well, causing increasing ecological risks.

For example for ecological risks in coastal zones, the increasing occurrence of anoxic zones could be taken (Rachor and Albrecht 1983, Niermann 1990). Risk aversion requires the reduced use of ecosystem services, thus possibilities to reduce the nutrient losses caused for example by different land use systems in the catchments could be studied. As these possibilities are either connected with lower yields or with higher technical efforts it is necessary to keep both economic and ecological risks as low as feasible. But as risk awareness of societies for economic and ecological risks is variable multiple combinations have to be analysed. In the Elbe case study of the EUROCAT project three scenarios, 'Deep Green', 'Business as usual, e.g. Global markets' and 'Policy Targets', were covered (Nunneri et al. 2002).

Here, ecological integrity is operationally defined as the guarantee that those processes at the basis of ecosystems self-organising capacity are protected and kept intact. Adaptation capacity and development potential (e.g. use of exceeding energy for building structures) belong essentially to self-organising capacity. The self-organising capacity of ecological systems is thereby based on multiple networks of processes as shown in Fig. 2.

The selection of the process "Exergy Capture" stems from the "Non equilibrium principle as formulated by Kay (2000) and Jørgensen (2000). These authors state that during their development ecosystems move further away from the thermodynamic equilibrium using incoming solar radiation (exergy e.g. usable energy) to build up as much dissipative structures (e.g. biomass) as possible. In the case of coastal zones not only energy stemming from solar radiation is available to support the photosynthesis, other energy flows for instance coupled with organic and/or inorganic nutrient inputs from the atmosphere or from adjacent regions have to be taken into account as well.

Another important process to enhance the self-organising capacity of ecosystems is their tendency to (re)cycle limiting substances - especially nutrients - in order to keep the ecosystems as efficient as possible. The contribution of this process to ecosystem development has been described in detail by Higashy et al. (1991) and Ulanowicz (2000). As a rule of thumb it can be assumed, that the cycling intensity increases with the complexity of the trophic network and with decreasing nutrient availability (oligotrophic situation). However, the availability of limiting nutrients and energy in ecosystems depends on storage capacity as well as on input to the system (Kutsch et al. 2001).

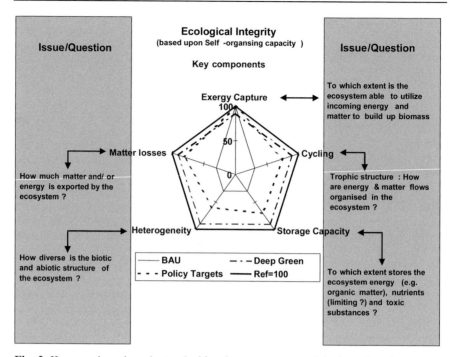

Fig. 2. Key questions, in order to elucidate key components of the integrity of ecological systems. The different diagrams indicated with BAU – Business as usual, Deep Green and Policy Targets represent possible ecological states caused by different intensities of the use of ecosystem services according to different socio-economic value settings. The ecological impact is indicated by a relative comparison with a reference situation, which could be pristine conditions

Taking the holistic perspective of this approach into account, the term storage capacity is understood to include matter that is stored in the sediments, even if these substances are stored for long periods. This forces us for example to analyse not only changing abiotic constraints, (e.g. currents in coastal waters have the potential to alter the accessibility of nutrients), but also the quantity of nutrients which could be mobilised. Via this perspective recent processes are taken into account while in parallel historical developments are valued in terms of their potential to become operative. Thus, the capacity and the exchange rate of the pools is decisive for the long-term availability of nutrients (e.g. nitrogen), energy (e.g. carbon), as well as the possibility to dampen or to buffer temporarily external inputs.

The extent to which ecological systems can utilise this storage capacity thereby depends on the biotic diversity of the system (Kay 2000, Holling and Gunderson 2002). In addition this heterogeneity is also a pool of possibilities represented by the species, which might become dominant under certain environmental conditions. The absence of certain species on the other hand, may therefore minimise the resilience of the whole ecological system (see citation in Box 1).

Finally, ecological systems have the tendency to minimise matter losses, because for instance lost nutrients or organic matter cannot be utilised anymore to build up biomass. Thus matter losses reduce the capacity of primary and secondary production, which are essential functions of ecosystems. Furthermore, matter losses from one ecosystem will be an input to adjacent ecosystems, this process informs about ecosystem internal processes and also about indirect effects on neighbouring ecosystems which are caused via changed matter losses. Thus, in summary, we argue that exergy capture, cycling of elements, storage capacity, heterogeneity (diversity) and matter losses are important elements of ecosystem functions although some are difficult to measure. Together these would serve as indicators of functional ecosystem integrity. In the next sections we will discuss the potential of suggested indicator systems and models to assess ecosystem functioning of a coastal area in the North Sea.

Box 1. Relationship between diversity and resilience of ecosystems

'When grappling with this broader relationship between diversity and resilience two hypotheses are commonly discussed: Ehrlich´s, (1991) "rivet" hypothesis and Walker´s, (1992) "driver and passenger hypothesis. Ehrlich´s hypothesis proposes that there is little change in ecosystem function as species are added or lost, until a threshold is reached. At that threshold the addition or removal a single species leads to a system re-organisation. This model assumes that species have overlapping roles, ands that as species are lost the ecological resilience of the system is decreased, and then overcome entirely. Walker proposes that species can be divided into "functional groups" or "guilds", groups that act in an ecologically similar way. Walker proposes that these groups can be divided into "drivers" and "passengers". Drivers are "keystone species", that control the future of an of an ecosystem, while the passengers live in but do not alter significantly this ecosystem. However, as conditions change, endogenously or exogenously, species shift roles. In this model, removing passengers has little effect, while removing drivers can ...' have a large impact. Ecological resilience resides both in the diversity of drivers, and in the number of passengers who are potential drivers'. From: Gunderson et al. (2000).

The indication of self-organising capacity of ecosystems

In order to meet the major objective of this paper, it is necessary to analyse whether the Water Framework Directive (WFD) presently 'demands' indicators capable of describing the functioning of ecosystems. The second question is whether we can use the quality elements for coastal waters listed in the WFD shown in Table 1. In this table most elements (1 to 16) are items that primarily describe the state of coastal waters, while no. 17, 18 and 19 are emphasising the functioning of the ecosystem. Because our perspective on self-organising capacity, or integrity, of ecosystems represents a top-down approach and is more aggregated than the WFD-elements presented in Table 1, it is assumed that the concept of ecological integrity used here (and following Barkmann and Windhorst 2000) has the potential to serve as an integrating approach, coupling structures and processes of ecosystems.

Table 1. Biological and pysico-chemical quality elements according to Annex V, chap. 1.2.5, Water Framework Directive, European Union (2000)

Quality Element	Criteria in coastal waters
Phytoplankton	1. Composition 2. Abundance 3. Biomass 4. Bloom frequency/intensity
Macroalgae	5. Sensitive taxa 6. Cover
Angiosperms	7. Sensitive taxa 8. Abundance
Benthic invertebrate fauna	9. Abundance 10. Diversity 11. Presence of sensitive taxa
Fish	Not yet determined
Tidal regime	12. Freshwater flow regime 13. Direction and speed of dominant currents
Morphological conditions	14. Depth variations 15. Structure and substrate of coastal bed 16. Structure and condition of coastal zone
General physico-chemical conditions	17. Nutrient concentrations 18. Temperature, oxygen balance and transparency 19. Values for (17) and (18) must permit functioning of ecosystems at good status
Specific synthetic and non synthetic pollutants: • All priority substances identified as being discharged into the water body • Other substances identified as being discharged in significant quantities into the water body	High status: 20. Synthetic: close to zero/below detection limits; non-synthetic: background levels Good status: 21. EQS

Another approach to the development of indicators of the ecological state of coastal waters has been undertaken by Kabuta and Laane (2003). The major distinction is the top-down approach, from broad policy themes like biodiversity and ecological functioning to the definition of measurable indicators, which are connected to policy and management topics of the coastal and marine ecosystems in the Netherlands. Kabuta and Laane (2003) recommend the selection of information about 13 indicator species to indicate species (groups), ecotopes and populations (groups) beyond the topic 'Biodiversity' and to select 10 indicator species to indicate the productivity of the ecosystem, it's feeding structure (types) and the hydro-morphodynamic situation. The indicators are placed under two categories according to the forces that influence them. Indicators that are autonomously influenced by the natural dynamics and processes of the ecosystem are grouped under the category system indicators. Those indicators that are strongly influenced by forces due to human utilisation are placed under the category utilisation indicators. Some of the indicators are placed under both categories. This approach ensures the quantitative estimation of the effects of both human and natural forces on the integrity of the ecosystem. The approach of Kabuta and Laane (2003) broadly overlaps with the strategy to indicate the ecological integrity of coastal zones chosen in this paper. By adding "Storage capacity" and "Matter losses" a full agreement could be achieved. A remaining question is whether it will be feasible to get reliable information with a suitable spatio-temporal resolution for these ecosystem processes.

Applying models to indicate the ecological state of ecosystems

Generally, models are useful instruments in surveys of complex systems, they can be used to reveal the level of interaction between the various properties of the system whilst revealing the weaknesses and the gaps in our knowledge about the system (Jørgensen 1988). Recently different simulation models have been developed to reflect the ecological dynamic of coastal ecosystems (OSPAR 1998). In this paper the European Regional Seas Ecosystem Model (ERSEM) will be used to study the applicability of ecosystem models as tool to describe ecological reference conditions. ERSEM has been developed within an EU-project between 1990 and 1996, focussing the knowledge of six marine institutes across Europa. An overview is given by Baretta et al. (1995). A spatially explicit variant has been elaborated by Lenhart (2001). We used ERSEM to assess ecological integrity for the Elbe plume within the larger German Bight.

Three nutrient loading reduction levels (1995=100) were selected, representing the three scenarios (Fig. 2) "Business As Usual" (80%), "Policy Targets (70%) and "Deep Green" (60%). According to Behrendt et al. (2002), a 10% level of the 1995 nutrient load of the Elbe represents pristine background conditions, corresponding to a forest cover in the whole Elbe catchment.

Case study: Application of the ecosystem model ERSEM to describe reference conditions in the Elbe plume

For our ERSEM calculations, we used forcing data for 1995, which serves as the reference year in this section. The Elbe plume area is shown in Fig. 3. First, modelled values for 1995 have been compared with measured values in the German Bight (Table 2). Deviations ranged between 0-100%, where winter DIP had the highest difference (60% on average). Still, these orders of magnitude are sufficiently satisfactory, and the spatial patterns were consistent.

Table 2. Comparison of model results and field measurements, assessed by Hesse (personal communication). More general benchmarking of ERSEM has been elaborated during the ASMO Workshop in 1996 (OSPAR 1998)

Box	68		69		78	
	Model	Field	Model	Field	Model	Field
Mean Winter DIN [mmol N m⁻³]	31.1	36.5	119.0	65.6	135.9	139.2
Mean Winter DIP [mmol N m⁻³]	1.2	0.9	3.4	1.7	3.6	2.5
Mean Winter DIN/DIP ratio	26.2	40.4	34.9	40.9	39.8	50.8
Mean Winter DIN/Si ratio	2.5	3.0	1.8	1.8	1.6	2.3
Mean Winter DIP/Si ratio	9.5	10.3	5.3	4.6	4.0	4.8

To demonstrate the sensitivity of the model, riverine nutrient loads were reduced in further model runs by 10% steps. These calculations are based on the initialisation of the model by running the simulation for 30 years with repeated forcing, depending on the scenario with reduced load or stable nutrient load for the standard year 1995. After a repeating annual cycle is generated after 30 years, the values for all state variables on January 1ˢᵗ were then used as initialisation for the actual simulations. To visualise the spatial gradient, we compare modelling outcomes for box 78, at the mouth of the river, with those for the larger plume area in the German Bight, i.e. boxes 58-78 (cf. Fig. 3).

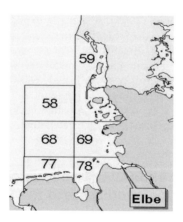

Fig. 3. ERSEM boxes in the coastal zone used for the Elbe case study. The pooled box represents the volume-based sum of the ERSEM boxes 58, 59, 68, 69, 77 and 78 taken together

While box 78, as the input box for the Elbe river load, is the most sensitive one of the analysed boxes, the results for the pooled, larger plume area show a rather limited response (Fig. 5), probably due to effects from adjacent boxes and/or to dilution.It should be realised here that for these scenario runs only the nutrient input of the river Elbe has been reduced, while that from other tributaries to the North Sea have been maintained at the 1995 level. This explains that even drastic reductions of the nutrient loads from the Elbe may cause comparatively small changes in the larger Elbe plume. The modelled drastic decline in nutrient loading of the Elbe plume thus probably would lead to susbtantial changes in winter nutirent concentrations in the river mouth, but not in the wider German Bight. Furthermore, responses in pelagic chlorophyll, net primary productivity and phytoplankton community composition appear to be comparatively small, even in the river mouth (box 78).

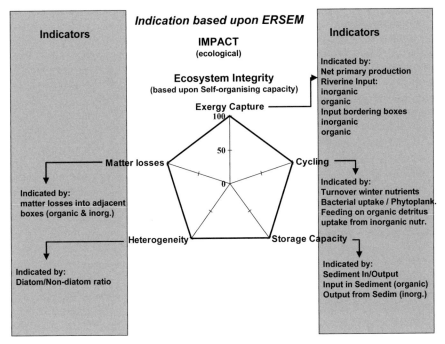

Fig. 4. Available ERSEM elements to describe ecosystem integrity (Nunneri et al. 2002)

The next step in the analysis is a comparison of the three scenarios within the framework of ecosystem integrity indicators (cf Fig. 4). We selected (1) primary production, (2) the annual turnover rate of winter nutrient stocks, (3) nutrient gain by the sediment, (4) the diatom/non-diatom ratio, and (5) nutrient losses out of each ERSEM box, as indicators of the five elements of ecosystem integrity (cf. Fig 3.4, see above). Calculations have been summarised for box 78 (Table 3).

The three scenarios did not lead to a substantial reduction in net primary productivity for box 78, when compared with the two outer bands, i.e. the 100% or 1995 loading and the 10% or pristine situation (Table 3). Also the changes in the other four indicators were limited. Overall, some non-linearity is present in the responses of three important variables, i.e. spring chlorophyll, net primary production and the diatom/non-diatom ratio (Fig 3.5), the latter two are also included as integrity indicators (Table 3) . Quantitative outputs from Table 3 were converted to relative change after scaling against the maximum range of change between the

1995 load and the 10% load of the perceived pristine condition, with the maximum set at 100%. Nitrogen and phosphorus were considered to be equally important. Their scaled changes were therefore added and the sum divided by a factor two to arrive at one compound indicator for nutrients.

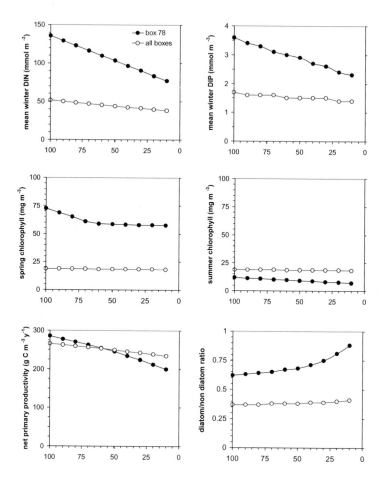

Fig. 5. Impact of a reduction in mutrient loading from the river Elbe on the ERSEM modelling results for areal boxes of the Elbe mouth (box 78) and the adjacent German Bight (all boxes, cf Fig. 3). Decrease of nutrient loading is plotted relative to that of 1995 (=100) from left to right. Presented are mean winter Dissolved Inorganic Nitrogen and Phosphorus (DIN and DIP), spring and summer chlorophyll, net primary productivity and the diatom/non-diatom ratio

Table 3. Impact of reduced nutrient loading according to three scenarios (80, 70, and 60%, see text) on five indicators of ecological integrity for the ERSEM box adjacent to the mouth of river Elbe (box 78, cf Fig. 4): These indicators are derived from Fig. 3. Also included are the 1995 loading as 100% and an assumed pristine condition (10%)

Load (1995=100%):	100%	80%	70%	60%	10%
1. *Exergy capture*: net primary production (g C m^{-2} y^{-1})	286	271	263	254	200
2. *Cycling*					
Turnover winter dissolved inorganic nitrogen (y^{-1})	3.1	3.3	3.4	3.6	4.5
Turnover winter inorganic phosphorus (y^{-1})	5.8	6.1	6.3	6.4	7.0
3. *Storage capacity:* sediment net nutrient gain (mmol m^{-3}-y^{-1})*					
Nitrogen input	571	533	513	492	374
Nitrogen output	540	504	483	463	354
Net sediment nitrogen gain	31	29	30	29	20
Phosphorus input	34	31	30	29	22
Phosphorus output	32	30	28	27	20
Net sediment phosporus gain	1.9	1.8	1.7	1.6	1.2
4. *Heterogeneity:* (diatom/non-diatom ratio)	0.62	0.64	0.65	0.67	0.88
5. *Matter losses from box 78* (mmol m^{-3}-y^{-1})					
Organic nitrogen losses	1109	991	929	870	578
Inorganic nitrogen losses	91	86	84	81	67
Sum nitrogen loss	1200	1076	1013	951	645
Organic phosphorus	21	19	18	17	13
Inorganic phosphorus	6	6	6	5	5
Sum phosphorus loss	27	25	24	23	17

* sediment input is sinking plus uptake by benthic filter feeders

The constructed diagram (Fig. 6) shows that in the selected case study positive effects for the ecological status can be achieved, but that even the "Deep Green" scenario remains quite far away from the assumed pristine conditions. Based upon the calculated relative value it is possible to see the extend to which the different reactions of the selected indicators mirror an overall change of the ecological quality of the coastal ecosystem. These outcomes can then be set against the context provided by with its economic and ecological risk perspective. The 100% scale for the ordinate was chosen because it allows the maximum distance between the reference year and the assumed pristine conditions to be shown and the relative change of the ecological status, which can be achieved by the selected reduction scenarios. However, the reduction of the Elbe nutrient load alone could lower ecological risks like for example the occurrence of anoxic zones.

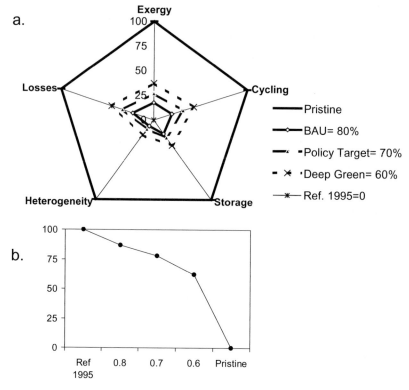

Fig. 6. (a) Relative distance of different ecological state indicators of the Elbe coastal zone compared to assumed pristine conditions according to different reduction scenarios of nutrients loads. The indicators from Table 3 were used, i.e. exergy = primary production, cycling = turnover of winter nutrients, storage = (sediment input − sediment output), heterogeneity = diatom/non-diatom ratio, losses = nutrient output out of the box. Relative distance was scaled between 0 (=1995 load, the heart of the radar plot) and 100 (=10% of 1995 load, the pristine condition), as indicated in the text. **(b)** averaged change of ecological risks in the Elbe box 87 for the different reduction scenarios of riverine nutrient loads (1995=100%, Pristine=10%). The average is taken over the five indicators of ecological integrity

Concluding remarks

As presented in the last section it is feasible to use an ecosystem approach to indicate the ecological state of coastal waters with models. Though only results for one case study and one model have been presented, it seems to be feasible to indicate the integrity of coastal ecosystems based on model results as well as with monitored data connected with policy targeted monitoring schemes (Kabuta and Laane, 2003). The presented case suggests that indicators of ecosystem integrity, though preliminary in nature, are feasible on the larger spatial scales required for coastal zone management. Also, these indicators of ecosystem functioning would be useful in addition to the indicators presently proposed in the WFD. Furthermore, it must be possible to interface this description of the ecological status with socio-economic evaluations, thus allowing one to investigate which economic efforts – in this case in the Elbe catchment – are capable of achieving a certain change of the ecological status.

Acknowledgements

This work was funded by the EUROCAT project of the European Union (EVK1/2000/00510).

References

Baretta JW, Ebenhöh W, Ruardij P (1995) An overview over the European Regional Sea Ecosystem Model, a complex marine ecosystem model. Neth. J. Sea Res. 33:233-246

Barkmann J (2000) Eine Leitlinie für die Vorsorge für unspezifische ökologische Gefährdungen, in Jax K (Hrsg.) Funktionsbegriff und Ungewissheit in der Ökologie. Peter Lang, Europäischer Verlag der Wissenschaften, Frankfurt, pp 139-152

Barkmann J, Windhorst W (2000) Hedging our bets: the utility of ecological integrity. In Jørgensen SE, Müller F (eds): Handbook of Ecosystem Theories and Management. Lewis, Boca Raton, pp 497-517

Baumann R (2001) Indikation der Selbstorganisationsfähigkeit terrestrischer Ökosysteme. Dissertation, Christian-Albrechts-Universität zu Kiel http://e-diss.uni-kiel.de/math-nat.html

Behrendt H, Bach M, Kunkel R, Opitz D, Pagenkopf GW, Scholz G, Wendland F (2002) Quantifizierung der Nährstoffeinträge der Oberflächengewässer Deutschlands auf der Grundlage eines harmonisierten Vorgehens. Umweltforschungsplan des Bundesministers für Umwelt, Naturschutz und Reaktorsicherheit, Forschungsvorhaben: 29922285

Colijn F, Kannen A, Windhorst W (2002) The use of indicators and critical loads, EUROCAT Deliverable 2.1 http://www.iia-cnr.unical.it/ EUROCAT/project.htm

Ehrlich PR (1991) Population diversity and the future of ecosystems. Science 254:175

European Union (2000) Water Framework Directive. In Official Journal of the European Communities L327

Fifth International Conference on the Protection of the North Sea, Bergen 2002: Progress Report

Gunderson LH, Holling CS, Petersen GD (2000) Resilience in ecological systems. In Jørgensen SE, Müller F. (eds): Handbook of ecosystem theories and management. Lewis, Boca Raton, pp 385-394

Higashy M, Patten B, Burns TP (1991) Network trophic dynamics: an emerging paradigm in ecosystems ecology. In: Higashy M, Burns TP (eds) theoretical studies of ecosystems – the network perspective. Cambridge University Press, Cambridge. pp 117-151

Holling CS, Gunderson LH (2002) Resilience and adaptive cycles. In Gunderson LH, Holling CS (eds) Panarchy – Understanding transformations in human and natural systems. Island Press, Washington. pp 25-62

Jørgensen SE (1988) Fundamentals of ecological modelling. Elsevier, Amsterdam, 391 p.

Jørgensen SE (2000) The Tentative fourth law of thermodynamics. In Jørgensen SE, Müller F (eds): Handbook of ecosystem theories and management. Lewis, Boca Raton, pp 161-175

Kabuta SH, Laane RWPM (2003) Ecological performance indicators in the North Sea: development and application. Ocean Coast Manage 46:227-297

Kay JJ (2000) Ecosystems as self-organising holarchic open systems: Narratives and the second law of thermodynamics. In Jørgensen SE, Müller F (eds): Handbook of ecosystem theories and management. Lewis, Boca Raton, pp 135-159

Kutsch WL, Steinborn W, Herbst M, Baumann R, Barkmann J, Kappen L (2001) Environmental indication: A field test of an ecosystem approach to quantify biological self organization. Ecosystems 4:49-66

Ledoux L, Vermaat JE, Bouwer LM, Salomons W, Turner RK (2005) ELOISE research and the implementation of EU policy in the coastal zone. In: Vermaat JE, Bouwer LM, Salomons W, Turner RK (eds) Managing European coasts: past, present and future. Springer, Berlin, pp 1-19

Lehnhart HJ (2001) Effects of river nutrient load reduction on the eutrophication of the North Sea, simulated with the ecosystem model ERSEM. In Kröncke I, Türkay M, Sündermann J (eds): Burning issues of North Sea ecology, Proc 14th int Senckenberg Conference North Sea 2000, Senckenb marit 31:299-311

Moschella PS, Laane RPWM, Back S, Behrendt H, Bendoricchio G, Georgiou S, Herman PMJ, Lindeboom H, Skourtous MS, Tett P, Voss M, Windhorst W (2005) Group report: methodologies to support implementation of the Water Framework Directive. In: Vermaat JE, Bouwer LM, Salomons W, Turner RK (eds) Managing European coasts: past, present and future. Springer, Berlin, pp 137-152

Niermann U. (1990) Oxygen deficiency in the south eastern North Sea in summer 1989. ICES C.M. 1990

Nunneri C, Windhorst W, Kannen A (2002) Scenarios and indicators: a link for pressures and impacts in the Elbe catchment, following the DPSIR approach. SWAP Conference Proceedings, Norwich, 2-4 September 2002

Nunneri C, Turner RK, Cieslak A, Kannen A, Klein RJT, Ledoux L, Marquenie JM, Mee LD, Moncheva S, Nicholls RJ, Salomons W, Sardá R, Stive MJF, Vellinga T (2005) Group report: integrated assessment and future scenarios for the coast. In: Vermaat JE, Bouwer LM, Salomons W, Turner RK (eds) Managing European coasts: past, present and future. Springer, Berlin, pp 271-290

OSPAR (1993) North Sea Quality Status Report, ISBN 1 872349 06 4

OSPAR (1998) Report of the ASMO modelling workshop on eutrophication issues, 5.-8. November 1996, OSPAR Commission, The Hague.

Rachor E, Albrecht H (1983) Sauerstoffmangel im Bodenwasser der deutschen Bucht. Veröff. Inst. Meeresforschung. Bremerhaven 19:209-227

Ulanowicz RE (2000) Ascendancy: a measure of ecosystem performance. In Jørgensen SE, Müller F (eds): Handbook of ecosystem theories and management. Lewis, Boca Raton, pp 304-315

Walker BH (1992) Biological diversity and ecological redundancy. Conserv Biol 6:18-23

4. Bathing water quality

Stavros Georgiou[1]

Abstract

This chapter conducts a multidisciplinary investigation into the public and scientific acceptability of coastal bathing water health risks and proposals to revise EC Bathing Water legislation in the context of UK coastal waters. The research incorporates physical/technical, economic, and public/social assessment components, which are deployed using a mixed methodological approach. It is found that although bathing water quality has been improving and the risks of gastrointestinal illness falling, a significant level of disease burden from this source may still exist across the population. A further tightening of standards and consequent clean up of bathing water may thus be possibly warranted. A cost-benefit analysis of possible proposals to revise the EC bathing water Directive suggests that the economic benefits of doing so would outweigh the costs incurred. These findings are qualified by a number of important lessons and insights regarding attitudes towards risk management and regulation, and issues such as trust, blame and accountability of the institutions and regulatory process involved in setting standards for bathing water quality.

Introduction

In the last few decades, both the general public and policy makers have become increasingly concerned about sewage discharges to coastal bathing waters in the European Union and the consequent risks to public health (House of Lords 1994-5, CEC 2000, CEC 2002). The public health risks of sewage discharged into coastal marine waters are derived from human population infections. The sewage contains various micro-organisms that have been shown to be pathogenic and the causative agents of several human diseases. The main risk faced by people bathing in sewage-contaminated water is in increases to minor morbidity such as gastrointestinal and

---

[1] Correspondence to Stavros Georgiou: s.georgiou@uea.ac.uk

J.E. Vermaat et al. (Eds.): Managing European Coasts: Past, Present, and Future, pp. 75–101, 2005.

upper respiratory tract ailments. The European Commission (EC) Bathing Water Directive of 1976 (CEC 1976) sets out standards for designated bathing waters which should be complied with by all member states. This has been one of the first and most important elements of European Water Policy. The 1976 Bathing Water Directive reflects the state of knowledge and experience of the early 1970's, in respect to its technical-scientific basis, the managerial approach and the involvement of the public. Recently changes in science and technology as well as in managerial experience have obliged the Commission to consider revision of EU environmental legislation where appropriate. Further legislation has thus been proposed on more than one occasion by the EC in the form of revisions to the 1976 Directive (CEC 1994, CEC 2000, CEC 2002). However, policy makers and regulators face a number of dilemmas in the area of coastal bathing water health risk policy. There is a question mark over the level of protection to be afforded against minor illness acquisition by EC standards. The costs of tightening these standards are considerable and the health gain associated with any tightening is likely to be measured in terms of self-limiting and minor illness, such that there is a question as to whether any expenditures on sewage cleanup represent effective and efficient use of resources. Regulators and governments have to balance the public desire for better environmental quality with the economic impact of policy changes on both water bill payers and the financial health of water companies. Furthermore, any new policy must be compatible with EU Water Policy, which has been completely restructured by the adoption of the Water Framework Directive and which provides a coherent managerial framework for all water related EU Legislation.

Given this public health, political, economic and water policy background the central purpose of this chapter is to conduct an investigation into the public and scientific acceptability of coastal bathing water health risks and the proposed revision to EC legislation in the context of UK coastal waters. The research incorporates physical/technical, economic, and public/social assessment components, which are deployed using a mixed methodological approach. The physical/technical assessment focuses on an epidemiological and disease burden analysis (Beaglehole et al. 1993) of the health risks from bathing in faecally contaminated UK coastal waters. The economic assessment focuses on an economic cost-benefit analysis of the EC Bathing Water Directive standards, whilst the public/social assessment focuses on a psychosocial analysis of the public's perceptions of health risks, environmental quality and behaviour regarding coastal bathing waters and related EC standards. The chapter brings together the insights and lessons from each of these components in order to offer a number of policy relevant recommendations regarding proposals to revise the EC Bathing Water Directive.

The mixed methodological approach

This investigation into the public and scientific acceptability of coastal bathing water health risks and legislation uses a mixed methodology involving both quantitative and qualitative elements that are able to generate different types of policy relevant information. Both elements have an essential role to play due to the diverse nature of the theoretical backgrounds that are being brought together. In general, qualitative approaches provide more in-depth information on fewer cases whereas the quantitative approaches provide more breadth of information across a larger number of cases. Quantitative research methods are premised on the assumption that the relevant constructs of interest can be expressed in meaningful numerical ways within a given context. However, they are often criticised for their reductionist nature in the face of real world complexity and diversity. In addition, due to their often technical nature, they may obscure 'proper' interpretation by the public. Qualitative research techniques are more flexible in this respect, being more able to explore the public's knowledge and understanding of the issues involved, and to provide insights into the process by which respondents answer questions the way they do. In addition, they can be used to discuss quantitative results with stakeholders and relate these to the conclusions made.

The quantitative and qualitative approaches tend to place the process of analysis in different scientific and social settings and so provide different kinds of information. In the context of public policy research, the type of approach used depends on the type of information that policy and decision makers are looking for in specific policy domains, as well as the type of information the public is able to deliver and their willingness to participate in the process. When considering research related to such policy decision-making as that undertaken here, it may be important to distinguish clearly between societal and researchers preferences about how public participation and decision making procedures are, or should be, organised. Given the different types of information and approaches, the mixed methodological approach is useful since it makes external validation of the results easier, as well as allowing more flexibility given the importance that any results may have for informing decision-making.

The quantitative element of the mixed methodology used in this paper incorporates questionnaire surveys and existing scientific data collections, whilst the qualitative element consists of focus groups. Each of the elements was considered in terms of what would be appropriate to satisfy the aims and objectives of the research. The existing scientific data used was provided by the UK Environment Agency. This contained microbiological compliance data for all UK bathing waters. The other questionnaire surveys and focus groups were specifically conducted for the purposes of this research. In total two separate face-to-face survey questionnaires were employed along with two sets of focus groups. One of the surveys was conducted on a regional geographical basis at locations in East Anglia. The regional case study nature of this survey was necessary for logistical reasons related to the need to interview beach visitors on site. The survey included questions related to both the economic and public/societal components of the investigation. The focus groups were conducted in the East Anglian city of Norwich and based solely on local residents. The focus groups also included questions related to both the economic and public/societal components of the investigation. The second survey questionnaire was nationally based and contained questions related solely to the physical/technical component of the investigation. The national and regional basis of data collection methods associated with the various components of the investigation to some extent reflects the three scales/levels of analysis associated with each type of component. The physical/technical perspective tends to focus on populations, whilst the economic and public/societal perspectives tend to look at information at the individual and social/cultural group level.

Physical/technical assessment

This section considers the epidemiological and disease burden component of the analysis. It reviews the microbiological state of UK coastal bathing waters and their compliance with the EC Bathing Waters Directive during the period 1999 to 2001, as well as projected compliance under the proposed provisions of the revised Directive. This feeds into the derivation and estimation of excess risk of gastrointestinal illness associated with the actual state of UK coastal waters. Finally, using this information as well as data from a survey of British beach use across the English and Welsh population, an estimate is made of the current absolute disease burden for gastrointestinal illness arising from bathing in faecally contaminated UK coastal waters and the change that may arise from various representative improvements to the current status of UK coastal bathing waters.

Table 1 shows descriptive statistics for the microbiological quality of UK coastal waters over the period 1999-2001. The data contained in the table was collated by the Environment Agency for the annual compliance assessment of bathing water at UK beaches (559 locations in 1999). The data contains records of faecal coliform and faecal streptococci counts (colony forming units per 100 ml – cfu/100 ml). The table shows descriptive statistics for the raw organism concentrations and the $\log_{10}$-arithmetically transformed organism concentrations[2]. A constant of one was added prior to logarithmic transformation ($\log_{10}$) of the variables, to allow for inclusion of zero values. The latter are included as statistical distributions of such organism densities in samples taken from beaches around the UK coast have been found to show a $\log_{10}$-normal pattern (Wyer et al. 1995).

Looking at the measures of central tendency and variability of the two microbiological parameters over the years 1999-2001 it can be seen that UK coastal bathing waters have been improving over this period. This improvement can also be seen in relation to compliance with the EC Bathing Water Directive Standards as shown in Table 2, for both the existing mandatory and guideline values, over the period 1999-2001[3].

---

[2] The arithmetic mean of the log-transformed variable is equal to the log of the geometric mean of a variable.

[3] It should be noted that although the geometric mean values of water quality are improving, the inherent variability in the distribution of the water quality data is not characterised by the use of this statistic (WHO 2001). This can be problematic in that it is such variability that produces high values at the top end of the distribution that are of most concern in relation to public health. A percentage compliance system will however be more reflective of any top end variability in the distribution of water quality data, though it is affected by greater statistical uncertainty and is a less reliable measure of water quality (WHO 2001).

Table 1. Microbiological quality of UK coastal bathing waters (1999-2001)

	Arithmetic mean ± standard deviation	Range	Geometric mean	$\log_{10}$ std. deviation	Number of observations
Indicator: faecal coliform (cfu/100ml)					
1999-2001	208±1187	0-68,000	35	0.69	33,324
1999	251±1390	0-68,000	39	0.72	10,963
2000	214±1207	0-60,000	38	0.69	11,259
2001	161±915	0-30,400	29	0.64	11,102
Indicator: faecal streptococci (cfu/100ml)					
1999-2001	130±1012	0-88,000	20	0.6	33,323
1999	149±939	0-45,000	22	0.71	10,963
2000	134±1178	0-88,000	21	0.67	11,259
2001	108±891	0-50,000	17	0.64	11,101

Note: Data includes inland bathing waters. Takes no account of abnormal weather waivers. Also includes waters for which, more than or less than the usual 20 samples were obtained. The limit of detection is 10 FC/100ml or 10 FS/100ml

Table 2 shows that for the 2001 bathing season, 530 of the designated bathing waters (95%) in the UK complied with the EC bathing water standard at the Mandatory level, whereas 365 bathing waters (66%) complied with the more stringent Guideline values. The compliance rate for the United Kingdom improved from 91 per cent in 1999 to 95 per cent in 2001. However going back even further, compliance has improved from 76% in 1991.

Table 2 also shows compliance rates of UK bathing waters over the period 1999-2001 with the various scenarios being proposed as revisions to the Directive. Whilst UK coastal waters are clearly improving according to the criteria under each revision scenario, the % of bathing waters complying varies considerably depending on the precise nature of the revision. Revision Scenario 1 shows a compliance picture that is roughly comparable with the current EC bathing water Directive. However, under Scenario 2, the number of compliant bathing waters falls to two-thirds of the total. Under the alternative Scenario 2a, which bases compliance on the 95th percentile approach advocated by the WHO (WHO 1998) compliance is worse still at just under 50% of the total. Under the even more stringent revision scenario 3, only one quarter of UK bathing waters pass the requirements. The excess risks of gastrointestinal illness associated with bathers' exposure to the quality of UK coastal bathing waters present over the period 1999-2001 are now assessed. It should be noted that this is not the only illness associated with faecal contamination of bathing waters. Nevertheless it has been the main focus of most of the epidemiological work, and the illness for which there is the most credible scientific evidence of a clear dose-response relationship with water quality. Furthermore, it is what most of the policy decisions undertaken in this area have been concerned with, and hence the analysis is restricted to these risks only.

Table 2. Compliance of UK Bathing Waters with EC Directive (1999-2001)

Directive standard	Parameters per 100 ml (% of samples to comply)		% of UK bathing waters complying with Standard in year[1] (n = total number of bathing waters sampled)		
	Faecal coliforms	Intestinal enterococci	1999 (n=546)	2000 (n=557)	2001 (n=557)
Existing mandatory	200 (95)	NA	91	94	95
Existing guideline	100 (80)	NA	50	54	66
Revision scenario 1	NA	200 (80)	87	90	92
Revision scenario 2	NA	200 (95)	55	61	66
Revision scenario 2a	NA	200 (95)[2]	41	46	50
Revision scenario 3	NA	50 (95)	18	22	26

[1] Includes all coastal bathing waters in England, Wales, Scotland and Northern Ireland, as well as nine inland bathing water sites. Takes no account of abnormal weather waivers. Also includes waters for which more than or less than 20 samples were obtained.
[2] Rather than basing compliance on a percentage of samples lying below the limit value, an alternative is to base compliance on a percentile approach in which one assesses whether the specified percentile (in this case 95th) value for the sample exceeds the limit value. This is the approach advocated by WHO (1998).

The risk of illness for a distribution of exposures to water of different qualities, such as that found around the UK coast, is not given by the relevant epidemiological dose-response function alone. Rather the dose response function, as derived from epidemiological studies, has to be used along with the statistical distribution of the related microbiological parameter densities of the relevant coastal waters. Such a distribution describes the exposure of the bathing population to the different qualities of water around the coast. In this way the proportion of bathers likely to suffer from gastrointestinal illness can be derived for the statistical distribution of UK coastal bathing water quality for any relevant period.

Based on the estimation procedure described in Wyer et al. (1999) and WHO (2001). The dose-response relationship from Kay et al. (1994) has been applied to the faecal streptococci probability density function for identified beaches around the UK coast for the period 1999-2001. This gives the expected excess rate of gastroenteritis (per 1000) for a beach with water quality described by the $\log_{10}$ mean and $\log_{10}$ standard deviation of the distribution. Of the 1000 persons assumed to be exposed, 621 experience water quality unlikely to produce any health effect (Fig. 1) . Of the 379 who experience water quality that might make them ill, 79 become ill with symptoms of gastroenteritis. Using this estimation procedure, the risks of gastrointestinal illness associated with bathers' exposure to the quality of UK coastal bathing waters over the period 1999-2001 were derived (Table 3). As can be seen the excess risks of illness have been falling as the quality of bathing water has improved over the period.

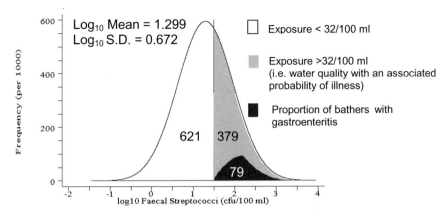

Fig. 1. Integration to calculate total excess gastroenteritis for a faecal streptococci exposure distribution based on the average UK bathing water quality data 1999.2001 (total curve area adjusted to 1000).

Whilst the estimates of excess risk of gastrointestinal illness are based on the actual $\log_{10}$ mean and $\log_{10}$ standard deviation values for UK bathing waters over the period 1999-2001, it is interesting to examine the effect of using a fixed $\log_{10}$ standard deviation for faecal streptococci, as was carried out by the WHO (WHO 1998) to derive the guideline values found in the proposed Revision (guideline value equal to 200 Intestinal Enterococci/Faecal Streptococci). This will lead to differences in the health risks for people exposed above the threshold value depending on how the true standard deviation of a beach varies from the fixed standard deviation (cf. Table 3).

Table 3. Excess risk of gastrointestinal illness associated with UK bathing waters (1999-2001)

year	$\log_{10}$ faecal streptococci concentration (cfu/100 ml)[1]		estimated excess risk - number ill/1000 exposures
	mean	std deviation	
1999-2001	1.299	0.672	79 (94^2)
1999	1.339	0.710	90 (100^2)
2000	1.322	0.665	82 (98^2)
2001	1.235	0.636	65 (86^2)

[1] Includes inland bathing waters. Takes no account of abnormal weather waivers. Also includes waters for which, more than or less than the usual 20 samples were obtained.
[2] WHO fixed Std. Deviation (0.8103) used to estimate excess risk (see discussion in text).

The use of the fixed $\log_{10}$ standard deviation of 0.8103 leads to an overestimate of the excess health risks for people exposed above the threshold level. This is because the actual $\log_{10}$ standard deviation is less than the fixed value and hence there is a more narrow spread of values and thus exposures. This has implications for the use of a single parameter value: local variations in standard deviation will mean that risks of illness will vary even though the same guideline value standard is in place. For example applying the 95th percentile guideline value contained in the Bathing Water Directive revision (guideline value equal to 200 IE/FS) to the $\log_{10}$ standard deviation for faecal streptococci concentrations associated with current (2001) UK coastal waters would result in an excess risk of gastrointestinal illness of about 68 per 1000 exposures (6.8%). This same guideline value applied to the $\log_{10}$ standard deviation for the waters from which it was derived (11,607 EU bathing waters) corresponds to an excess risk of gastrointestinal illness of 50 per 1000 exposures (5%). In order to achieve equal reductions in risk across different waters, this would require the use of differential guideline values across the different waters. So for the case of the current (2001) $\log_{10}$ standard deviation of UK coastal waters, one would have to have a 95th percentile guideline value of about 150 IE/FS in order to achieve a 50 per 1000 exposures (5%) excess risk of gastrointestinal illness.

The application of the gastrointestinal illness risk estimates to data on British[4] beach use/bathing behaviour amongst the English and Welsh population in order to estimate the absolute disease burden for gastrointestinal illness arising from bathing in faecally contaminated UK coastal waters is now considered. Although previous estimates of relative disease burden have been made (Kay et al. 1997), this is the first time that actual beach and bathing water usage rates have been used to calculate absolute disease burden. The beach use/bathing behaviour data was taken from a questionnaire survey undertaken as part of a project commissioned by the Department of the Environment, Food and Rural Affairs (DEFRA) whose objective was to find out people's preferences for changes in various beach attributes (EFTEC 2002). The survey provided data on coastal recreation bathing behaviour (an in particular, respondents bathing water related exposures) for a representative sample of 809 people from the English and Welsh population, which could be combined with the epidemiological risks of gastrointestinal illness established earlier.

Applying the exposure to coastal water figures for the sample found in the bathing behaviour survey to the excess risk of gastrointestinal illness estimates derived earlier and, multiplying by the number of people in the English and Welsh population, it is possible to establish the gastrointestinal illness disease burden for England and Wales arising from faecal contamination of UK coastal waters. Table 4 above shows the relevant calculations in order to estimate the gastrointestinal illness disease burden under a number of different assumptions regarding the excess risk of suffering gastrointestinal illness, and according to swim/dip and combined categories of bathing associated water activity. The total number of exposures for the survey sample can be estimated using either the mean or median number of exposures per person from the bathing behaviour survey. This total number of exposures figure is then divided by the total number of people in the survey sample (809) to give the exposure (to risk) rate for the total sample (rather than for just those undertaking the activity). This is then multiplied by the excess risk of gastrointestinal illness and the population of England and Wales (52.9 million) to give the disease burden for England and Wales arising from bathing in faecally contaminated UK bathing waters.

---

[4] It is assumed that UK beaches and British beaches are synonymous since there are only 13 beaches in Northern Ireland.

Table 4. Gastrointestinal illness disease burden for the English and Welsh population. The best guess overall estimate is highlighted in bold.

Activity	Total no. of exposures in 2001 (95% confidence interval)	Total sample exposure rate [=(1)/809] (95% confidence interval)	Excess risk of gastro-intestinal illness[2] (prob. per person)	Disease burden: number of excess cases of gastro-intestinal illness per year [=(2)x (3)x 52.9 million] (95% confidence interval) (all x million persons)
Exposure calculated on the basis of mean				
Swim/dip	460 (313-607)	0.57 (0.39-0.75)	0.065	1.97 (1.35 - 2.57)
			0.050	1.52 (1.04 – 1.98)
			0.043	1.30 (0.89 - 1.70)
			0.020	0.60 (0.41 - 0.79)
All bathing associated water activities	1011 (568-1454)	1.25 (0.64-1.86)	0.065	4.29 (2.20 - 6.39)
			0.050	3.3 (1.69 – 4.92)
			0.043	2.84 (1.46 - 4.23)
			0.020	1.32 (0.68 - 1.97)
Exposure calculated on the basis of median				
swim/dip	260	0.32	0.065	1.10
			0.050	0.85
			0.043	0.73
			0.020	0.34
All bathing associated water activities	413	0.51	**0.065**	**1.76**
			0.050	1.35
			0.043	1.16
			0.020	0.54

[1] In order to calculate the total number of exposures for use in the grossing up exercise, use can be made of either the mean or median exposures per person from the bathing behaviour survey. The median is used since it is less susceptible to outliers in the sample, whose effect will be greatly multiplied when grossing up estimates to the population level.

[2] The figures relate to risks related to swimming/dips only, and may or may not be correct for the other bathing associated water activities. Epidemiological evidence relating to the other high exposure activities such as surfing, etc., is currently inadequate for a parallel figure to be established for these activities (WHO, 2001). The single risk value is thus applied for all bathing associated water activities. Note also that the probability on each exposure is assumed additive (see later section on immunity).

Whilst there is some uncertainty over the precise current excess risk of suffering gastrointestinal illness (due for example to uncertainty about the $\log_{10}$ normality of the bacterial probability density function), the figure derived in Table 3 for the year 2001 (65 per 1000 exposures) is nevertheless used[5]. In addition, three other estimates associated with the revised EC Directive/WHO guideline values are shown (50, 43 and 20 per 1000 exposures). These estimates cover the likely range of risks associated with compliance of bathing waters in the UK to the proposed EC/WHO Guideline Values, and are based on calculations from an Environment Agency analysis undertaken for DEFRA (Environment Agency 2002).

---

[5] Whilst the possibility of non-$\log_{10}$ normality of the bacterial probability density function was considered, the evidence was inconclusive.

The predicted 'baseline' gastrointestinal illness disease burden resulting from bathing in faecally contaminated UK coastal waters for the year 2001 ranges between 4.29 and 1.1 million cases (Table 4), depending on the category of bathing associated water activity and whether the mean or median number of exposures per person is used as the basis of the total number of exposures calculation ('best guess' estimate is 1.76 million).

The table also shows the predicted gastrointestinal illness disease burden associated with the possible risk figures that represent improvements from the current status of UK coastal bathing waters. These range between 3.3 million cases and 0.34 million cases, depending again on the category of bathing associated water activity and the basis of the total number of exposures calculation, as well as the estimate of risk used (p=0.050, 0.043 or 0.020).

Although there is some degree of uncertainty associated with the disease burden figures, principally due to issues regarding the shape of the bacterial probability density functions associated with UK coastal waters and with the effect of prior population immunity impacts on illness (Hunter 2000), a figure somewhere in the region of 1.75 million cases of gastrointestinal illness per year may be considered to be a central approximation for current disease burden.

The disease burden associated with improved levels of coastal water quality in the UK varied considerably, again depending on the assumptions used to generate the estimate, and specifically on the improvement level being considered. Hence, the estimated disease burden reduction, ranges from about 0.4 million to 1.2 million cases per year depending on the specific excess risk reduction considered.

Economic assessment

This section considers the economic investigation of the EC Bathing Water Directive revision. In particular it seeks to consider the question of whether the revision is worthwhile in terms of the economic benefits of coastal bathing waters complying with it, or whether the resources required to afford compliance would be used more efficiently to achieve other societal goals. The economic benefits are estimated using a contingent valuation study (Mitchell and Carson 1989), which considers a bathing water quality improvement scenario based on a revised Directive. The focus is on the public's willingness to pay (WTP) for particular bathing waters to comply with such legislation, and by implication on the public health benefits afforded to individuals and society[6]. These economic benefits are compared to the costs of implementing changes to bring bathing waters up to the required standard.

---

[6] The main focus of the EC Bathing Water Directive Standards is with public health concerns, though it is recognised that there will nevertheless be additional benefits from bathing waters complying with the standards in terms of recreational/amenity, aesthetic, ecological and non-use considerations. The studies undertaken in this chapter likewise have as their primary focus the public health benefits, though alternative motivations, stemming from the additional benefits mentioned above, may also find some expression in the WTP values being expressed.

The contingent valuation (CV) study was designed to estimate the economic benefits associated with improvements in water quality at <u>all</u> beaches in the Anglian water region (37 beaches in total) such that they all comply with a revised EC Bathing Water Directive. The study comprised of an identical CV survey questionnaire undertaken at two coastal and one urban locations in East Anglia, and was, wherever possible, designed to correspond to the NOAA 'Blue Ribbon' panel guidelines (Arrow et al. 1993) on conducting CV studies.

A contingent valuation survey requires that the change in the provision of the good that respondents are being asked to value is communicated and understood by them. A procedure to elicit respondent's values is then required (elicitation method), as well as a mechanism by which respondents are told that they will have to pay for the change in provision (payment vehicle). One needs to be confident that respondents are actually valuing the specific change in provision and not some other more general change. These elements are usually contained within an information statement, a valuation scenario and questions, and debriefing questions. The elicitation method used in this study was a referendum style payment principle, followed by an open-ended WTP question. The payment vehicle used was an increase in water rates per year, which although problematical (due to the fact that visitors to the coastal location may be from outside the charging area) was nevertheless considered to be the most likely way of financing any bathing water improvements.

Survey respondents were informed about, sewage contamination of bathing water and the subsequent possible health risks from bathing, as well as the existing EC bathing water standards. In this respect they were informed of the current status quo regarding the standard of bathing water quality and associated risks of illness associated with most beaches in the region. This information stated that although most beaches in the region pass the existing Directive, the health risks associated with beaches which satisfy the standard is as follows: 'out of every 1000 bathers, 51 will suffer from vomiting, diarrhoea, indigestion or nausea accompanied by fever; 20 will suffer from respiratory illness such as sore throat, runny nose, coughing; 54 will suffer ear ailments, and 24 will suffer from eye ailments. Some bathers may suffer more than one of these illnesses at the same time.'

Respondents were then asked to consider the introduction of a new standard, which should result in further reductions in risks to health at those beaches that satisfy the new standard. They were told that in order for all beaches in the Anglian region to achieve compliance with the new standard, extra expenditure in the form of higher water rates may be required. Respondents were then asked a payment principle question, in which those agreeing to the principle were asked a further WTP amount question. A budget constraint remainder was given prior to the payment principle and WTP amount questions. In addition, prior to the WTP amount question, a reminder was given that respondents already pay for sewage treatment in order to ensure compliance with the existing directive, and therefore the benefit of the new standard is in terms of further reductions in risks to health at those beaches that comply with the new standard.

In describing the proposed new EC Directive standard, it was not possible to define the specific health risk probability reductions associated with compliance (since scientific evidence was limited). In this respect the contingent commodity being offered was implicitly framed in terms of a change between two perceived 'publicly acceptable' health risk levels. The first associated with the existing directive and the second with the revision. Hence although the framing of the contingent commodity is very much in terms of public health concerns, the reliance on respondents perceiving the changes in health risks means that there is scope for them to incorporate additional benefit motivations (other than just public health risk reductions) into their valuations. Given the use of a change in perceived 'publicly acceptable' health risk levels, it was decided to explicitly examine the variation in people's perceptions regarding this change. Prior to the valuation questions therefore, respondents were asked to state what they themselves expected in terms of proportional health risk reductions (in terms of incidence of illness) from the new EC standard relative to the existing EC standard.

The survey was administered using in-person interviews. The sample of respondents were chosen at random amongst the population of visitors to Great Yarmouth and Lowestoft beaches, and a partially stratified sample was chosen amongst the population of household residents in the city of Norwich. No particular claims are made in terms of representativeness of the sample with respect to any particular population of interest. In fact, the sampling strategy was such as to obtain a varied sample rather than a true cross section. Data were obtained from a total of 616 respondents.

Table 5 presents a summary of the mean WTP amounts found for each of the three site samples, as well as the combined sample, according to respondent's expectations regarding the reductions in number of illnesses achieved by compliance with the revised Directive. These mean WTP values are aggregated for the English and Welsh population using 2002 prices and converted to net present values using a 25 year time frame and discount rates of 6% and 3.5%. The benefit aggregations make the assumption throughout, that the WTP values are representative of the WTP values of the English and Welsh population at large. It is acknowledged that the various samples may not be highly representative of the population. In addition, it should be noted that the CV studies used to generate the benefit estimates only covered improvements at a small proportion of the total number of bathing waters in England and Wales, and hence the estimates are possibly underestimates of possible countrywide improvements. In order to work out the aggregate WTP for the English and Welsh population per year the relevant mean WTP value is multiplied by the number of number of households in England and Wales, equal to 24 million, at the time of this study.

Turning now to the costs of controlling bathing water pollution to a level where water quality complies with the standards laid out by the EC Bathing Water Directive, unlike the benefits estimates, for which there were no previous figures available, two previous estimates of pollution control costs exist and are considered. The first set relate to the cost compliance assessment (CCA) that was commissioned by the UK Department of the Environment and given in evidence to the 1995 House of Lords Select Committee on the European Communities Enquiry, which considered the EC's 1994 proposal to revise the 1976 EC Bathing Water Directive (HOL 1994-5). The CCA required the evaluation of costs associated with four possible scenarios. Scenario A_{1994} is the Commission's 1994 proposal, which introduces a mandatory standard for faecal streptococci, and an enterovirus

Table 5. Summary of mean WTP and net present value of benefits of revised EC bathing water directive (aggregate for English and Welsh households, based on a survey with 616 respondents)

Variants depending on the expected benefits due to an expected reduction in illnesses	Study location and sample[1]	mean WTP per household/pa (£) Study year prices	2002 prices[2] (1)	aggregate WTP/pa[3] = (1) * 24 million (£2002 million)	total net present value of benefits (£2002 million) (over 25 years at 6% discount rate)	total net present value of benefits (£2002 million) (over 25 years at 3.5% discount rate)
Revised directive (irrespective of expected reduction in illnesses)	Great Yarmouth	27.77	30.51	732	9923	12492
	Lowestoft	31.58	34.70	833	11285	14206
	Norwich	49.13	53.98	1296	17556	22101
	Gt Y + L + N	**36.14**	**39.71**	**953**	**12914**	**16258**
Reduction by 100%	Gt Y + L + N	50.69	55.70	1337	18114	22803
Reduction by 75%	Gt Y + L + N	42.10	46.26	1110	15044	18939
Reduction by 50%	Gt Y + L + N	**28.00**	**30.77**	**738**	**10006**	**12596**
Reduction by 25%	Gt Y + L + N	26.56	29.18	700	9491	11948

[1] The study sample relates to which site location sample set of WTP figures are being considered.
[2] Figures from year that original WTP derived (either 1995 or 1997) are adjusted by GDP deflators (UK Treasury figures) to give 2002 prices
[3] Aggregate WTP for England and Wales is found by multiplying the household WTP figures by the number of English and Welsh households = 24 million.

standard. Scenario B_{1994} is the existing Directive made more stringent by making mandatory the standards that are presently the optional *Guideline* standards. Scenario C_{1994} is the Commission's 1994 proposal except for the omission of the more stringent enterovirus requirement. Finally Scenario D_{1994} is the existing directive plus a new mandatory standard for faecal streptococci.

The second set of cost compliance figures relate to a second cost compliance assessment report commissioned by the UK Department of the Environment, Food and Rural Affairs in response to the EC's 2000 proposal to revise the 1976 EC Bathing Water Directive (Cascade Consulting 2002). The assessment examines the costs of three scenarios for upgrading bathing water quality. These are all based on increasingly stringent levels of faecal streptococci that correspond to WHO's microbiological assessment categories for bathing waters (WHO 2001). Scenario C_{2000} is equivalent to the current mandatory EU standards, while Scenario B_{2000} is roughly equivalent to the current *Guideline* EU standard. Finally Scenario A_{2000} is the strictest standard in the WHO's classification categories shows the indicative parameters and their respective limit values associated with each of the seven revision scenarios.

The two sets of cost compliance figures relating to the 1994 and 2000 proposals for revising the EC Bathing water directive are shown in Table 7, using net present values, based on 2002 prices, a 25 year time frame and discount rates of 6% and 3.5%[7]. As can be seen the figures vary considerably depending on the particular scenario considered. As expected the strictest scenarios under each set of revision proposals (scenario A for the 1994 and 2000 revisions) are the most costly. Two of the scenarios (B_{1994} and B_{2000} - shown in bold) from each set of revision proposals, both relate to the same Guideline standard of the current Directive and hence serve as a cross check of the credibility of the two cost compliance assessments. It is interesting to note that, although the individual capital cost and operating cost figures for scenario B appear to diverge somewhat between the 1994 and 2000 figures, the net present cost figures are very similar (the figure for B_{2000} is about the mid point of the range given for B_{1994}.

The cost estimates for the various revision scenarios in Table 7 can now be compared with the various benefits estimates for the different estimation scenarios in Table 5. It would appear that the benefits of a revised Directive outweigh the costs of even the most stringent of the revision scenarios, irrespective of respondents' expectations regarding reductions in the number of illness from compliance Given the fact that the benefit estimates may even be conservative underestimates (since they may only cover improvements at a small proportion of the total number of bathing waters in England and Wales with certainty), then it seems likely that the benefits will outweigh the costs even allowing for any sources of imprecision in their estimation. It is acknowledged that there may be problems over the representativeness of the samples in the two CV studies, such that the benefits estimates are somewhat biased, though on balance it is felt that this is unlikely to make any material difference to the finding of positive net economic benefits associated with bathing water pollution control.

---

[7] At the time of writing 6% is the rate of discount used by the UK Treasury in its 'Green Book', though it is thought that this is likely to change to 3.5% in the next revision.

Table 6. Cost compliance assessment for different directive revision scenarios

Proposed directive revision scenario	Indicative parameter	Limit values
A_{1994}	Faecal streptococci	400 cfu/100 ml
	enterococci	0 pfu/10 l
B_{1994}	Total coliform	500 cfu /100 ml[1]
	Faecal coliform	100 cfu/100 ml[1]
	Faecal streptococci	100 cfu /100 ml[3]
C_{1994}	Faecal streptococci	400 cfu /100 ml
D_{1994}	Total coliform	10000 cfu/100 ml[2]
	Faecal coliform	2000 cfu /100 ml[2]
	Faecal streptococci	1000 cfu /100 ml
A_{2000}	Faecal streptococci	<40 cfu/100 ml[4]
B_{2000}	Faecal streptococci	40-200 cfu/100 ml[4]
C_{2000}	Faecal streptococci	201-500 cfu/100 ml[4]

[1] 80% of samples should not exceed this level
[2] 95% of samples should not exceed this level
[3] 90% of samples should not exceed this level
[4] 95 Percentile
Note: The cost figures relate to the eight affected companies (excluding Northern Ireland and Scotland), and it is thought that the impact in some water company areas might be twice the national average (HOL 1994-5).

Table 7. Net present costs of EC Bathing Water Directive revision scenarios (aggregate for English and Welsh bathing waters)

Proposed directive revision scenario	Capital cost £2002 million[1]	Operating cost £2002 million/pa[1]	Total net present cost - £2002 million (over 25 years at 6% discount rate)	Total net present cost - £2002 million (over 25 years at 3.5% discount rate)
A_{1994}	1,971-5,096	84-180	3,111-7,539	3,406-8,171
B_{1994}	1,370-3,173	60-120	2,184-4,802	2,395-5,223
C_{1994}	529-1,322	24-48	855-1,974	939-2,142
D_{1994}	24-48	0	24-48	24-48
A_{2000}	590	500	7365.18	9119.18
B_{2000}	280	230	3396.58	4203.42
C_{2000}	2.9	0.5	9.68	11.43

[1] Costs adjusted where necessary by UK Treasury GDP deflators to give 2002 prices

Public/social assessment

Background

The focus of this final component is on examining the public acceptability of coastal bathing water health risks and the revision of the EC bathing water Directive standard from a social/public perception standpoint. In particular it examines social/public perceptions of environmental quality, health risks and health risk regulation and management, in the context of specific UK beach sites and the proposal to revise the directive. The aim is to broaden the scope and understanding of the coastal bathing water problem provided by the previous two components in a number of ways. The analysis seeks to better understand how people achieve, justify and sustain particular evaluations and actions towards the bathing water issue, and to better incorporate the social/public perception perspective in the development and implementation of coastal bathing water health risk policies and standards. Psychological and economic instruments for assessing the importance of the bathing water risk issue are compared, whilst setting the findings in social, institutional and cultural perspectives. In particular the analysis seeks to explain the motivations behind perceived magnitude of risk, environmental quality and stated intentions to behave, such as willingness to pay for bathing water improvements. The analysis considers attitudes towards risk management and regulation, and issues such as trust, blame and accountability of the institutions and regulatory process involved in setting standards for bathing water quality. The context is always in terms of informing the real-life policy debate over EC coastal bathing water standards.

Approach

The approach taken in undertaking this investigation draws on a number of diverse and eclectic theoretical sources, which are considered useful to informing the real life policy debate over EC coastal bathing water standards. These range from various types of social cognition models (Fishbein and Azjen 1975, Rotter 1954, Bandura 1977, Wallston et. al. 1978) to psychometric risk perceptions analysis (Slovic 1992) and cultural theory (Dake 1992). Given the disciplinary perspective of this section, the methodological approach made use of a combination of a questionnaire survey of the general public and smaller focus group meetings.

The survey questionnaire for the public/social assessment was undertaken as part of a combined survey along with the contingent valuation study questions discussed in the previous section. Design of the questionnaire was thus influenced by elements from economic, psychological and sociological models. The survey questionnaire contained questions representing key variables from six categories defined by reference to the different theoretical approaches mentioned above. The six categories are outlined below.

Views of Nature (world views). These questions attempted to ascertain respondents' underlying beliefs about the environment, and their worldviews in general. Respondents' views of nature, or 'myths of nature' as proposed by cultural theorists were elicited. This led to the construction of variables describing how respondents viewed the natural world from the point of view of it being adaptable to pressures (ADAPT), controllable by expert management (EXPMAN), fragile and vulnerable to pressures (FRAGILE), and unpredictable in the way it responds to pressures (UNPRED). Respondents were asked to assign values, on a five point Likert Scale, indicating the extent to which they agree or disagree with these views (1= disagree strongly, 3 = neither agree or disagree, 5 = agree strongly).

Knowledge and Experience. These questions enquired about respondents perception of their awareness of risks to health from polluted bathing waters (AWARENESS), whether they had heard of the current EC standard (HEARDSTD), and whether they themselves or a member of their family had been ill as a result of swimming in polluted bathing waters (ILLNESS).

Self Efficacy. Respondents were asked if they felt personally capable of making a decision about the new EC standard (CAPABLE), whether the decision should be left to experts (EXPERTS), and whether public consultation should be courted on the issue (PUBCON).

Expectations. Respondents were asked about whether they believed that their participation in the survey would have an important input into the decision making process (IMPINPUT), if the implementation of a new EC bathing water standard was realistic in practice (REALISTIC), whether the success or failure of a new EC standard would be largely a matter of chance (CHANCE). Participants were asked if they trusted the Government to implement the new EC standard (TRUSTSTD). Respondents were then asked to estimate what decrease in health risks (as a proportion of existing risks) they would expect from a new EC standard (EXPRED).

Values. These questions related to the importance to the respondent of the new EC standard, both personally (IMPPERS) as well as to the nation (IMPNAT), and whether the trustworthiness of government in implementing EC directives was an important issue to the individual (TRUSTIMP). Participants were also asked if the proposed EC standard was something that particularly interested the respondent (INTEREST). Finally, respondents were asked to rate on a Likert-type scale (1= not important, 5 = very important) how important it was in terms of their health that the bathing water at beaches in the Anglian Water region should pass the new EC standard (IMPHEA), as well as how important they thought action on a set of coastal environmental problems was (ISSUES).

Personal context and characteristics. Each individual was asked a set of questions about their sex (SEX), age (AGE), income (INCOME), level of education (EDU>16, i.e. educated beyond age sixteen), whether they had young children or not (CH<10), or were members of various environmental groups (ENVGROUP), and leisure interest groups such as the "Surfers against Sewage" pressure group (INTGROUP).

The variables measured in the survey were treated as either continuous, i.e. measured on a scale, or dichotomous, i.e. respondents answered 'yes' or 'no' to a question. The following four response variables were analysed simultaneously:

1. Perceived magnitude of health risk from polluted bathing water, measured in terms of how serious a risk they thought pollutants in coastal bathing waters were generally to people in the UK, on a five point Likert-type scale (continuous);

2. Perceived current water quality rating, measured on a seven point Likert-type scale from –3 (very poor) to +3 (very good);
3. Willingness to pay, in principle, for an increase in water rates to achieve implementation of a new EC Bathing Water Directive which reduces risk to a level determined by participants (dichotomous);
4. For those who stated they would, in principle, be willing to pay some amount, a further open-ended question was asked about the amount they would be willing to pay as an annual increase in water rates per year (continuous).

Due to the mixture of binary and continuous response variables, and the fact that not all participants gave responses to all questions, the data were analysed using a mixed response multivariate model of the form described by Langford et al. (1999b). In addition, a set of four focus group interviews (each 6 members) were undertaken to provide further insights, and interpretation of survey findings. Using a sample of respondents from the questionnaire surveys, four focus groups were established according to respondents scores regarding their cultural theory solidarities (Thompson et al. 1990, Marris et al. 1998, Langford et al. 1999a). Full details of the selection procedure and logistics of each of the four group meetings are given in Georgiou, (2003). Briefly described, the four solidarities are characterised as being:

- *Hierarchists*: belief in the smooth running of society on prescribed guidelines, framed in legislation and institutional classifications, with control being vested in formal, hierarchical systems of authority, associated with the belief that expert management can solve environmental crises;
- *Egalitarians*: in common with hierarchists, there is a strong sense of society, but not along institutionalised guidelines. Individuals are not granted authority because of their position, and decisions are reached through negotiation. The environment is seen as potentially fragile, and easily damaged by human actions;
- *Fatalists*: tending to have low social associations, but a strong sense of social distinctions ('us and them'). Like hierarchists, fatalists autonomy is controlled by institutional systems, but these are believed to be corrupt and self-interested, excluding the fatalists from meaningful involvement. The environment is believed to be unpredictable and uncontrollable;
- *Individualists*: having low social associations, and no belief in formal institutions, or responsibility towards society as a whole. Individualists believe power and resources are allocated by competition, rather than position and status. The environment is viewed as being adaptable to changes resulting from human actions.

The group discussion protocol focused on: public perceptions of bathing water health risk information; possible solutions to coastal bathing water pollution problems; the extent to which WTP reflects individual preferences; the appropriateness of weak and strong sustainability criteria for the setting of standards; trust and accountability in the agencies and groups concerned with bathing water issues, and how this influences/affects WTP.

Survey results

Turning now to the results, a summary of the multilevel modelling results for magnitude of risk, current water quality, willingness to pay in principle, and willingness to pay amounts is given in Table 8. The explanatory variables were modelled simultaneously, and hence the results are for a complete model including information on world views, personal characteristics, self efficacy, expectations, values and knowledge and experience.

Table 8. Multilevel modelling results for magnitude of risk, current water quality, willingness to pay in principle, and willingness to pay amounts. Number of respondents was 616.

Category	Variable	Magnitude of risk	Water quality rating	Payment principle	WTP amounts
Views of nature	ADAPT	---			
	EXPMAN		++	++	
	FRAGILE	+			
	UNPRED				--
Personal	SEX				
characteristics	AGE	---	++++		
	INCOME				
	CH<10			++	
	EDU<16			+++	++
	INTGROUP			+	
	ENVGROUP			+	
Self efficacy	CAPABLE		-	--	
	EXPERTS	-			--
	PUBCON	++++	++		--
Expectations	EXPRED			---	++++
	IMPINPUT		++	+++	
	TRUSTSTD		+		
	REALIST			++++	
	CHANCE				
Values	INTEREST	++			
	TRUSTIMP		-		
	IMPPERS		-	+++	
	IMPNAT	++			+++
	IMPHEA			++	++
	ISSUES	++++			
Knowledge and	ILLNESS	++	--		
experience	AWARENESS	++	+++		
	HEARDSTD	-	+++		

Note : +/- = p < 0.10, ++/-- = p < 0.05, +++/--- = p < 0.01, ++++/---- = p < 0.001; + = positive correlation, - = negative correlation

Considering views of nature, magnitude of health risk from polluted coastal bathing waters was negatively associated with a belief in the adaptability of nature, and positively with a view that nature is fragile, and prone to damage by human actions. This can be related to cultural solidarities, with adaptability of nature being associated with individualism, and fragility with egalitarianism (Marris *et al.*, 1998). Younger people were more likely to estimate a higher risk, although this was the only significant personal characteristic variable. A higher perception of risk was also strongly associated with a desire for public consultation, and that experts and policy makers should not be allowed to take decisions on the public's behalf. Interestingly, none of the behavioural expectations variables were significant predictors of risk perception, but several measures of importance value were, including personal interest in the issue and a belief that coastal bathing water quality was an important national issue. This cluster of predictors outlines a belief in the importance of general environmental protection, associated with interest and importance of the particular issue, in line with interest in coastal environmental issues generally (such as waste disposal, preservation of natural heritage and coastline protection). In addition, higher perceived risk was also associated with previous perceived illness of self or a family member attributed to exposure to polluted seawater and perceived awareness of the risks. However, there was a weak negative association between perceived risk magnitude and knowledge of the EC standard.

A positive perception of current water quality rating was associated with a belief in the efficacy of expert management of environmental problems, associated with the hierarchy solidarity, which views institutional solutions and regulatory approaches as the best way to tackle environmental pollution (see focus groups discussion below). Older people were also more likely to perceive the water quality as better than younger people, and this was again the only significant personal characteristic variable. A belief in public consultation as a means of developing better water quality measures was positively associated with water quality rating, but feeling currently capable of making a decision was negatively associated. Two expectation variables, namely trusting the implementation of the new standard, and belief that answering the questions in our survey would be an important input to the decision making process (respondents were informed that a report on the work would be sent to the House of Lords Select Committee) were positively associated with water quality rating. In contrast, personal importance of the issue, and feeling that trustworthiness of the Government in implementing the standard was important were negatively associated with high water quality rating. This cluster of predictor variables suggests that people who believe in expert management and trust the current institutional approaches tend to perceive the current water quality as being higher. There was also a positive association between higher water quality rating and perception of awareness of the risks and knowledge about the current EC standard. However, there was a negative association with previous perceived illness experience.

For the payment principle question, there was a positive association with belief in expert management, and several of the personal characteristic variables were significant, namely, having higher education, having children, and being a member of an environmental or interest group. Belief in being capable of making a decision was interestingly negatively correlated with a positive response, suggesting that those who refused to pay were certain of their beliefs. Three expectations variables were significant, with belief that the respondent was having an important input into the decision making process and belief that implementation of a new standard was realistic being positively correlated with a positive response. This is important, as it suggests that saying "yes" to the payment principle is to a degree dependent on belief in the action being offered and the perceived importance of the contingent valuation study in determining benefits. Those who were willing to pay something had lower expectations of the reduction in risk, suggesting that those who wanted a greater reduction in risk were objecting to the payment principle question. High personal importance value was also associated with saying "yes", as was importance to personal health, suggesting that more immediate personal concerns were determining the response to this economic payment question. None of the variables to do with knowledge or previous experience were significant predictors.

Unpredictability of nature was the belief associated with lower WTP amounts (of those who were willing to pay anything at all). This supports other results (Langford et al. 1999c, Marris et al 1998) that those with a more fatalistic outlook, believing that industry and government act out of largely self-interested motives, are less willing to commit themselves to institution-based improvements. Out of the personal characteristic variables, only higher education was associated with higher WTP (income was not significant). However, higher WTP amounts were positively associated with the size of the expected reduction, suggesting that WTP amounts were more based around what people would like for their money than with income constraints in this case. WTP amounts were also negatively correlated with both public consultation and experts taking decisions. These two explanatory variables were not highly correlated in the model (r = -0.20), perhaps surprisingly, but negative associations with both may suggest a preference for the status quo, rather than potentially expensive public consultation or further expert analysis. Importance to personal health was a predictor of higher willingness to pay, as was perceived importance to the nation. Again, none of the knowledge and experience variables were significant.

Focus group results

The focus group discussions produced interesting differences between the cultural theory defined categories. Members of the individualist, hierarchy and egalitarian focus groups expressed surprise that the current estimates of health risk from polluted coastal bathing waters were so high (100 – 150 minor illnesses per thousand bathers, Fleischer et al. 1998). However, there were differences in the interpretations put on this information. Egalitarians, who were least surprised by the high figures, acknowledged that the illnesses were minor, but clearly stated they were still serious because they were unnecessary, and used the information to further justify their belief that the current standard was very inadequate. In contrast, individualists immediately doubted the validity of the statistics, and questioned how people could be certain that illnesses arose from exposure to pollution. There was a belief that illness was largely due to individual weakness (e.g. poor immune systems) or the overuse of antibiotics and lack of exposure to germs in society in general. Illnesses were therefore interpreted as being minor, and exposure to risk largely a matter of individual choice, or common sense. Hierarchists also commented that some people may be more susceptible, but were more concerned with having a yardstick to judge the figures against, for example, health risks from unpolluted beaches or beaches in other countries, such as around the Mediterranean. Risk was therefore interpreted in a relativistic way, needing comparison in a wider arena area to determine acceptability. Fatalists accepted the figures readily, and described them as being shockingly high. They used the information to validate their beliefs that the present standard was useless, and that it would be difficult to do anything to alleviate the current bad situation.

Regarding risk management and regulation, hierarchists focused on public awareness and education, and placed importance on identification of the sources of pollution, so that technology and legislation could be used to reduce pollution from problem industries. In contrast, egalitarians interpreted risk management to mean the removal of risks to the public. There was a desire for re-nationalisation of utility industries, such as water, and concern over profiteering from the privatised water companies. Health risks were also framed in more general environmental terms, with human health being one of a number of important concerns for society. Individualists were concerned about pollution, but looked to external causes, such as waste and oil from ships, and the importance of agricultural and industrial effluent from the rest of Europe. Importance was attached to the efficiency and accountability of the privatised water companies in providing less expensive and more effective ways of tackling the problem, but it was accepted that "sewage is a fact of life". Fatalists agreed that water companies should be responsible for taking action to solve the problem, but wouldn't do so unless forced to. However, they also doubted that legislation would be introduced effectively to achieve this. They perceived government and industry to be quite separate entities, both divorced from the public interest. Risk management was framed in terms of institutional failure due to self-interest and lack of motivation for change.

The focus groups also brought out important differences in interpretation of an economic solution to the problem, and the use of willingness to pay to measure benefits. Individualists concluded that willingness to pay was a good measure of how serious people were about their concerns – "putting their money where their mouths are". They also supported the use of an economic solution as being the most realistic and effective, whilst expressing unease about 'expert opinion'. Hierarchists, on the other hand, saw an economic solution as a necessary evil on the road towards a better standard of quality, which could be achieved through effective legislation and application of new technology. Egalitarians had reservations about the payment vehicle, namely an increase in water rates, as they believed that the privatised water companies were more interested in profits than providing services. They wanted to know what they were being asked to pay for, and focused discussion on the provision of better information, including an estimate of the costs involved – a few doubted that environmental goods could be allocated meaningful economic values. They were also concerned with people's ability to pay, rather than willingness to pay. The fatalists favoured tougher environmental standards, but believed these would never be achieved, and that people in general would not be willing to pay for them. They therefore interpreted the willingness to pay question in terms of an unfavourable view of society's motivation to change, and in particular, pay for change. They also believed that people realistically had very little choice, and so doubted the usefulness and validity of estimating willingness to pay, but concluded that an economic solution was more feasible than fixed standards because it was less expensive.

The issue of trust in the government was very important to the questionnaire respondents, though a minority felt that the government could be trusted to implement the new standard. Individualists and hierarchists were least concerned about trust and governance, believing that regulatory change, technological innovation and market mechanisms could be combined to provide workable solutions. Hierarchists were generally satisfied with the agencies involved in regulation and sought action by parliament to ensure proper regulation of the Water Companies. Individualists focused on the perceived inefficiency and insensitivity of the EU, commenting negatively on their inflexibility and financial profligacy. Egalitarians, in contrast, were concerned about the amount of bureaucracy and lack of democracy in the EU, and fatalists expressed both ignorance and mistrust of the operation of the EU or the Government.

The results of the above analyses suggest that people distinguish between their perception of risk, risk management and environmental quality, and economic measures of commitment, such as willingness to pay. Respondents also seemed to assume the dual role of consumers and citizens, depending on the issue discussed. Participants recognised that risks may be hidden, such as bacteria, but still used sensory data to define their judgements of polluted water. The action associated with awareness of risks was avoidance behaviour, such as refusal to go into the sea and not eating seafood. Lack of complete knowledge of the risks was bolstered by anecdotal evidence and media coverage of illnesses that has been attributed to pollution in seawater. Some participants reacted against this, believing that health risks are exaggerated. In contrast to this individual-based definition of risk and risk avoidance, collective or public responsibility for pollution was accepted in the light of the need to identify specific main polluters, and participants took a pragmatic view of risk management. Nevertheless, different cultural solidarities provided different justifications for both their concern and choice of management option.

The questionnaire survey also highlighted differences between risk and willingness to pay. Magnitude of risk to society was associated with importance value, general environmental concerns and previous knowledge and experience. In contrast, willingness to pay in principle was associated with personal characteristics, expectations and personal importance to health. This shows that different ideas are being used to construct an answer to magnitude of risk compared with a stated intention to provide funds to reduce the risk. In particular, how realistic the policy proposed was, and how important the participant's involvement in the survey was perceived to be were important predictors of being willing to pay something. Beyond this, willingness to pay amounts was associated with expected reductions in risk and importance to personal health, which are consistent with economic theory. Perceptions of current water quality were more closely linked to magnitude of risk, than willingness to pay, with knowledge and experience again being important explanatory factors. Perception of good quality was generally associated in a belief that 'everything is basically OK' in terms of risk exposure and management. This finding was backed up in the focus groups where hierarchists, believing in the efficacy of expert management, defined risks in relative terms and decided upon education of the public and improved legislation to reduce exposure to risk.

Willingness to pay was generally accepted as a pragmatic way of measuring public commitment to reduce health risks, given a number of caveats. For individualists, efficiency was the key issue, whilst for egalitarians equity and general environmental concern featured prominently. Fatalists believed an economic solution was making the best of a bad job, whilst hierarchists looked forward to an 'ideal' situation where a 'gold standard' could be imposed.

Recommendations and conclusions

A number of conclusions and recommendations stem from the analysis. Firstly, it is undoubtedly the case that coastal bathing water quality in the UK has been improving and consequently the risks of gastrointestinal illness falling over the last decade as sewage cleanup has taken place. However, whilst most bathing waters currently comply with the minimum requirements of the European Directive on bathing water (CEC, 1976), it is unclear to what extent UK bathing waters will be compliant with the provisions of a revised Directive that is more stringent and relevant to public health concerns. Furthermore, the analysis has indicated that a single parameter value to define a revised guideline value across all waters in the EU may not be appropriate.

Using data on bathing related water exposures alongside relevant WHO disease burden methodology, a 'ball park' estimate around 1.75 million cases of gastrointestinal illness per year was found for disease burden in the English and Welsh population associated with current (2001) levels of coastal water quality in the UK, whilst there would be a reduction from this figure of between 0.4 million to 1.2 million cases per year, depending on the specific water quality improvement undertaken. Given the uncertainties associated with both current and possible future guideline values, it is believed that there needs to be a move away from a reliance on a single guideline value across all EU waters. Instead EU legislation should aim towards a system that provides for a comprehensive and flexible approach to the control of recreational water environments that better reflects health risks and provides enhanced scope for effective management intervention. Such an approach is enshrined in the approach known as the "Annapolis Protocol" (WHO, 1999), and would involve an extended implementation of the WHO methodology used to derive the proposed EC Guideline Values.

With respect to the economics of revising the bathing water Directive, the cost-benefit analysis undertaken here, whilst based on a number of assumptions with a significant degree of uncertainty attached to them, indicates that a further tightening of standards and consequent cleanup of bathing waters is appropriate. This finding is qualified by a number of important lessons and insights. The great majority of respondents accepted an economic approach to the mitigation of the problem, though individuals with different viewpoints arrived at this decision for different reasons. Willingness to pay was generally seen as a reasonable way of assessing public commitment to reducing risks to health. However this needs qualification with respect to proper apportionment of blame and responsibility to those who pollute the sea, the distribution of impacts across different sectors of society, and the setting of the issue of health risks in the context of wider environmental issues. People's perceptions, preferences and behaviour are based on very different criteria and concerns. The reasons why people will or won't pay, and how much they pay are based on very different factors. This means that policy makers need to be informed not only of the economic costs and benefits of bathing water cleanup, but also about the motivations and expectations that people convey in relation to a reduction in risks. Individuals use social and institutional frameworks to define their responses to bathing water risk, and make clear distinctions between their personal risk avoidance strategies and societal management of risk issues.

The analysis undertaken found that the various factors affect to a varying degree the public's faith in any proposed new standard, and their expectations of the health benefits that the standard will deliver. Overall, it can be concluded that individuals across all four identifiable cultural categories (egalitarian, fatalist, hierarchist and individualist) are concerned either with the current level of risk, the current practices for implementing the standard, and with issues of efficiency and accountability of the Water Companies. Any attempt at redrafting and successfully implementing a new standard must take account of these concerns.

References

Arrow KJ, Solow R, Portney PR, Leamer EE, Radner R, Schuman EH (1993) Report of the NOAA panel on contingent valuation. Fed Regis 58:4602-4614

Bandura A (1977) Social learning theory, Prentice-Hall, Englewood Cliffs.

Beaglehole R, Bonita R, Kjellström T (1993) Basic Epidemiology, WHO, Geneva

Cascade Consulting (2002) Revision of the Bathing Water Directive: evaluation of the draft WHO beach classification methodology, (Final report submitted to Department for Environment, Food and Rural Affairs)

Commission of the European Communities (1976) Council Directive of 8th December 1975 concerning the Quality of Bathing Water (76/160/EEC), Off J Eur Community, L31/1, Brussels

Commission of the European Communities (1994) Commission Proposal for a Council Directive concerning the Quality of Bathing Water, Communication from the Commission to the European Parliament and the Council, COM (94) 36 final, Brussels

Commission of the European Communities (2000) Developing a New Bathing Water Policy, Communication from the Commission to the European Parliament and the Council, COM (2000) 860 final, Brussels

Commission of the European Communities (2002) Proposal for a Directive of the European Parliament and of the Council concerning the quality of bathing water, Communication from the Commission of the European Communities, COM (2002) 581 final, Brussels.

Dake K (1992) Myths of nature: culture and the social construction of risk. J Soc Iss 48:21-37

EFTEC (2002) valuation of benefits to England and Wales of a revised bathing water quality directive and other beach characteristics using the choice experiment methodology (Final Report submitted to Department for Environment, Food and Rural Affairs), Economics for the Environment Consultancy Ltd, London.

Environment Agency (2002) WHO standards for bathing waters, Mimeo for DEFRA Water Quality.

Fishbein M, Ajzen I (1975) Belief, Attitude, intention and behaviour: An introduction to theory and research, Addison-Wesley, Reading

Fleisher JM, Kay D, Wyer M, Godfree AF (1998) Estimates of the severity of illnesses associated with bathing in marine waters contaminated with domestic sewage. Int J Epidemiol 27:722-726

Georgiou S (2003) Coastal Bathing Water Health Risks: Assessing the Public and Scientific Acceptability of Health Risk Standards, PhD Thesis, University of East Anglia

House of Lords (1994-5) Select Committee on the European Communities, 1st Report, Bathing Water, HL Paper 6-I; and 7th Report, Bathing Water Revisited, HL Paper 41, HMSO, London

Kay D, Fleisher JM, Salmon RL, Jones F, Wyer MD, Godfree AF, Zelenauch-Jacquotte Z, Shore R (1994) Predicting Likelihood of Gastroenteritis from Sea Bathing: Results from Randomised Exposure. The Lancet 344:905-909

Kay D, Falconer RA, Salmon R (1997) Modelling Disease Burden Attributable to Recreational Water Exposure and 'Health Gain' from Waste Water Treatment, Final Report – Costs and Benefits of Urban Waste Water Treatment - EPSRC Project GR/K64150/01, Engineering and Physical Sciences Research Council

Langford IH, Georgiou S, Bateman I, Day RJ, Turner RK (1999a) Public perceptions of health risks from polluted coastal bathing waters: a mixed methodological analysis using Cultural Theory. Risk Analysis 20:691-704

Langford IH, Marris M, McDonald A-L, Goldstein H, Rasbash J, O'Riordan T (1999b). Simultaneous analysis of individual and aggregate responses in psychometric data using multilevel modelling. Risk Anal 19: 669-677

Langford IH, Marris C, O'Riordan T (1999c) Public reactions to risk: social structures, images of science and the role of trust. In Bennett PG and Calman KC (eds.) Risk Communication and Public Health: Policy, Science and Participation. Oxford University Press, Oxford.

Marris C, Langford I, O'Riordan T (1998) A Quantitative Test of the Cultural Theory of Risk Perceptions: Comparison with the Psychometric Paradigm. Risk Anal 18: 635-647

Mitchell RC, Carson RT (1989) Using Surveys to Value Public Goods: The Contingent Valuation Method. Resources for the Future, Washington

Rotter JB (1954) Social Learning Theory and Clinical Psychology, Prentice-Hall, Englewood Cliffs

Slovic P (1992) Perceptions of Risk: Reflections on the Psychometric Paradigm, in S Krimsky and D Golding (eds), Social Theories of Risk, Prager, Westport

Thompson M, Ellis R, Wildavsky A (1990) Cultural Theory, Westview Press, Boulder, Co.

Wallston KA, Wallston BS, DeVellis R (1978) Development of the Multidimensional Health Locus of Control (MHLC) Scales, Health Educ Monographs 6:160-170

World Health Organisation (1998) Guidelines for Safe Recreational Water Environments: Coastal and Fresh Waters, Draft for Consultation ref. EOS/Draft/98.14, Geneva

World Health Organisation (1999) Health based monitoring of recreational waters: the feasibility of a new approach (*The 'Annapolis Protocol'*), WHO/SDE/WSH/99.1, Protection of the Human Environment, Water, Sanitation and Health Series, Geneva

World Health Organisation (2001) Bathing Water Quality and Human Health, WHO/SDE/WSH/01.2, Protection of the Human Environment, Water, Sanitation and Health Series, Geneva

Wyer MD, Fleisher JM, Gough J, Kay D, Merret H (1995) An Investigation into Parametric Relationships between Enterovirus and Faecal Indicator Organisms in the Coastal Waters of England and Wales, Water Res 29: 1863-1868

Wyer MD, Kay D, Fleisher JM, Salmon RL, Jones F, Godfree AF, Jackson G, Rogers A (1999) An Experimental Health Related Classification for Marine Waters, Water Res 33:715-722

5. Establishing coastal and marine reserves – with the emphasis on fisheries

Han Lindeboom[1] and Saara Bäck

Abstract

Marine reserves or protected areas with certain restrictions are formed for several reasons: protection of species or specific life stages, protection of habitats such as spawning, resting or feeding areas, and creation of more natural age composition in populations. Areas are established to prevent continuous impacts of human actions such as certain disturbance of fishing techniques. For scientific research and monitoring purposes marine reserves are indispensable. It is recommended that in marine reserves where fisheries, other destructive human activities and local pollution are forbidden or very limited, scientific research is carried out in order to reveal trends in species composition, abundance and age distribution. These data should be used for comparative studies with non-protected areas, and then be applied to obtain more sustainable use of resources, including optimal production and optimal nature preservation. For successful marine reserves it is necessary to define clear objectives for the closure, to include the stakeholders in the planning process from the beginning, to design proper, manageable and legally controllable boundaries, and to raise awareness and education. The EU Water Framework Directive (WFD) includes regulations for establishing monitoring programmes also for protected areas. Regular monitoring and evaluation programs should be executed to see if the objectives are met and to renew the management plans and redesign the areas if necessary.

Introduction

The coastal zone forms a productive boundary between ocean, coastal systems and land with characteristically high amounts of energy and nutrients that stimulate both high biological productivity and a wide diversity of habitats and species. The

---

[1] Correspondence to Han Lindeboom: hanl@nioz.nl or han.lindeboom@wur.nl

J.E. Vermaat et al. (Eds.): Managing European Coasts: Past, Present, and Future, pp. 103–117, 2005.
© Springer-Verlag Berlin Heidelberg 2005.

coastal zone is also facing human induced changes like pollution, eutrophication, urbanization, land reclamation, over fishing and exploitation of other living populations and other natural resources. In modern marine management plans sustainable use and protection and the precautionary principle are high on the agenda. It is a major challenge to manage and use the coastal areas and find the best sustainability compromise between use and protection.

The European Union Water Framework Directive (2000/60/EC) includes actions that deal with Protected Areas and their management. All Member States should establish a register of areas that require special protection under Community legislation e.g. for conservation of marine habitats and species. For example, Natura 2000 marine and coastal sites where the maintenance or improvement of the water quality is an important factor in their protection.

Intensive fisheries are one of the biggest threats to marine ecosystems. There are many signals that fishing activities affect the marine ecosystem on local and sometimes regional scales. Stocks of economically important species and biodiversity are declining. There is evidence that in the Dutch sector of the North Sea at least 25 species have decreased significantly in numbers or have totally disappeared (Bergman et al. 1991). On the other hand some opportunistic species have increased in numbers. On the Dutch Continental Shelf, fishing is now so intensive that every square meter is trawled on average once to twice a year. Bradshaw et al. (2002) showed that in the Irish Sea the negative change of benthic populations was correlated with the time period over which the site was trawled rather than intensity of activity. In comparison with other possible causes like pollution, eutrophication, or climatic changes, the results from fisheries impact studies led to the conclusion that changes observed in the North Sea ecosystem over the past 100 years can, to a great extent, be attributed to fisheries (Lindeboom and de Groot 1998).

Coastal and marine reserves and Marine Protected Areas (MPA) are important tools to sustain these coastal ecosystems (Parrish 1999, Boersma and Parrish 1999). Reserves have been established as a management tool to compensate for the over fishing on coastal populations (Garcia-Charton and Perez-Ruzata 1999) and to support sustainable fisheries (Mangel 2000). It is postulated that establishing marine reserves could improve the conservation of exploitated fish species (Boersma and Parrish 1999) and thus increase the annual catches of fish in surrounding areas (Pezzey et al. 2000). Recently, Myers and Worm (2003) investigated the development of large ocean fish such as tuna, swordfish, sharks, marlins and cod, which decreased worldwide by 90% in the last half century. To restore the populations they suggested, apart from lower fishing intensity, the establishment of reserves in the open oceans.

However, marine reserves are not isolated from long-term and long-range chemical pollution or effects of extensive nutrient concentrations in ambient waters. Proper reserves may need substantial buffer zones (Simberloff 2000). And to sustain the water quality, environmental management actions that are carried out outside the reserve boundaries are often necessary (Murray et al. 1999).

In this chapter we will address the question, why and how marine reserves are established and their role in future management of marine systems, with the emphasis on fisheries management.

Marine reserves for conservation purposes

Most marine management documents that have appeared in recent years start with the concept of a sustainable use of marine resources. The term itself is rarely defined. According to Agardy (1997): "It is now touted the world over as the solution to real and prospective global, regional and local environmental problems. Prolonged economical gain, ecological sound development, low-level use of renewable resources or parity among all resource users are terms often expressed. The most common meaning of ecological sustainability has to do with the ecosystem function. For an activity to be sustainable the activity must not cause environmental degradation in the systems sense. Removing organisms from an ecosystem or interfering with its critical processes can only be sustained over time if the system's functioning is not adversely impacted".

Many international agreements and EU Directives like the habitat Directive (92/43/EEC) and Water Framework Directive (2000/60/EC) include ideas of protection of habitats and species and sustainable use of marine resources. The combination of these policies recalls the same idea as in Rosenberg et al. (2000) on healthy habitats, which are also productive. They also brought the concept of healthy and productive ecosystem functioning to traditional fishery management, including the concept of essential fish habitats (EFH). In many fisheries agreements the importance of habitat protection is manifested (Turner et al. 1999). Depending on the past and present day status, and on the desired local ecosystem functioning, the sustainability needs to be defined for specific areas. This includes clear definitions of sustainable protection of non-target species and the definition of thresholds beyond which the risk of changes in the ecosystem are considered unacceptable. One of the great challenges is to set these definitions for the marine environment on local, regional and global scales. Then effects on both target and non-target fish species must be limited to levels that do not cause a decline and eventually collapse of the defined ecosystem properties.

Management of sustainable use of ecosystems requires information on the functioning of the system, on the actual and potential uses of its components and the effects of exploitation. This is true especially because ecosystems are not static, unchanging entities, but rather a complex and dynamic web of interactions that are affected by cumulative impacts (Agardy 1997, Lindeboom 2002). In order to effectively tackle the substantial marine conservation problems, clear questions need to be asked on how to sustain ecosystem functions and biodiversity, how to continue a sustained use of living resources, and how to modify our behaviour to reach that goal.

Part of that goal may be reached by establishing marine areas including no take zones or closed areas where the constant pressure of human activities is minimized (Wallace 1999, Mangel 2000). The so called "precautionary principle" implies that actions that produce irreversible change to ecosystems (e.g. extinctions and permanent restructuring of food webs) must be avoided, and risks and uncertainties must be taken into account. As long as we are not certain about the long-term effects of fisheries, the maintenance of relatively non-impacted areas may be an important part of a precautionary approach. Following the approach on land, the time has come to seriously consider the creation of real nature conservation areas in the open sea, where the marine ecosystem may develop without continuous human induced pressures.

The question of the size of the marine reserve is complicated. Reserves protect animals completely if they never leave the reserve area. The size of such an area depends upon the species to be protected. For species with low mobility as sessile benthos, e.g. the long-living shellfish, the area can be rather small, but considerable areas are needed for animals with long home ranges (Kramer and Chapman 1999). In general, Edgar and Barrett (1999) showed that the effectiveness of marine reserve corresponds with reserve size.

Reasons to create protected areas

Protection of specific species or groups of species

Species for which it may be important to establish protected or closed areas include: species in imminent danger of extinction; species that play a central role in ecological communities, often called 'keystone species'; species that may serve as an indicator of the ecological condition; and species that may help to raise public awareness (Agardy 1997).

For the Dutch North Sea, rays are a good example. These organisms disappeared from the coastal zone entirely, most likely mainly due to fishing (Walker 1998). Using marker experiments, Walker (1998) showed that these animals do not wander all through the North Sea but remain mostly within 20 km of their place of release. She recommends closed areas with the size of ICES rectangles (50x50 km) where local ray populations may re-establish themselves. Other animals that might return in such areas include oysters and lobsters.

Another example of marine reserves that are established to protect specific species are Seal Protection Areas along the Finnish Baltic coast. In the Finnish classification code of threatened species grey seals are a vulnerable species. The EU species directive demands the establishment of conservation areas for grey seals. The population of grey seals nearly disappeared in the last century. Until the 1960s overexploitation and harmful substances such as DDT and PCB affected the reproduction. There has been and still is a conflict between the fisheries and conservation and protection actions of grey seals. Fishermen state that grey seals eat the salmon directly from nets leaving only remains. Some compensation is paid to fishermen by the state. In state owned sea and coastal areas seven reserves were established in 2001. These areas are also included in the EU Natura 2000 network. The aim is to protect grey seals and their habitats and these areas will be used for research and monitoring. Some parts of the reserves are closed the whole year around and there are some restrictions, which deal with access to the area and fishing. Hunting is completely forbidden in these reserves.

Protecion of juvenile fish from early destruction

Habitat requirements change during species life cycles and thus one single marine reserve may not cover all life stages (StMary et al. 2000). There are only a few examples when reserves are recommended or are already established for certain life stages e.g. spawning sites of gag (*Mycteroperca microlepii*) and scamp (*M. phenax*) (Koenig et al. 2000) or plaice (Piet and Rijnsdorp 1998).

Along the Netherlands, German and Danish coast, an area called "plaice box" was established in 1989 in order to diminish mortality of fish juveniles. The "box" was intended to cover the major distribution area of the main commercial fish species such as plaice, sole and, to a lesser extent, cod. At first, the area was closed from 1 April till 30 September for beam and otter trawlers exceeding 300hp (221 kW). In 1997 the area was closed for trawlers exceeding 300hp for the whole year. Comparing the "plaice box" with a reference area, Piet and Rijnsdorp (1998) showed that the overall size structure of the commercially exploited fish species increased due to the change in trawling effort whereas that of non-target species did not change. The species composition was not significantly affected. Other trends that were observed both within and outside the "box" were: a general increase of species richness due to the influx of southerly species, and a decrease of the relative abundance of plaice. The latter led to the fishermen's opinion that the "plaice box" does not function as protection of fish stocks. However, it is likely that other causes such as, natural variation led to a decrease of plaice in the ten-year period that the "box" has existed. Lindeboom (2002) indicated large changes in the Wadden Sea and North Sea ecosystem in the late 80's, leading to smaller biomasses of shellfish in the Wadden Sea and possibly plaice in the North Sea. Lessons to be learned from the "plaice box" so far are that excluding fishing pressures leads to measurable changes in the marine ecosystem. But temporarily the 'positive' effects may be completely overshadowed by other trends in the natural system. To overcome this problem long periods of closure and continuous monitoring of both the ecosystem and the behaviour of the stakeholders are needed.

Creation of a more 'natural' age composition within fish populations

One of the features of overfished fish populations is a shift in the age distribution towards younger specimen (Daan 1989). In the past, fish like cod could grow to an age of 40 years or more, more recently specimen older than six or seven years are very rare. These age shifts have also been recorded in non-target species (van der Veer et al. 1990). These age shifts may influence the capability of populations to sustain sudden collapses caused by, for example, cold winters or diseases. In closed areas fish that stay in that area can grow until their natural death, thus increasing the mean age, rendering the populations less vulnerable to natural variations. In addition model results indicate that marine reserves could play a beneficial role in the protection of marine systems against overfishing (Gerber et al. 2003).

Protection of certain habitats, such as reefs, seagrass beds, maerl grounds, stony areas

Specific habitats listed in EU Habitat Directive like reefs, sandbanks, seagrass beds and maerl grounds can easily be damaged physically by movable fishing gears, oil or gas extraction and coral, sand or gravel mining. This damage may lead to a decrease in the natural functioning of these areas and in the long term even to their disappearance. Marine reserves closed to these activities are an adequate instrument to protect these vulnerable habitats. In all European coastal areas, sea grass fields, kelp beds, and reef forming organisms like shellfish and worms and specific stony areas need protection. The EU is establishing an extensive inventory of the most threatened areas, and Natura 2000 conservation areas, where different activities are regulated. In these areas, specific activities will only be allowed as long as their sustainable conservation status is maintained.

Prevention of the continuous impact of certain fishing techniques which change the ecosystem

Maybe in many areas we have to give up our traditional preoccupation with conserving structures or specific species, and instead direct ourselves towards safeguarding the critical ecological processes and properties that are responsible for maintaining the desired habitat and ecosystem functioning. In this approach we take the direct impact of the fisheries as starting point. Depending on the fisheries intensity and the direct effects on target and non-target species, managers may decide that this is not tolerable *at infinitum* in fast marine areas. As part of a 'precautionary approach', the creation of areas where the impact of fisheries on structures and non-target organisms is negligible may be a good conservation option. Special fishing techniques, like dedicated long-lining, may then be developed and applied.

Protection of areas for scientific research and monitoring purposes

There are various reasons for establishing protected areas for marine research such as comparative research between reserves and non-protected areas (Edgar and Barrett 1999, Kelly et al. 2000). An example comes from the Dutch sector of the North Sea where a biological monitoring program was started in 1988 (Duineveld, 1992). The aim is to establish possible trends in the development of benthic fauna during a period of 5-10 years. The research on the direct effects of fisheries (Lindeboom and de Groot 1998) indicates that the infauna and epifauna are easily influenced by fisheries on a short-term scale. Thus fisheries may influence the data collected in monitoring programs, rendering these data useless for establishing possible eutrophication or pollution trends. If trends, caused by actions other than fisheries, are to be monitored, the sampling sites should be off-limits to the fisheries. There could be more distorted data sets if beam trawlers ploughed the sampling area an unknown amount of times prior to the sampling. Studies of the settlement and survival of benthic organisms, studies of sediment-water exchange or the transport of suspended matter, and even the benthic mapping executed by ICES-members in 1986 (Künitzer et al. 1992) are possible examples of programmes whose results have been affected by fishing activity.

Another purpose for protected areas is to study long-term effects of fisheries. Long-term changes in the underwater ecosystem have often been observed, and many of these seem to be related to the fisheries (Bergman and Lindeboom, 1999). But so far it has been impossible to indisputably identify cause-effect relationships. For example, rays have disappeared from the Dutch coastal zone most likely due to fisheries. But incontestable evidence is still needed. Comparisons between large fished and relatively unfished areas may provide such evidence, or give results which clearly reject the hypothetical relationship between fisheries and the occurrence of rays. Such research will even yield more conclusive results if a large fishing-free zone is created in a previously heavily fished area. The size of such an area depends upon the species to be studied. For sessile benthos, such as the long-living shellfish, the area can be rather small, but considerable areas need to be closed off for migrating animals.

Comparing the effects of fisheries with the effects of other anthropogenic influences will be a major task of applied scientific research. However, it is almost impossible to quantitatively estimate the individual effects of fisheries, eutrophication and pollution in a certain marine area. The establishment of a protected region in such an area may provide the practical means to study the effects of different anthropogenic activities. To investigate optimal future management strategies one could then execute different experimental management options in sub-areas of this closure, e.g. allowing different fishing intensities or gears, in connection with a proper monitoring programme.

An example: The Dutch North Sea

At present, there is consensus about the view that the North Sea is heavily overfished (Daan 1996, Bergman et al.1991, Bergman and Lindeboom 1999). A reduction of fishing effort of about 20-40% will enable the commercial stocks to build towards a more natural population structure (numeric as well as qua age distribution). Such populations are less vulnerable to natural fluctuations, which will result in less sudden changes in licensed quota and thus in higher economical profits. A reduction in fishing effort will also lead to a reduction of the impact. Furthermore, alternative gears have to be designed to catch target fish more selectively and to minimize the by-catch and mortality in undersized and non-target fish and invertebrates. As even the most selective trawling gears will have bottom contact by means of the groundrope, direct mortality will still be induced in epifauna species living on the seabed (e.g. bivalves, sponges), slowly swimming fish, and egg capsules of rays and whelks. Habitat structures build by tube building worms or bivalves will be destroyed as well. Therefore, a significant reduction of fishing effort and the development of selective gears will not result in a sufficiently low fishing mortality to enable the recovery of populations of sensible animals as rays, long lived bivalves, sedentary epifauna species (sponges, anemones, hydroids), whelks, and structure building fauna species. For the conservation of these species the designation of areas closed to harmful fisheries is needed.

In the early 1990s a study was contracted on the necessity and feasibility of the designation of protected areas in the Dutch sector of the North Sea. This was planned as a contribution to the conservation and, where possible, rehabilitation of a natural diversity of ecologically valuable areas (Bergman et al. 1991). The objectives of such a designation would be:

1. To preserve, rehabilitate and develop natural values by limiting the effects of human activities that cause detectable changes;
2. To protect animals which are an integral part of the Dutch sector of the North Sea.

First, four criteria were developed that may be used for the designation and selection of areas which qualify for a protected status:

1. The extent to which specific activities have developed into a threat to the existence or normal functioning of groups of animals or species;
2. Whether a prohibition or restriction of certain human activities would reduce this threat;
3. The use of ecological criteria, such as diversity, representativeness, integrity and vulnerability to identify the areas most suitable for a protected status;
4. The question whether there are adequate legal instruments to ensure effective protection of the selected areas.

Taking into account the effects of different human activities and the above criteria, it was concluded that an area directly northwest of the Frisian Islands qualifies for a protected status. In this area, containing coastal waters, sandy bottoms, the Frisian Front area, muddy areas and restricted stony areas, it will be possible to protect different types of benthic communities, including invertebrates and fish.

The following protective measures have been proposed for the area: 1) close the area for all types of fisheries throughout the year; 2) prevent or minimize oil containing discharges from offshore mining installations; 3) take area specific measures with respect to offshore mining, shipping, military activities, sand extraction, dumping and the laying of cables and pipelines whenever the situation in the area calls for such measures; 4) consider additional measures if the area is to be used as a reference area for scientific research.

Following the publication of Bergman et al. (1991) the Dutch government considered initiatives to establish a protected area in the Dutch sector of the North Sea and to study the actual protective effects of such an area. However, due to very strong opposition from the fisheries sector, the political judgement not to create a new trouble area when agriculture was already causing so many problems, and the lack of support at a European level, the idea was temporarily abandoned. Recent EU policies and Directives now provide an appropriate context to reopen the discussion.

Box 1. The Humber and nature conservation (by Tony Edwards)

Background. The Humber Estuary with its deep water navigation and its industries plays a vital role in the UK's economy. The estuary, where over 300 000 people live, has the country's largest complex of ports and one of its biggest clusters of the chemicals and oil refining industry, all protected by tidal flood defences. It is also of outstanding value for wildlife conservation, particularly waterfowl. Sea level in the estuary during the twentieth century rose at a rate of 2-3 mm per year, increasing to a predicted 6mm per year as a result of global warming. The Humber is a dynamic estuary with a tidal range of up to 7m, and the channels and sandbanks are continually moving. The inter-tidal area has been greatly reduced since large-scale reclamation commenced early in the seventeenth century. The habitats of particular interest are coastal lagoons, Atlantic salt meadows, reedbeds, mudflats, sandbanks and the estuary itself, which has populations of the endangered river and sea lampreys. There is also a thriving grey seal colony at the estuary mouth. A review of the wildlife designations is in progress by the UK Government's nature conservation agency. The review includes the possible extension of the existing RAMSAR site and the Special Protection Area (SPA) of the European Birds Directive to include all of the inter-tidal habitat, and a Special Area for Conservation (SAC) under the Habitats Directive.

Protection of the European Marine Site. The quality of the Humber has improved greatly over the last 10 to 15 years as pollution in the inland industrial catchments has been cleaned up, and treatment provide for sewage and industrial effluents discharged directly to the tidal waters. The Environment Agency is developing the Humber Estuary Shoreline Management Plan (HESMP) to provide a sustainable strategy for the long-term investment in flood defences to counter rising sea level and other risks of flooding. The Objectives of the project are to develop a coherent and realistic plan for the estuary's flood defences that is: compatible with natural estuary processes and adjacent human developments; sustainable, technically feasible; economically viable; environmentally appropriate; and socially acceptable. The broad strategy is to maintain a line of defence around the Humber and to assess if the alignment can be improved in some places. In front of the main urban and industrial areas there is no alternative to raising the defences on their existing line. Flood defences will be set back in some rural areas to compensate for habitat losses. Setback can also result in greater stability of embankments by having foreshores to dissipate the erosive energy. The first two Humber managed realignment sites are being developed and another eleven potential sites are being evaluated. In most cases the Environment Agency will buy the farmland.

The Habitats Regulations require "relevant authorities" to produce collectively a single management scheme covering day to day activities, rather than "plans and projects". There are 39 Humber Relevant Authorities – Environment Agency, English Nature, local authorities, drainage boards, sea fisheries committees, water companies, navigation and harbour authorities, and the Ministry of Defence (for the weapons range at the estuary's mouth). Matters being examined are flood defence maintenance works, land drainage, diffuse pollution, maintenance dredging, fisheries management, physical barriers to fish movement, and tourism and recreation impacts. The management scheme, including a prioritized and costed action plan for each partner organisation, should be published in 2004. The implementation of the scheme and the condition of the site will be monitored and other actions developed if necessary.

Further information:

A.M.C. Edwards (2001) River and estuary management in the Humber catchment. in D. Huntley, G. Leeks and D. Walling (eds.) A land-ocean interaction study: measuring and modelling fluxes from rivers to the coastal ocean. IWA Publishing, pp. 9-32

A.M.C Edwards, R. J. Freestone and C. P. Crockett, (1997) River management in the Humber catchment. Sci Total Environ, 194/195:235-246

Environment Agency (2000) Planning for the rising tides: the Humber Estuary Shoreline Management Plan

Participation and involvement of stakeholders

It is suggested that stakeholders should be identified in the beginning of the project and during the establishment process of marine reserves consultative and community based procedures should be followed. Local views should be given significance alteration in the consultancy process and state departments should be flexible and willing to, negotiate and follow a "bottom – up" approach rather than being dogmatic and dictatorial via "top-down" procedures (Hughey 2000, Nickerson-Tietze 2000).

A possible mistake that was made in Netherlands in designation of protected areas in the Dutch sector of the North Sea in the early 1990s was not to involve all stakeholders in the discussion from the start. The lack of proper consultation and local politics and press coverage created a very hostile fisheries community. Although one may wonder if the politicians would have reacted at all because conflicts of interests between governmental departments also were playing a crucial role. However, involving the fishing community from the outset could have avoided many antagonistic reactions. A dialogue about future measures between politicians, managers, scientists and fishermen should start from the planning phase of the project.

Established marine reserves

There is a lot of experience concerning the establishment of reserves especially in tropical areas. Some of the most famous examples are the Great Barrier Reef in Australia, the Galapagos Islands, Manado in North Sulawesi in Indonesia and the Saba Natural Reserve in the Dutch Antilles. Often these parks are multi-user protected areas where certain functions, like fisheries, anchoring, diving etc., are allowed or completely forbidden in parts of the areas. Such multi-user protected areas are or should be created all around the world. However, Craik et al. (1990) state that "the selection of sites usually owes more to the fact that they are not in demand for more obvious economic priorities than the intrinsic nature of the ecosystem".

Many more or less successful marine reserves have been established in the tropics where the visibility and the attractiveness of coral reefs for tourists is a major drive for protection. Also sanctuaries for birds, turtles or sea-mammal protection have been successfully established in coastal areas. However, there is a lack of information on the values of submarine nature conservation and thus this is hindering the process of establishing marine reserves further offshore. Fortunately, more and more sophisticated technical methods are being developed for surveying the under water world, both for geological and biological values. The pressure and need to establish reserves also in EEZ (Exclusive Economic Zone) is increasing in Europe (Andrulewicz and Wiegat 1999) and thus in future more offshore marine reserves could be established. It is much more difficult to sell the natural values of fish, benthos or permanently submersed habitats, than that of breeding birds or dolphins. So far, only if the protection leads to quantifiable increases in harvests or profits, will the creation of this type of reserves be considered.

Halpern and Warner (2002) reviewed 112 independent measurements of 80 reserves and showed that the higher average values of population density (91% higher), biomass (192% higher), average organism size and diversity (20-30% higher) inside reserves (relative to controls) reach mean levels within a short (1-3 y) period of time and that the values are subsequently consistent across reserves of all ages (up to 40 y). These values were independent of reserve size, indicating that even small reserves can produce high values and the authors offer evidence that marine reserves of all sizes can engender biological responses.

Numerous studies indicate that establishing marine reserves increases biodiversity and stocks of commercial species. Here we present only a few examples: increased biomass of reef fish (Tupper and Rudd 2002), increased condition and reproductive potential of white seabreams in the northwestern Mediterranean (Lloret and Planes 2003), increased density, mean size and egg production of snappers (Willis et al. 2003), increased abundance, size, biomass and reproductive output of spiny lobster, *Jasus edwardsii* (Kelly et al. 2000), increased abundance and biomass of reef fish (Koenig et al. 2000), and increased reproductive output of northern abalone, *Haliotis kamtschatka* (Wallace 1999).

Procedures to establish marine conservation areas

Closed areas or multiple use marine protected areas are two possible tools that contribute to sustainable use of marine resources. An active involvement of a spectrum of stakeholder groups moves marine management from ineffective governmental sector control towards conservation that benefits both humans and nature.

There are several principles for the successful establishment of marine protected areas. The following are after Agardy (1997), Hughey (2000), and Lindeboom (2000):

1. Clearly define specific objectives and criteria for marine protected or closed areas at the outset;
2. Identify the stakeholders in each case. Get as much input from stakeholders as possible. The involvement of the stakeholders, in many cases users of the area e.g. the fishermen, is crucial for different reasons. Stakeholders have traditional (historic) knowledge about resource dynamics and ecosystems that will be important to determine levels of sustainable use. Also, stakeholders can increase the public awareness and promote good marine stewardship, including use, responsibility and protection;
3. Make the planning process truly participatory and consultative, as opposed to allowing user groups to comment on a plan developed by a single stakeholder (usually a government agency);
4. Design zoning to maximize protection for ecologically critical areas, while allowing sustainable use in less sensitive, vulnerable, or important areas. If non-destructive fishing techniques are available they could be allowed in (part of)

the area. It may even be possible that more environmentally friendly fishing techniques that at present are not economically feasible (e.g. long-lining) become profitable if destructive techniques, like beam trawling are banned in larger areas;

5. Design marine protected area boundaries based on field surveys so that they reflect ecological reality as much as possible (avoid squares and other 'unnatural' shapes, encompass estuaries and landward sides of coastal zones, etc.). people should also easily be aware when they are inside or outside the area. Boundaries along well-established lines, e.g. the edges of the ICES rectangles, may facilitate both awareness and control;

6. Be prepared to alter the design or the management as more ecological and sociological information becomes available;

7. Design the marine protected area and develop its management plan with feasibility in mind- and look for ways to self-finance management operation from the onset;

8. Obtain international recognition of the protected area, and assure a worldwide-adopted legal status. Important instruments in this context include: EU habitat directive NATURA 2000 network, United Nations Convention on Law of the Sea (UNCLOS); United Nations Environment Programme (UNEP); UNESCO's Biosphere Reserve Programme; Agreements from Agenda 21 of the Rio Meeting; and the Ramsar Convention on Wetlands of International Importance;

9. Develop monitoring and evaluation methodologies that are appropriate to the specific objectives and include these in design criteria. Hereby, both the monitoring of biological, economical and social parameters and the prioritization of research needs should be closely linked to the management objectives (FAO 1998);

10. Form an independent and multi-user group to manage the marine protected area and monitor its effectiveness using established benchmarks;

11. Undertake valuation exercises under a broader public periodically to ensure that the full value of the marine protected area is being realized;

12. Use the marine protected area as a way to raise awareness and stimulate education;

13. Use individual marine protected areas as a starting point for more effective marine policies overall-either to begin a representative network of MPAs on a national or international scale, or to draw attention to larger scale environmental problems such as land-based sources of pollution, regional overexploitation, or habitat destruction.

Conclusion

There are good reasons to create protected areas or marine reserves in the marine environment. Nature conservation calls for them, scientific research desperately needs them, and even fisheries might benefit from them. However, since the establishment of such areas in the open seas of Europe will demand the approval of the

European Community, and because economics may be affected by the creation of fishing-free areas, a long and difficult political process lies ahead during which socio-cultural aspects will also have to be taken into account (Fiske 1992). Only an approach that integrates the needs and possibilities of all managers, exploiters, and scientists involved will facilitate the successful creation of real marine reserves in our coastal zones and the open sea.

References

Agardy TS (1997) Marine protected areas and ocean conservation. R.G. Landes Company and Academic Press, Inc., Austin, Texas

Andrulewicz E, Wiegat M (1999) Selection of southern Baltic banks- future marine protection areas. Hydrobiologia 393:271-277

Bergman MJN, Lindeboom HJ (1999) Natural variability and the effects of fisheries in the North Sea: towards an integrated fisheries and ecosystem management? In: Biogeochemical Cycling and Sediment Ecology, J.S.Gray et al. (eds). Kluwer, Dordrecht, pp.173-184

Bergman MJN, Lindeboom HJ, Peet G, Nelissen PHM, Nijkamp H, Leopold MF (1991) Beschermde Gebieden Noordzee -noodzaak en mogelijkheden-. NIOZ Report 1991-3: 195 pp

Bradshaw C, Veale LO, Brand AR (2002) The role of scallop dredge disturbance in long-term changes in Irish Sea benthic communities: re-analysis of an historical dataset. J Sea Res 47:161-184

Boersma PD, Parrish JK (1999) Limiting abuse: marine protected areas, a limited solution. Ecol Econ 2:287-304

Craik W, Kenchington R, Kelleker G (1990) Coral-reef management. In Ecosystems of the world 25; Coral reefs. (Ed. By Z. Dubinsky). Elseviers Science Publishers, Amsterdam

Daan N (1989) The ecological setting of North Sea fisheries. Dana, 8, 17-31

Daan N (1996) Desk study on medium term research requirements in relation to the development of integrated fisheries management objectives for the North Sea. RIVO-DLO report C054/96: pp. 83

Duineveld GCA (1992) The macrobenthic fauna in the Dutch sector of the North Sea in 1991. NIOZ Report 1992-6: 19 pp

Edgar GJ, Barrett NS (1999) Effects of the declaration of marine reserves on Tasmanian reef fishes, invertebrates and plants. J Exp Mar Biol Ecol 232:107-144

FAO (1998) Integrated coastal area management and agriculture, forestry and fisheries. Food and Agriculture Organization of the United Nations, Rome, Italy

Fiske FJ (1992) Sociocultural aspects of establishing marine protected areas. Ocean Coast Manage 18:25-46

Garcia-Charton JA, Perez-Ruzafa A (1999) Ecological heterogeneity and the evaluation of the effects of marine reserves. Fish Res 42:1-20

Gerber LR, Botsford LW, Hastings A, Possingham HP, Gaines SD, Palumbi SR, Andelman S (2003) Population models for marine reserve design: A retrospective and prospective synthesis. Ecol Appl 13:47-64

Halpern BS, Warner RR (2002) Marine reserves have rapid and lasting effects. Ecol Lett 5:361-366

Hughey K (2000) An evaluation of a management saga: The Banks peninsula Marine Mammal Sanctuary, New Zealand. J Environ Manage 3:179-197

Hutchings JA (2000) Collapse and recovery of marine fishes. Nature 6798:882-885

Kramer DL, Chapman MR (1999) Implications of fish home range size and recolonisation for marine reserve function. Environ Biol Fish 1-2:65-79

Kelly S, Scott D, MacDiarmid AB, Babcock RC (2000) Spiny lobster, *Jasus edwardsii*, recovery in New Zealand marine reserves. Biol Conserv 3:359-369

Koenig CC, Coleman FC, Grimes CB, Fitzhugh GR, Scanlon KM, Gledhill CT, Crace, M (2000) Protection of fish spawning habitat for the conservation od warm-temperate reef-fish fisheries of shelf-edge reefs of Florida. Bull Mar Sci 3:593-616

Künitzer A, Basford D, Craeymeersch JA, Dewarumez J-M, Dörjes J, Duineveld GCA, Eleftheriou A, Heip C, Herman P, Kingston P, Niermann U, Rachor E, Rumohr H, De Wilde PAWJ (1992) The benthic infauna of the North Sea: species distribution and assemblages. ICES J Mar Sci 49:127-143

Lindeboom HJ, De Groot SJ (Eds) (1998) The effects of different types of fisheries on the North Sea and Irish Sea benthic ecosystems. NIOZ Report 1998-1 / RIVO-DLO Report C003/98, 1-404

Lindeboom HJ (2000) The need for closed areas as conservation tools. In: Kaiser MJ, De Groot SJ (eds.) Effects of fishing on non-target species and habitats. Blackwell Science Ltd., Oxford, pp 290-302

Lindeboom HJ (2002) Changes in coastal zone ecosystems. Wefer G, Berger H, Behre K-E, Jansen E in: Climate development and history of the North Atlantic realm. Springer, Berlin, pp 447-455

Lloret J, Planes S (2003) Condition, feeding and reproductive potential of white seabream Diplodus sargus as indicators of habitat quality and the effect of reserve protection in the northwestern Mediterranian. Mar Ecol Prog Ser 248:197-208

Mangel M (2000) Trade-offs between fish habitats and fishing mortality and the role of reserves. Bull Mar Sci 3:663-674

Myers RA, Worm B (2003). Rapid worldwide depletion of predatory fish communities. Nature 423:280-283

Murray SN, Ambrose RF, Bohnsack, LW, Carr, M, Davis, GE, Dayton PK, Gotshall, D, Gunderson, DR, Hixon, MA, Lubchenko, J Mangel, M, MacCall, A, Ardle DA (1999) No-take reserve networks: Sustaining Fishery populations and marine ecosystems. Fisheries 11:11-25

Nickerson-Tietze DJ (2000) Community-based management for sustainable fisheries resources in Phang-nga bay, Thailand. Coast Manage 1:65-74

Parrish R (1999). Marine reserves for fisheries management: Why not. Reports of Californian Cooperative oceanic fisheries Investigations, October, 1999, pp. 77-86

Pezzey JCV, Roberts CM, Urdal BT (2000) A simple bioeconomic model of a marine reserve. Ecol Econ 1:77-91

Philippart CJM (1998) Long-term impact of bottom fisheries on several by-catch species of demersal fish and benthic invertebrates in the south-eastern North Sea. ICES J Mar Sci 55:342-352

Piet GJ and Rijnsdorp AD (1998) Changes in the demersal fish assemblage in the south-eastern North Sea following the establishment of a protected area ("plaice box"). ICES J Mar Sci 55: 420-429

Rosenberg, A, Bigford, TE, Leathery, RL Bickers, K (2000). Ecosystem approaches to fishery management through essential fish habitat. Bull Mar Sci 3:535-542

Simberloff D (2000) No reserve is an island: Marine reserves and nonindigenous species. Bull Mar Sci 3:567-580

St.Mary CM, Osenberg CW, Frazer TK, Lindberg WJ (2000) Stage structure, density dependence and the efficacy of marine reserves. Bull Mar Sci 3:675-690

Tupper M, Rudd MA (2002) Species specific impacts of a small marine reserve on reef fish production and fishing productivity in the Turks and Caicos Islands. Environ Conserv 29:484-492

Turner SJ, Thrush SF, Hewitt JE, CummingsVJ, Funnell G (1999) Fishing impacts and the degradation or loss of habitat structure. Fish Manage Ecol 5:401-420

Van der Veer HW, Creutzberg F, Dapper R, Duineveld GCA, Fonds M, Kuipers BR, Van Noort GJ, Witte JIJ (1990) On the ecology of the dragonet *Callionymus lyra* L. in the southern North Sea. Neth J Sea Res 26:139-150

Walker PA (1998) Fleeting images: Dynamics of North Sea ray populations. PhD thesis, University of Amsterdam

Wallace SS (1999) Evaluating the effects of three forms of marine reserve on North Abalone populations in British Columbia, Canada. Conserv Biol 4:882-887

Willis TJ, Millar RB, Babcock RC (2003) Protection of exploited fish in temperate regions: high density and biomass of snapper *Pagrus auratus* (Sparidae) in northern New Zealand marine reserves. J Appl Ecol 40:214-227

6. Valuing Coastal Systems

Mihalis S. Skourtos[1], Areti D. Kontogianni, Stavros Georgiou, and R. Kerry Turner

Abstract

Integrated coastal zone management involves an assessment of development needs and economic inequality, pressures from population growth and mass tourism as well as social and cultural conflicts. In this context, economic valuation of coastal functions that provide goods and services is an important tool. Its applications and caveats are reviewed. Published value estimates range widely (i.e. 0.05-200,000 US\$ ha^{-1} y^{-1}) depending on function valued, method used and local welfare (f.x. expressed as GDP). An argument is made that, despite this variation and despite imperfect knowledge of ecosystem complexity, societal preferences or the 'real value' of nature, decisions on coastal development will be made and thus be helped best by the rational provision of scientific knowledge, from both natural sciences and socio-economics. A mixed methodological approach is therefore suggested to be most useful in practical, and multidisciplinary, situations. Three cases are presented of European valuation exercises at different spatial scales on the coast.

Introduction

Over the 1990s the European Union has gradually but fundamentally changed the structure and objectives of its environmental programmes for coastal protection. By 1995, however, the Dobris assessment of the European environment still noted that 'it is perhaps surprising that at present no comprehensive coastal zone management (CZM) scheme exists for Europe' (Stanner and Bourdeau 1995). Since then there has been a gradual shift in focus towards a more coherent framework for research on coastal functions and values, which could support integrated management. The now generally accepted scoping platform DPSIR has proved its worth (Turner et al. 1998, Turner, this volume). Such an approach needs to facili-

---

[1] Correspondence to Mihalis Skourtos: mskour@aegean.gr

J.E. Vermaat et al. (Eds.): Managing European Coasts: Past, Present, and Future, pp. 119–136, 2005.

tate communication within multidisciplinary research teams; it needs to recognise the functional continuity from watersheds to the coasts thereby helping to locate the scale of intervention less on the basis of traditional jurisdictions, and more towards appropriate ecosystem scales; it must encompass participatory management schemes which promise a substantive change in the exploitation of local knowledge.

It is worth noting that these policy shifts have in turn generated a growing body of economic research. For example, future progress towards the consolidation of a European integrated coastal zone policy will take place within the framework of the recent Water Framework Directive's evolving 'legacy'. The new Directive will provide a much more integrated and strategic (river-basin) approach to European water policy, explicitly recognising the interdependencies between ecological and socio-economic realities. Economic methods and tools are particularly relevant to this task, as water resource allocations must be guided by full cost recovery and cost-effectiveness criteria and therefore be in line with polluter pays principle. A similar approach has been taken by the recent proposal for a Directive concerning the quality of bathing water, presented by the Commission to the European Parliament and the Council (CEC 2002).

Starting from the 'wise use' sustainability imperative, the management of coastal zones: should take account of: (a) developmental needs and economic inequality; (b) pressure from population growth, immigration and mass tourism; and (c) social and cultural conflicts. Given this context what role should economic valuation play in integrated coastal zone management (ICZM); and will cost and benefit estimation (CBA) of changes in the coastal environment pass the test of analytical robustness, political usability and communicative adequacy? Use of CBA in European environmental agencies is sparse, though the idea of applying economic techniques when balancing environmental with developmental trade-offs has gained a certain credibility in UK, Germany and, to a lesser extent, Scandinavian countries (Bonnieux and Rainelli 1999). The policy relevance of benefit assessment for coastal management is aptly demonstrated in UNEP/MAP's explicit attempt to introduce 'resource consciousness' in its Regional Strategic Environmental Action Plan. Since, within the wider context of ICZM, 'raw cost information is insufficient to support investment decisions' what is needed is an investment plan where 'benefits [..] derived from the reduction or avoidance of pollution impacts on resources of social, economic and environmental value' are demonstrated. Moreover, in order for benefit estimates to be of relevance to prospective investors, their definition should include 'the conservation of resource for their existence (or non-use) value' [UNEP 1999, p. 67-69). Lastly it is worth mentioning that an increased academic interest in economic valuation in the last years is documented (Humphries et al. 1995, Pimentel et al. 1997, Bower and Turner 1998, Daily et al. 2000, Balmford et al. 2002).

This chapter reviews the current state of knowledge concerning the use of economic valuation techniques within the wider framework of integrated coastal zone management and presents some conclusions and interpretations of the available evidence, along with suggestions for further research. The chapter is structured as follows: we start with a concise description of the problem setting; then follow economic valuation methods, and the empirical content and applicability. The chapter ends with a number of practical examples from the published literature.

Facts and values in sustainable coastal management

There must be something special about coasts: throughout modernity, an ever-increasing number of people are continuously inhabiting the coastal or near-coastal part of the Earth. Their historical importance in the development of human civilization is therefore obviously beyond doubt. As meeting points of land, water and air, coasts have served to provide food and security, industrial and commercial development and, lately, leisure and conservation. As the process of industrialization and economic expansion, has accelerated, coastal zones have come under heavy pressure from human activities. The pace of human relocation from inland towards the coast has been described as 'one of the greatest human migrations of modern times' (Tibbetts 2002). The ensuing problems include physical modifications and habitat loss through coastal erosion, contamination and coastal pollution and depletion of fisheries. As a consequence, approx. 85% of the European coast is at high or moderate risk form development-related pressures (Bryant et al. 1995).

The problem is illustrated by the fate of coastal wetlands in the Mediterranean, a valuable source of natural capital that has been destroyed and degraded to a great extend. Their loss and/or degradation in this century amounts to 73% of the marshes in Greece, 86% of the most important wetlands in France, 60% of wetlands in Spain and 15% of lakes and marshes in Tunisia (MedWeT 1996). The situation is, as expected, crucial for island states and/or nations with a long shoreline. In Greece, for instance, a handful of indicators aptly demonstrates the importance of the coast and its vulnerability to human pressures: Coastal areas represent 72% of total territory, 86% of population, 88% of employment in manufacture, 90% of tourist activities and 90% of energy consumption (OECD 2000).

. Though the loss of valuable assets, such as the coastal resources, is well documented, this is not the case with the consequent, indirect or second order losses in economic values that this process entails. Efforts to highlight the economic or value side of the process of coastal change are scattered in a number of reports and studies addressing predominantly North American and to a lesser extend European and Third World coastal resources (David et al. 1999, Spurgeon, 1999, Dunn et al. 2000, Turner et al. 2001, Ledoux and Turner 2002). In a recent meta-review of economic valuation studies of coastal wetlands, Brander et al. (2003) observed a wide range in estimated values, with GDP being the most important factor to explain the variance in the data set. Their findings (Table 1) confirm the notion of high private and social values generated by the coastal resources put forward by Ledoux and Turner (2002).

Table 1. Range of estimated values of coastal resources for four important functions generating goods and services (as US$2000 ha^{-1} y^{-1})

Coastal resource function	Median value (range; number of observations for each function)
Recreation	491 (5-200,086; 52)
Water quality	288 (2-102,300; 30)
Fisheries (commercial)	201 (0.05-55,861; 72)
Biodiversity	214 (8-200,086; 12)

Source: Brander et al. 2003

Coast: A complex system

Coasts are notoriously complex systems encompassing highly variable biotic and abiotic components. What does the recognition of ecosystem complexity entail for the economic approach to valuation? From the beginning of economic thought, economists have taken for granted the analytical legitimacy of simplification when investigating capitalist production and the consequences of economic behaviour. Today's environmental crisis reveals the fact that such purposeful abstractions can prove problematic when used for policy prescriptions. It is this contradiction between the notion of abstract, quantifiable economic value and the concrete, limited and qualitative *physis* that went, with rare exemptions, unnoticed in the development process of economic tools and techniques. Classical political economy, with its main representatives Adam Smith, David Ricardo and Thomas Robert Malthus, has always been striving to keep the memory of pre-industrial, concrete wealth of *oeconomia natura* alive but could not resist the sweeping force of market reality and neoclassical formalism (Skourtos 1998).

Meanwhile, fundamental changes have occurred in our understanding of the functions and values of coastal ecosystems, and these have prompted many recent international efforts to protect and sustainably use them. Thanks to joint efforts with natural scientists, our 'production functions' linking natural and engineering processes with economic goods and services are far better understood. In spite of scientific advancement though, the gaps in our knowledge remain considerable. Integrated approaches to environmental planning with proper stakeholder involvement offer a possible way forward (Harremoes and Turner 2001).

With respect to economic valuation, two main conclusions can be drawn: Firstly, since we are forced to act in the face of potentially irreversible ecosystem change we have to be proactive and, consequently, conservative in our management plans. A sustainable use of resources has to take into account the existence of thresholds and other irregularities in the functioning of ecosystems under what is widely known as 'safe minimum standards' (SMS) approach (Randal and Farmer 1995, Crowards 1996).

The above recognition enhances the relevance of *ex ante* economic valuation studies. However, these involve societal uncertainty with respect to of future preferences, needs and incomes. It is plausible to assume that present societal preferences and needs are fuzzy and lack articulation. The act of eliciting present preferences is therefore criticised as blurring the process of *eliciting* existing preference structures with that of *constructing* them (Sagoff 1988, O'Neil 1997). The fact remains that complexity of both ecosystems and societies does not cancel out the need for hard choices in the face of both natural and societal uncertainties.

Trade-offs at the coast and at the margin [2]

'It is best to view the coast as a common resource, available to all. However, we need to apply certain standards of resource allocation and use to the coast, in order to sustain its attractiveness' (Carter 1988, p. 2). The ideal of a natural resource common to all, echoing in the above paragraph, is in obvious contrast with the realities of current environmental planning. The knowledge of ecosystem complexity and systemic interrelationships prompts the scientist to attach an *absolute* importance at the specifities of ecosystem processes and structures confounding the notions of functions and values per se (Toman 1997). A second thought though reveals the fact that modern ecological science does accept the notion of relative and therefore hierarchical importance referring to single species and ecosystem types (Perrings 1995). Notions like 'keystone species', 'critical biotopes' or 'critical functions' reveal this fundamental fact, which within a management perspective allows someone to think about trade-offs when designing effective and efficient conservation priorities.

Ecologists argue that the main benefit of preserving natural ecosystems is related to the maintenance of critical ecosystem services and the integrity of the life-support systems (e.g. Costanza et al. 1997). Landscape functions change on several spatial and temporal scales that are of concern to conservation planning, particularly at the coast. Conservation in hotspot areas, where fast-moving changes are driven by spatially undifferentiated economic development policies, as is the case in the coastal zones of the European South, needs compelling conservation planning processes to enforce its claims and arguments. Traditionally, conservation planning has to cope with the rigorous selection of 'spatial conservation objects' (habitat remnants, natural areas) featured by biotic communities, landscapes or species, representative of the ensemble of biodiversity. Furthermore, the 'natural areas' selected should form a coherent and stable nature reserve system. However, one should keep in mind that, even within the previously described framework, the prioritisation of conservation effort remains of the highest importance since *ad hoc* procedures of allocation of nature, human resources and funds may seriously jeopardize the efficiency of conservation planning. At large- to medium- spatial scales, ecological strategies to establish priorities about what to conserve are mainly based on the identification of biological richness, rarity or complementarities of biota among conservation units (i.e. sites, remnants, biotopes, grid cells etc.). Species lists of various taxa (e.g. groups such as birds, plants or mammals, indicator species, threatened species, etc.) are used to identify hotspots and/or threatspots (Troumbis and Dimitrakopoulos 1998), or mega-diversity countries (Mittermeier and Werner 1990). However, as Dinerstein and Wikramanayake

---

[2] This section is based on joint work with our colleague A. Troumbis

(1993) have already underlined, prioritisation approaches based only on species valuation fail to incorporate potential threats for biodiversity, which is the very essence of both conservation effort and prioritisation. Thus, at a realm scale, they have proposed the use of 'conservation potential/threat index' (CPTI), which has been used to forecast the effects of mid-term deforestation on conservation in the Indo-Pacific. On similar lines, Turner et al. (2001) have recognized the importance of the ecosystem value (primary value) as opposed to the value of the parts (secondary values) and the inability of the traditional economic valuation procedures to capture it. The problem of the cost of biodiversity (i.e. species) conservation has been addressed in recent approaches to optimise the selection and allocation of nature reserves (e.g. Lombard et al. 1997, Ando et al. 1998). When land prices are included in the selection procedure, the constraint of 'value for money' imposes two alternative approaches: the first seeks to minimize costs by taking into account land prices while including a fixed number of species. The second maximizes the number of species protected for a given cost (Pimm and Lawton 1998). Whatever the perspective is, it is well established that biodiversity conservation planning should advance by combining ecological patterns with practical and political considerations.

Scaling down to an area that is typical of many European coasts (i.e. coastlines of 100s of km), requires testing the validity of general conservation planning protocols and methods. For instance, the major challenges for conservation planning at that scale are (1) the identification of 'core natural areas', and (2) the identification of areas or landscape elements which facilitate the control of abiotic, biotic and social-economic conditions for biodiversity within the 'core areas'. Buffer zones and ecological corridors are the best-known types of measures. Their function for a certain habitat is to supply, buffer, extract and retain water, nutrients, energy, organisms and man. A tract of land can have more than one function for a habitat. For example, an organism-supply area should guarantee, by its design and location, optimal movements between core habitats of the catchment/coastal zone continuum. Movements can serve different functions for organisms such as dispersal of youngsters, seasonal or daily migration from/to resting or foraging places.

Nevertheless, a decade after the Convention on Biological Diversity (CBD, UN 1992), the core issue of definition of biodiversity conservation priorities at a global scale remains unsettled. One of the main contentious issues is specifically related to differing perceptions and the operational definition of the intrinsic value of biodiversity. CBD includes the statement that *'ultimately, all ecosystems should be managed for the benefit of humans'*. It also includes the principle of *"benefit-sharing"*. According to these assumptions, the objectives of management of land, water and living resources are a matter of societal choice. Almost inevitably, conflicts arose over whether any framework policy text, such as the CBD, could legitimately say that all ecosystems must be managed, and if so, whether that should always be for human benefit or, on the contrary, whether it is ever legitimate to deny the right of humans to use living resources.

On the one hand, many stakeholders (countries, land owners, producers, etc.) see the alleviation of poverty as the central issue for their societies and therefore view the prime function of natural resources (including coastal zones) as a means to reduce human misery. On the other hand, other stakeholders accept legislation that forbids any human activity in designated pristine lands and some others insist on the intrinsic value of biodiversity. In the international political arena, these conflicts are overcome through formalistic compromises such as: *'Ecosystems should be managed for their intrinsic values and for the tangible or intangible benefits for humans, in a fair and equitable way'* (CBD/COP V Decision V/6 2001).

Unfortunately, such compromises can never resolve practical problems at the local scale, where evaluation of alternative development paths depend on what is meant by *'choice'* - and, for that matter, by *'society'*. Societies rarely really choose which way they will develop. Individuals make choices that have a proximate effect on their lives. Many of these choices produce externalities that individuals do not know about, or prefer to ignore, and that have a major long-term influence on the 'choices' that society drifts into.

ICZM aims at preserving coastal resources, their ecological functioning and ultimately their values, by applying adequate land use planning within a social, institutional and economic context. So far, several categories of values have been defined: economic, aesthetic, ethical, scientific, evolutionary and ecological. The multidimensionality of the coast adds to the confusion about the different values of its components. The various actors may have different or even conflicting perceptions on the significance of these values because of cultural differences, difficulties in calculating benefits or placing a monetary value on living entities, products or services. Rational arguments that would strengthen the perception of the public and policy makers about the seriousness of threats to coastal resources need to be established through new methodologies to valuate and evaluate the various forms of coastal goods and services from the perspective of all societal actors.

What then can scientific method offer to the resolution of such conflicts, especially at local scales and within ecosystem entities that mediate multiple functions? To help "society" make informed decisions in using space and resources we have to start by quantifying ecosystem functions and identifying needs. We can then ask, what kind of knowledge and information does the policy-making process need in order to comply with a sustainable use of spaces and resources. It is evident that besides data and predictive ability regarding changes in ecological parameters, there is a fundamental need to prioritise alternative uses by means of both 'objective', ecological scores <u>and</u> 'subjective' economic values. It is through the combined use of both scoring systems that ecosystem values can practically 'speak truth to power' (House and Howe 1999). No wonder that economic valuation is, after all, a complex approach!

Valuation: A mixed methodological approach

As generally understood, environmental evaluation of projects and policies is a generic term relating to the identification, measurement and assessment of environmental impacts. Evaluation is a complex and multifaceted process involving a mixture of scientific and non-scientific approaches, a multitude of criteria and metrics . Evaluation is both a cognitive process as well as an institutional practice. It consists of a prior, analytical phase and a consequent synthetic phase. Analysis here means scientific identification and quantification of natural trends and impacts whereas synthesis is reserved for socio-economic and policy assessment of the impacts (McAllister 1980). The term valuation on the other hand is usually reserved for comparisons between objects and where *economic* valuation refers to assigning *relative* values to mutually *exclusive* objects. Economic values are relative, because they assess the importance of objects/policies always in relation to forgone possibilities for alternative objects/policies. Economic objects/policies valued in this context are mutually exclusive because they are scarce, i.e. you cannot have all of them at the same time. Accordingly, economic values are practically trade-off coefficients denoting the quantity of a good a person is willing to give-up (usually income) in order to secure the consumption of another (environmental quality). The process of economic valuation encompasses both the process of 'market valuation' proper as well as its complement 'non-market valuation' and can be cast in one of three forms (in ascending order of completeness):

- Cost-effectiveness analysis: Ranks the alternatives for achieving a certain physical target on the basis of the monetisation of costs incurred. Offers the least-cost options.
- Cost-benefit analysis: Ranks the alternatives on the basis of the sum of fully monetised net present values of costs and benefits. Offers the optimal (most efficient) choice.
- Multicriteria analysis: Ranks the alternatives on the basis of an explicit set of choice criteria and weighted preferences of stakeholders involved. Offers qualitative and quantitative solutions to complex choice problems.

In the realm of environmental assets, the magnitude of non-market values (accounting or shadow prices) depends upon four factors (Dasgupta 2001):

1. The conception of social welfare being adopted
2. The size and composition of existing stocks of assets
3. Production and substitution possibilities of existing stocks of assets
4. The way resources are allocated in the economy

These four factors imply that the economic valuation of natural assets transcends the narrow borders of conventional economic analysis. For example, factor 1 opens the possibility of deliberating alternative notions of social welfare measures and brings in notions of equity and fairness in the allocation of resources. Factor 4 invokes the importance of social institutions as allocative mechanisms and links their performance with over- and undervaluation of goods and services. Thus, economic valuation methodology gradually 'slides' to addressing old questions with a multitude of quantitative and qualitative approaches from the fields of cognitive psychology, cultural theory and philosophy. By doing so it inevitably turns to the wider context of socio-cultural setting wherein values are articulated, expressed and used. A prominent example of such methodological 'mixing' refers to the use of qualitative social research tools (i.e. focus groups, deliberative techniques) with which an in-depth understanding of perceptions, attitudes and motives towards natural assets is articulated. When coupled with a serious qualitative investigation, a number of 'anomalies' and puzzles in economic valuation studies emerge in a different light: protest bidders are better understood as a form of rejection of implied property rights in the valuation exercise, cultural, aesthetic and moral considerations are better viewed not as values per se but as motives behind stated or revealed choices; temporal instability and spatial differentiation of monetary estimates for similar goods could be anchored in differences in the framing of the overall valuation context; preference inconsistency may imply both cognitive constraints on the part of individuals as well as unfamiliarity with the good and its attributes, etc. Last but not least, we learn to appreciate the dynamics of values formation and expression since the numbers we get are 'snapshots' of a complex reality (O'Riordan 2001).

The analytics of economic valuation of natural resources

What is to be valued?

The first question we address in this section refers to the object of valuation. Since coastal zones are not yet unambiguously defined, we may as well start by counting goods and services that usually form the object of coastal zone valuation studies (Table 2).

Having identified the goods and services to be valued in a coastal zone and the relevant population/stakeholders to be addressed, the next step is to determine to nature of the uses of these goods and services. Uses refer to a spectrum of institutional settings from pure private (fishing) to pure public ones (critical habitat protection), as well as the intermediate cases of common pool and toll goods. Common pool resources are resources that are divisible but not excludable (beaches) whereas toll goods are excludable but not divisible (controlled entry to marine reserves). The importance of specifying exactly the nature of uses while designing the valuation of a coastal good or service lies with the sense of fairness and justice that the implied property rights invoke to users: where free access to beaches is considered to be a 'fundamental right' of users (as it is the case, for example, in Greece). A valuation scenario with entrance fees as payment vehicle is bound to produce a considerable number of protest bidders. It seems that the stronger the element of public use, the more reluctant the people are to express private values through surveys.

We notice here that in principle, and in accordance with the relative character of economic value, what we should value are less absolute stocks of goods and services but changes in their availability. A proper definition of the object of valuation in an ICZM context starts with a definition of the baseline and the alternative (policy-oriented) scenario (Bower and Turner 1998). Since scenarios are not predictive but conditional (what-if) exercises, such a framing of the valuation process addresses in a more suitable manner the inherent uncertainty about physical impacts of policies and projects on the natural environment.

Table 2. Typical goods and services produced in the coastal zone

	Renewable resources	Non-renewable resources
Goods, mainly used	Fish Shellfish Kelp and other species of seaweed Fresh water (desalinated sea-water) Energy from waves, tides and thermal or salinity gradients	Oil Gas Minerals Sand and gravel
Services, mainly consumed	Transport Defence Recreation (bathing, boating, fishing, skin-diving, observing wildlife) Disposal of degradable wastes and other residues	Disposal of waste and other non-degradable residues Facility sitting: on-shore/offshore, fixed or mobile industrial operations, e.g. material processing, marine terminals, ports, seabed pipelines and cables, power plants.

Source: CEC 1995

A number of benefits can thus be addressed including:

- Mitigation benefits composed of damage reductions and restoration benefits;
- Enhancement benefits resulting from increased flows of services;
- Preservation benefits flowing from marine reserves;
- Indirect economic benefits due to positive multiplier effects;
- Options benefits from keeping future use options intact.

Pareto-relevant welfare changes

Imagine a coastal zone with a variety of structural and functional characteristics. These characteristics indicate the potential of the coastal zone for supplying a specific function (i.e. to function as wintering biotope of wildfowl); they do not though guarantee automatically the supply of the said function (i.e. the final attraction of wildfowl for wintering purposes). Moreover, even if it is assured that the function is provided, it does not follow that the ecosystem is seen as providing the relevant services to humans (i.e. bird watching, research and educational activities). Finally, the supply of the services, though a necessary condition, is not by itself sufficient to determine the magnitude of the relevant economic values (King and Wainger 1999).

The above paragraph illustrates the anthropocentric approach to valuing nature. Taken literally, an anthropocentric approach does not concern itself with impacts on the environment *per se*, but only as long as these impacts impinge on human welfare. Attention is drawn to the fact that welfare changes include a differentiated palette of use as well as non-use aspects of environmental impacts. Individual

do care about other individuals (present or future) and an increasing number of valuation studies reveal non-selfish motives for protecting the environment. Often such motives are consolidated arguments in individual utility functions blurring the borders between citizens and consumers, as they have been traditionally defined (Sagoff 1988).

Impacts that people care about are causes of Pareto relevant welfare changes. Impacts that people do not care about are termed Pareto irrelevant welfare changes. Economic valuation studies are concerned with the former. The philosophical question whether Pareto irrelevant welfare changes should count in decision-making process is often debated in the literature under the heading of relative versus absolute, instrumental versus intrinsic, or anthropocentric versus ecocentric stances. Insisting on such dividing lines makes the philosophical debate on values rather sterile and fails to offers a practical guide to everyday resource management dilemmas (Turner 1999). An important consequence though of such a framing of the debate is to acknowledge that the concept of Total Economic Value capturing the sum of Pareto relevant welfare changes is only a partial and incomplete picture of the real value of nature. Primary or 'glue' ecosystem value, beyond individual preferences escapes our economic calculus but needs to be taken into consideration in one or another way (e.g. via imposed safe minimum standards).

The mechanics of preference elicitation

When goods and services exchange on the markets it is usually easy to observe and evaluate choices. Connecting choices to preferences and values is also observable in other forms of social contacts such as voluntary participation in the provision of public goods. The task becomes a complex one when we deal with environmental resources that are only partially present in market-similar settings: from elicited valuation statements contingent valuation experiments can only gauge as deep as the attitudinal level into human behaviour.

We can think of three possible ways to understand underlying preferences for environmental goods and services (Smith 1990): First, through observed choice, second, through verbal expressions and conversation, and third through observed adaptations due to learning. All three options have been to a lesser or greater degree utilised in the literature, spawning a variety of methodologies (Balmford et al. 2002; Bateman et al. 2002). Analysts have investigated a wide range of valuation problems and contexts including, for example, the mismatch between expert and public perceptions of environmental quality in coastal areas (Goodman et al. 1998); the differences between perceived and actual quality levels and their links to actual policy making and objectives setting (Hanley et al. 2003); and the potential to combine quantitative and qualitative data using stakeholder focus groups (Kontogianni et al. 2001).

Some practical examples

Three European cases were selected to span a range of spatial scales and illustrate the complexities that arise when attempts are made to value real world coastal functions in terms of goods and services, and include direct use, indirect use and non-use value estimates to somehow approximate total economic value.

Case 1. Combining publicly perceived and expert-judgment-based 'scientific' values of conservation quality. Source: Goodman et al (1998)

Stated-preference-techniques of non-market environmental valuation, such as CVM, rely on subjective, individual values among respondents, while scientists develop their own, possibly more objective measures of ecological values. In an attempt to bridge the gap, and thus enhance the policy relevance of stated preference techniques, the effort was undertaken to evaluate whether public preferences for conservation quality agreed with conservationists' assessment of the conservation value of coastal resources. It would then be possible to benchmark non-use values to conservation quality levels. The problem was conceived as a test for part-whole biases in evaluating goods with different extend of environmental protection.

The site under evaluation was the entire English coast; respondents were asked about their willingness to pay (WTP) additional taxes to avoid a loss of conservation quality. Those who answered positively were then asked how much of this additional tax they wanted to be spent in a specific group of coastal areas representing 10% of the entire coastline of England and Wales. Two specific groups were presented in a split sample; the first group representing areas with a relatively high level of conservation quality (Group I) and the second one representing areas with a low level of conservation quality (Group II). Data were collected through personal interviews following the relevant 'good practice' code of the NOAA Panel. A total of 806 questionnaires were administered of which 766 were usable.

The survey showed that a substantial portion of stated values related to non-use values. Overall, protest behaviour in the form of strategic bidding and free-riding does not appear to have been a significant problem within the study. Respondents reported a mean WTP of 48.36 pounds in additional annual household taxes for a coastal conservation programme for the entire coast. Respondents also appropriately distinguished the extend of environmental benefits provided by varying levels of the conservation programme's inclusiveness suggesting that perfect embedding did not occur at this level of analysis. The lack of significant difference between mean WTP for a conservation programme in the groups I and II though indicates that, overall, respondents did not express an economic preference for higher, rather than lower, levels of conservation quality. Additional analysis showed that in general respondents preferred higher levels of conservation quality, but that it may have been difficult for some to express their preferences in monetary terms. The modelling of valuation (bid) functions showed also that WTP was positively correlated with respondents' a) income, b) membership in an environmental organization, and c) use of the coast.

The authors conclude that coastal management policies based on the CVM may not coincide with the ecologically most preferable management strategy for coastal habitats.

Case 2: Combining revealed and stated preferences. Source: Hanley et al. (2003)

The hypothetical character of stated preference techniques and the consequent criticism that this methodological characteristic has provoked lies at the heart of the recent effort to

combine actual (that is, revealed) and contingent (that is, stated) information on individual choices. Two main approaches to using stated and revealed preferences data seem to exist. These are Random Utility Models combining stated and revealed preference data, and the Contingent Behaviour approach relating to either price or environmental quality changes. In this case, the benefits of coastal water benefits improvements were estimated by combining revealed preference data on actual visits to beaches, with contingent behaviour data relating the number of trips taken when hypothetical quality improvements occur.

The case study area was Scotland's south-west coast where bathing water quality has been problematic for many years, due to bacteriological contamination as measured by coliform counts. In collaboration with the Scottish Environmental Protection Agency the main bathing beaches which sampling should focus on were identified. On-site sampling based questionnaire was used which asked people i) about their trips to the beach where they were sampled (ii) about their activities they generally participate in on the beach (iii) about their trips to other beaches in the area, and (iv) about their perceptions of water quality at the beach where they were sampled. A proposed improvement in bathing water quality was then described and respondents were asked again whether this would cause them to visit the beach where they were being surveyed more frequently.

A first result of the survey concerns the subjective valuation of bathing water quality by respondents: there was an imperfect match between perceived water quality and biological monitoring results. Ranking beaches by subjective ratings of water quality would not therefore give the same picture as ranking by monitoring results. On a second stage of analysis, travel costs were estimating by using a figure of 10p per mile to represent the marginal costs of motoring. Applying a negative binomial random effects model a trip generation function was estimated where all the variables have the expected sign: Travel costs exert a strongly negative influence on trips as does also perceived water quality. The effect of the willingness to swim variable is positive and significant. Hypothetical improvements in water quality yielded only a 1.3 % increase in predicted trip frequency. The change in consumer surplus associated with enhanced water quality was 0.48 pounds per trip or 5.81 pounds per person. A rough guess of aggregate benefits amounts to 1.25 million pounds per annum.

Two interesting points arose from the study: The first is that perceived water quality is, as it is expected, a better measure in valuation models than actual ones. The second point though relates to the difficulties of using the results of studies based on perceived water quality from a policy point of view: Policy objectives and achievements relate to actual water quality rather than perceptions. This leads us to a related problem familiar within the stated preference literature: are intended trips a robust indicator of actual trips, in case the improvements described to respondents actually occur? The present study shows that combined revealed preferences-contingent behaviour models do not suffer from the hypothetical market bias often associated with contingent valuation.

Case 3. Nutrient pollution in the Baltic drainage basin – cost and benefits. Source: Turner et al. (1999), Markowska and Zylicz (1999)

A concerted attempt was made by a consortium of European researchers to estimate the costs and economic benefits of environmental improvements in the Baltic drainage basin. A 50% reduction in nutrient loading was adopted by the Helsinki Commission as a policy target in 1992.

A cost-effectiveness analysis was first carried out, to determine how to reach reductions in the nutrient load to the Baltic sea specified by international conventions. Measures A concerted attempt was made by a consortium of European researchers to estimate the costs and economic benefits of environmental improvements in the Baltic drainage basin. A involved the agricultural sector, sewage treatment plants, wetland restoration and traffic and other nitrogen oxides emissions sources. Marginal costs of these measures for

nitrogen and hosphorus reductions were calculated for all countries in the drainage basin that had coastal zones coincident with the Baltic sea. The relationship between possible nutrient reduction targets and associated minimum costs for their achievement was thus derived.

Although the results relied on some simplistic assumptions and suffered from missing information such as the retention and leaching of nitrogen and phosphorus, three important lessons could be learnt: (a) there are rapid increases in costs at reduction targets exceeding 40-45% reductions; (b) beyond 30% reduction, nitrogen reduction costs were estimated to be much higher than those for phosphorus for the same percentage reductions; (c) the cost of simultaneous reductions in both nitrogen and phosphorus loads would be less than the cost of separate reductions.

The cost-effective allocation of measures for a 50% reduction reveals that for nitrogen reductions, sewage treatment plants in the entire Baltic sea drainage basin account for about 33% of the reduction, wetland restoration contribute 33%, and the agricultural sector contributes mainly by reduction in nitrogen fertilisers, the cultivation of other crops, and changed practices for manure treatment. For phosphorus, sewage treatment accounts for 80% of reductions, and wetland restoration for 15%. A uniform 50% reduction also implied that the highest burdens would have to be carried out by Poland, Latvia, Lithuania, Estonia and Russia, raising compensation issues.

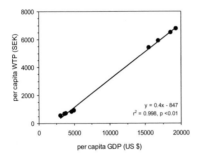 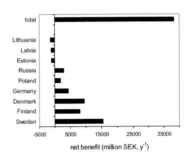

Fig. 1. Willingness-to-pay estimates from the Baltic as a function of GDP and net national benefits to riparian countries from a 50% nutrient load reduction to the Baltic Sea.

On the benefit valuation side, a total of 14 empirical valuation studies were carried out in three countries, Poland, Sweden, and Lithuania. These approximately addressed the 'total economic value' of reducing the effects of eutrophication, as well as sub-components of this total value such as: beach recreation benefits; existence and option values of preserving species and their habitats; and the benefits from preserving and restoring wetlands. The willingness to pay (WTP) data thus obtained allowed aggregate estimates for the three countries. Also, more controversially, they were aggregated across the two groups of economies around the Baltic Sea, i.e. transition and market economies, to give total basin wide benefit estimates. The project showed that WTP was strongly dependent on national GDP (Fig. 1).

In order to calculate basin wide benefit estimates, the values for the different activities carried out had to be added up, taking care not to double-count, and using the relevant correct populations. Since there are benefit estimates available for the same valuation scenario in only two of the 14 countries that are included in the Baltic Drainage Basin, any aggregation to the whole basin had to rely on strong assmptions.

The costs of pollution abatement and related economic benefit estimates were then brought together in a cost-benefit analysis framework. The results showed that there is considerable merit in the adoption of a basin-wide approach to pollution abatement policy in the Baltic and therefore in the implementation of an integrated coastal zone management strategy. Despite the pioneering nature (i.e. in the 'transition' economies) of some of the economic benefits research, there seems to be little doubt that a cost-effective pollution abatement strategy roughly equivalent to the 50% nutrients reduction target adopted by the Helsinki Commission would generate positive net economic benefits (benefits minus costs; Fig. 1). Results also indicated that a policy of uniform pollution reduction targets is neither environmentally nor economically optimal. Rather, what is required is a differentiated approach with abatement measures being concentrated on nutrient loads entering the Baltic proper from surrounding southern sub-drainage basins. The northern sub-drainage basins possess quite effective nutrient traps and contribute a much smaller proportionate impact on the Baltic's environmental quality state. The market economy countries such as Sweden, within whose national jurisdiction some of the southern sub-basins lie are also the biggest net economic gainers from the abatement strategy (Fig. 1). Finally, an important policy implication is that nutrient reduction measures in the Polish and Russian coastal zone areas would be disproportionally effective, but the financing of such measures would remain problematic if only 'local' sources of finance are to be deployed.

References

Ando A, Camm J, Polasky S, Solow A (1998) Species distributions, land values, and efficient conservation. Science 279:2126-2128

Balmford A, Bruner A, Cooper P, Constanza R, Farber S, Green RE, Jenkins M, Jefferiss M, Jessamy V, Madden J, Munro K, Myers N, Naeem S, Paavola J, Rayment M, Rosendo S, Roughgarden J, Trumper K, Turner RK (2002) Economic reasons for conserving wild nature. Science 297:950-953

Bateman IJ, Willis KG (1999) Valuing environmental preferences. Theory and practice of the contingent valuation method in the US, EU and developing countries. Oxford University Press.

Bateman IJ, Carson RT, Day B, Hanemann M, Hanley N, Hett T, Jones-Lee M, Loomes G, Murato S, Ozdemiroglu E, Pearce DW, Sudgen R, Swanson J (2002) Economic valuation with stated preference techniques. A manual. Cheltenham: Edward Elgar Publishing

Blamey RK (1995) Citizens, consumers and contingent valuation: An investigation into respondent behaviour. PhD thesis, Australian National University, Canberra

Bonnieux F, Rainelli P (1999) Contingent valuation methodology and the EU institutional framework. In: Bateman and Willis (eds) Valuing environmental preferences. Theory and practice of the contingent valuation method in the US, EU and developing countries. Oxford University Press, pp 586-612

Bower BT Turner RK (1998) Characterising and analysing benefits from integrated coastal management (ICM). Ocean Coast Manage 38:41-66

Brander LM, Florax RJGM, Vermaat JE (2003) The empirics of wetland valuation: a comprehensive summary and a meta-analysis of the literature. IVM report W-03/30 Institute for Environmental Studies, Vrije Universiteit, Amsterdam, The Netherlands

Bryant D, Rodenburg E, Cox T, Nielsen D (1995) Coastlines at risks: an index of potential development-related threats to coastal ecosystems. WRI Indicator Brief. Washington DC: World Resources Institute

Carter RWG (1988) Coastal environments. London: Academic Press

Commission of the European Communities (CEC) (1995) Communication from the Commission to the Council and European Parliament on the integrated management of coastal zones. COM (95) 511 final, Brussels

Commission of the European Communities (CEC) (2002) Proposal for a Directive of the European Parliament and of the Council Concerning the quality of Bathing Water. COM (2002) 581 final, Brussels

Constanza R, d'Arge R, de Groot R, Farber S, Grasso M, Hannon B, Naeem S, Limburg K, Paruelo J, O'Neil RV, Raskin R, Sutton P, van den Belt M (1997) The value of the world's ecosystem services and natural capital. Nature 387:253-260

Crowards T (1996) Addressing uncertainty in project evaluation: The costs and benefits of safe minimum standards. CSERGE Working Paper GEC 96-04, Norwich

Daily GC, Soderqvist T, Aniyar S, Arrow K, Dasgupta P, Erlich PR, Folke C, Jansson AM, Jansson BO, Kautsky N, Levin S, Lubchenco J, Maler KM, Simpson D, Starret D, Tillman D, Walker B (2000) The value of nature and the nature of value. Science 289:395-396

Dasgupta P (2001) Valuing objects and evaluating policies in imperfect economies. Econ J 111:1-29

David SD, Baish S, Morrow BH (1999) Uncovering the hidden costs of coastal hazards. Environment 41:10-19

Davos CA, Jones PJS, Side JC, Siakavara K (2002) Attitudes towards participation in co-operative coastal management: Four European case studies. Coast Manage 30:209-220

Dinerstein E, Wikramanayake ED (1993) Beyond hotspots – How to prioritize investments to conserve biodiversity in the Indo-Pacific region. Conserv Biol 7:53-65

Dunn S, Friedman R, Baish S (2000) Coastal erosion: evaluating the risk. Environment 42:37-45

Goodman SL, Seabrook W, Jaffry SA (1998) Considering conservation value in economic appraisals of coastal resources. J Environ Plann Manage 41:313-336

Hanley N, Bell D, Alvarez-Farizo B (2003) Valuing the benefits of coastal water quality improvements using contingent and real behaviour. Environ Res Econ 24:273-285

Harremoes P Turner RK (2001) Methods for integrated assessment. Reg Env Change 2:57-65

House ER. Howe KR (1999) Values in evaluation and social research. Thousand Oaks: Sage Publications

Humphrey S, Burbridge P, Blatch C (2000) US lessons for coastal management in the European Union. Mar Pollut 24:275-286

Humphries CJ, Williams PH, Vane-Wright RI (1995) Measuring biodiversity value for conservation. Annu Rev Ecol Syst 26:93-111

King, DM Wainger LA (1999) Assessing the economic value of biodiversity using indicators of site conditions and landscape context. Available at: http://www.ecosystemvaluation.org/

Kontogianni A, Skourtos MS, Langford IH, Bateman IJ, Georgiou S (2001) Integrating stakeholder analysis in non-market valuation of environmental assets. Ecol Econ 37:123-138

Ledoux L, Turner RK (2002) Valuing ocean and coastal resources: a review of practical examples and issues for further action. Ocean Coast Manage 45:583-616

Lombart AT, Cowling RM, Pressey RL, Mustart PJ (1997) Reserve selection in a species-rich and fragmented landscape on the Agulhas Plain, South Africa. Conserv Biol 11:1101-1116

Markowska A, Zylicz T (1999) Costing an international public good: the case of the Baltic Sea, Ecol Econ 30:301-316

McAllister DM. (1980) Evaluation in environmental planning. Assessing environmental, social, economic and political trade-offs. MIT Press, Cambridge, MA,

MedWet (1996) Mediterranean Wetland Strategy 1996-2006. Available at: http://www.iucn.org/

Mittermeier RA, Werner TB (1990) Wealth of plants and animals unites 'megadiversity' countries. Tropicus 4:4-5

O'Neill J (1976) Managing without prices: On the monetary valuation of biodiversity. Ambio 26:546-550

O'Riordan T (2001) On participatory valuation in shoreline management. In: Turner et al (eds) , Economics of coastal and water resources: valuing environmental functions. Kluwer, Dordrecht, pp 323-340

OECD (2000) Environmental performance reviews. Greece. Paris, France

Perrings C (1995) Ecological and economic values. In: Willis KG, Corkindale JT (eds), Environmental valuation, new perspectives. Cab International, Oxon pp 56-66

Pimentel D, Wilson C, McCullum C, Huand R, Dwen P, Flack J, Tran Q, Saltman T, Cliff B (1997) Economic and environmental benefits of biodiversity. Bioscience 47:747-756

Pimm SL, Lawton JH (1998) Planning for biodiversity. Science 279:2068-2069

Randall A. Farmer MC (1995) Benefits, costs, and the safe minimum standard of conservation. In: Bromley D (ed) The handbook of environmental economics. Blackwell, Oxford, pp 26-44

Sagoff M (1998) The economy of the earth. Cambridge

Skourtos M (1998) Nature. In: Kurz HD, Salvadori N (eds), The Elgar companion to classical economics Elgar, Cheltenham, pp 156-161

Smith VK (1996) Can we measure the economic value of environmental amenities? In: author?? (ed) Estimating the economic value for nature. Edward Elgar Publishing, Cheltenham pp 42-55

Spurgeon J (1999) The socio-economic costs and benefits of coastal habitat rehabilitation and creation. Mar Pollut Bull 37:373-382

Söderqvist T (2000) The Benefits of a less eutrophicated Baltic Sea, in Gren IM, Turner K, Wulff F (eds), Managing a sea: the ecological economics of the Baltic, Earthscan, London

Stanners D, Bourdeau Ph (eds) (1995) Europe's Environment. The Dobris Assessment. European Environment Agency: Copenhagen

Tibbetts J (2002) Coastal cities, living on the edge. Environ Health Perspect 110:674-681

Toman M (1997) Ecosystem valuation: An overview of issues and uncertainties. In: Simpson RD,Christensen NL (eds), Ecosystem function and human activities: reconciling economics and ecology. Chapman and Hall, New York

Troumbis AY, Dimitrakopoulos PG (1998) Geographic coincidence of diversity threatspots for three taxa and conservation planning in Greece. Biol Conserv 84:1-6

Turner RK (1999) The place of economic values in environmental valuation. In: Bateman and Willis, Valuing environmental preferences. Theory and practice of the contingent valuation method in the US, EU and developing countries. Oxford University Press, pp 17-41

Turner RK, Lorennzoni I, Beaumont N, Bateman IJ, Langford IH, McDonald AL (1998) Coastal management for sustainable development: analysing environmnetal and socio-economic change on the UK coast, Geogr J 164:269-281

Turner RK, Georgiou S, Gren IM, Wulff F, Scott B, Soderqvist T, Bateman IJ, Folke C, Langaas S, Zylicz T, Maler KG, Markowska A (1999) Managing nutrient fluxes and pollution in the Baltic: an interdisciplinary simulation study. Ecol Econ 30:333-352

Turner RK, Bateman IJ, Adger WN (2001), Economics of coastal and water resources: valuing environmental functions. Kluwer, Dordrecht

UNEP (1999), Strategic action programme to address pollution from land-based activities. UNEP Athens

7. Group report: Methodologies to support implementation of the water framework directive

Paula S. Moschella[1], Remi P.W.M. Laane, Saara Bäck, Horst Behrendt, Giuseppe Bendoricchio, Stavros Georgiou, Peter M.J. Herman, Han Lindeboom, Mihalis S. Skourtous, Paul Tett, Maren Voss, and Wilhelm Windhorst

Abstract

This chapter outlines the basic features and innovative aspects of the Water Framework Directive (WFD). Particular emphasis is given to problems and issues arising from the technical implementation of the environmental objectives set by the Directive. The difficulties of interpretation of key concepts such as ecological status, indicators and reference conditions are discussed in detail, and recommendations are given in order to avoid erroneous evaluation of these terms, leading to serious misclassification of the aquatic ecosystems. The consequences of the implementation of the WFD are also examined within the more specific context of water quality (bathing water quality, pollution by priority substances) and marine protected areas (MPAs). Tools and analyses to achieve environmental objectives and support the integrated management of water resources promoted by the Directive are discussed. It is suggested that models can be a powerful tool for prediction of reference conditions, the ecological classification of ecosystems and operational monitoring. The paper concludes with the importance of using an integrated approach for the implementation of the WFD, which can be achieved only by promoting communications and closer collaborations between scientists, economists and other stakeholders, particularly during the decision making process.

---

[1] Correspondence to Paula Moschella: pmos@mba.ac.uk

J.E. Vermaat et al. (Eds.): Managing European Coasts: Past, Present, and Future, pp. 137–152, 2005.

Introduction

After almost 10 years of scientific and political discussions the Water Framework Directive (EC 2000) came into force on 22nd December 2000. Before its creation, the EU legislation in respect of water resources was directed to specific issues (see also Ledoux et al., this volume) such as the control of dangerous substances discharged into the water environment (Dangerous Substances Directive, EC 1976), the monitoring and improvement of natural waters used for bathing (Bathing Water Directive, EC 1976) and the control of drinking water (EC 1998). At regional scale, each European country elaborated legislative instruments to deal with specific problems at local and national level. For example, separate national regulations were created for rivers, lakes and coastal waters. Despite the fragmented but consistent number of laws and regulations, two important aspects related to the water environment were still missing. First, aquatic ecosystems were not taken in sufficient consideration in the management of water resources. For example, the effects of a pollutant discharged in a river were mainly assessed on the basis of water quality standards, whilst the ecological implications for the ecosystem depending on that water body were not generally taken into account. Secondly, the ecological and societal needs were treated as separate issues, these being often a matter of conflict in the management of water resources.

For the first time in the history of European legislation, the Water Framework Directive adopted a holistic approach to the water environment. The overall aim of the Directive is to achieve a good ecological status of the aquatic environment and promote a sustainable use of the water. The policy adopted in the water management will therefore take into account not only water resources, but also the ecosystems depending on them, the human activities and needs. For surface waters, which include all inland surface waters, transitional and coastal waters, the most innovative aspects of the Water Framework Directives in respect of the previous legislations can be summarised in five points (EC 2003):

- *A river basin management approach*, where the different water compartments starting from the water sources, through the river systems to coastal waters are integrated in one single environment, regardless of whether part of it falls in different countries or regions. This new approach implies that water management need not be constrained by administrative or political boundaries, thus a common water management plan will have to be elaborated at both national and international level;

- *The concept of "good ecological and chemical status" of waters*, which must be achieved at the latest 15 years from the date of entry into force of the Directive. The ecological status is defined by biological, physical and chemical characteristics of the water environment. The chemical status is defined by quality standards and in relation to priority substances. In this new definition of good status of waters the Directive highlighted the importance of the ecological state of water resources in relation to the health of their ecosystems (see also Windhorst et al., this volume);

- *A combined approach for pollution control*, which sets emission limit values and quality standards under the same legislation. Before the Water Framework

Directive separate laws and regulations applied for the pollution point sources and diffusion sources. The introduction of a combined control of pollution links the causes (emission sources) with the effects for the water quality of the entire river basin district, bringing important changes in the pollution control management, particularly for the regulation of discharge of polluting substances from industries. This approach will have major consequences for the protection of coastal ecosystems, which are often indirectly affected by pollution sources located in other compartments of the river basin;

• *The concept of a more active public participation* in the river basin management. In this Directive the citizens and collective societal needs will have a greater influence in the decisions and actions to be taken for the implementation of the legislation (see De Bruin et al., O'Riordan, and Lise et al., this volume);

• *The concept of adequate water pricing* for a sustainable use or water resources. In this Directive, the price of water must reflect its true economic value, thus including also the costs of water used for leisure, navigational transports and the ecological value. This is one of the most innovative features of this legislation, as in several European countries these aspects of water usage have never been considered as specific costs to be included in the water price.

These key features have important consequences for the technical implementation of the Water Framework Directive and will involve great changes in the current approach, methodology and analysis of water management individually by European countries. The principles and the methodology of the new approach for the protection, management and sustainable use of the water environment are explained in detail in the Technical Annexes, particularly in Annexes II and V (EC 2000). These provide "instructions" on how the criteria and classification schemes must be established in order to achieve the environmental objectives. Despite the clear principles on which the whole Directive is based, the technical requirements and tasks described in the Annexes are quite complex and often controversial.

The aim of this chapter is to analyse the major implications derived from the implementation of the Water Framework Directive and to propose some suggestions, criteria and tools that might help in achieving the environmental objectives set in future. The main tasks of the directive, detailed in the Annexes for surface waters, are outlined alongside with the potential difficulties and specific issues arising from their implementation. These considerations will focus mainly on transitional and coastal waters.

Technical requirements for implementation of the WFD

The first step in the implementation of the Directive is to identify the river basin district, which is defined as "...*the area of land from which all surface run-off flows through a sequence of streams, rivers and possibly lakes into the sea at a single river mouth, estuary or delta*". This requirement sets the basis for the new integrated management approach adopted by the Directive. In practice, translating this definition into a classification is not so straightforward. For the first time coastal waters, as defined by the Directive, will be under the same management as all the other components of a river basin. The Directive requires coastal waters be assigned to a specific river basin. However, coastal waters are not discrete volumes of water like lakes or rivers and defining appropriate boundaries in relation to a river basin can result in a difficult task. Several coasts are affected to varying degrees by the output of different rivers, which are often under the control of different (inter)national authorities. Such geographical and administrative constraints will be solved only by intensifying active collaborations between river basin districts and promoting common actions in the implementation of the Directive at national and international level (Chave 2001).

Surface waters in the river basin districts are subdivided in water bodies such as rivers, lakes, estuaries and coastal waters. For each of those the Directive has set three main environmental objectives in relation with each other:

- To prevent deterioration of the status of all water bodies;
- To protect and restore all water bodies to a good ecological status and to a good ecological potential for heavily modified or artificial water bodies;
- To reduce pollution of water bodies from priority substances and eliminate the presence of priority hazardous substances.

Assessing the status of each water body is therefore the main task if these environmental objectives are to be achieved.

Classification of status of surface waters and implications for water management

The Water Framework Directive (WFD) has introduced a new approach in the assessment of the status of waters. The water quality standards used by the previous legislation will not be sufficient to define the status of surface waters. The Directive requires that the status of a water body be assessed by its ecological status and chemical status. The ecological status describes the condition of flora and fauna that live or depend on aquatic ecosystems and provides an index of the effects of human activities on the water environment. The chemical status describes the quality of waters which is defined by the presence / absence of pollutants exceeding the environmental quality standards listed in the Dangerous Substances Directive (EC 1976a) and priority substances set by the Water Framework Directive.

Ecological status and reference conditions

The ecological status of surface waters is defined in five different classes or levels (Table 1, see also Windhorst et al. this volume). Each class represents a certain level of disturbance to the aquatic ecosystem caused by human activities. The degree of disturbance is measured by how much the ecosystem changed from its original, undisturbed conditions, where *no or only very minor alterations* due to human activities occurred (EC 2000). These conditions represent the principal reference point around which the ecological scheme is built. Reference conditions have therefore a central role in the implementation of the Directive, which aims at restoring all water bodies to a good status, and where possible improving them to a high status. Despite its central role, reference conditions are not clearly defined in the WFD. If reference conditions are not correctly scientifically assessed (due to lack of data and personal views on the good status of an ecosystem), there could be a risk of misclassifying the water body, for example by placing it at lower ecological status class. As a consequence, human and financial resources would be unnecessarily used for restoration and monitoring programmes (Environment Agency 2002). It is therefore fundamental to establish criteria and tools that allow a correct definition of reference conditions, on the basis of which each water body will be assigned to a certain status class.

The Directive requires that the ecological status of the aquatic system be defined by a subset of biological indicators, called quality elements. These elements are components of the aquatic ecosystem (e.g. phytoplankton, fish etc.) that can be measured using parameters such as species composition or biomass (Table 2). The ecological classification will be based therefore on reference conditions of these quality elements. This implies that it is necessary to identify what these elements would be if the water body was affected by none or only very minor alterations resulting from human activities. Once physical and chemical reference conditions have been defined, these need to be translated into real reference biological values that will be used for developing monitoring systems (Environment Agency 2002).

Table 1. Ecological status classes and relative definitions for surface waters (Directive 2000/60/EC).

Status class	Definition
High status	The values of hydro-morphological, physico-chemical biological quality elements are similar to those determined for reference conditions, that is no or very minor anthropogenic impacts.
Good status	The values of biological quality elements show low deviation from those established for reference conditions. Values for physico-chemical quality elements fall within the range of environmental quality standards.
Moderate status	The values for biological quality elements show moderate changes from their reference conditions as a result of human activities.
Poor status	The values of biological quality elements deviate considerably from their reference conditions and the whole biological communities associated with the water body under undisturbed conditions is heavily modified by human activities.
Bad status	The biological quality element values show severe alterations and several components of biological communities associated with the water body under undisturbed conditions are absent.

Table 2. Biological quality elements in each water body and the parameters used to quantify them (Directive 2000/60/EC).

Biological quality element	Phytoplankton	Aquatic plants	Benthic invertebrates	Fish
Rivers	Composition Abundance	Composition Abundance	Composition Abundance	Composition Abundance Age structure
Lakes	Composition Abundance Biomass	Composition Abundance	Composition Abundance	Composition Abundance Age structure
Transitional waters	Composition Abundance Biomass	Composition Abundance	Composition Abundance	Composition Abundance
Coastal waters	Composition Abundance Biomass	Composition Abundance	Composition Abundance	

Feasibility of the ecological classification scheme and criteria for implementation

The description of the ecological status and the principles on which the ecological classification are based on has caused great concerns in the scientific world. The definitions provided by the Directive are rather generic; this is to allow each Member State to develop their own assessment criteria and adapt the classification scheme to the particular ecological characteristics and needs of the water bodies (Environment Agency 2002). The Directive, however, has set specific indicators, the biological quality elements, to assess the ecological status of the water body. The status of the water bodies is judged upon the reference conditions of phytoplankton, aquatic plants, invertebrates and fish. For coastal waters this set of indicators is further reduced, as fish are not included. Furthermore, the variables for which reference values will be established are generally restricted to the abundance and composition of the community; for instance, only phytoplankton communities are assessed also in terms of biomass. In this approach the complexity of an ecosystem is reduced to mechanistic processes and the generation of detailed knowledge of single, isolated parts. In this way, the exploitation of different functions of an ecosystem can be continued.

Evaluating the real status of the whole aquatic ecosystem by means of three, or four components poses a serious risk the correct classification of the water body will not be achieved. The Directive seems to adopt for biological communities the same approach used for monitoring water quality, which is assessed through periodical measurement of parameters such as the concentration of a pollutant or nutrients. In contrast with the relatively constant physico-chemical characteristics of the water body, biological communities vary greatly in space and time. It is therefore unlikely that the dynamic nature of an ecosystem can be represented adequately by the only by few variables as indicated in the Technical Annexes (Directive 2000/60/EC). The Directive also does not contemplate ecosystem functioning. For example, measuring the variation in composition and abundance of macrofauna and microphytobenthos in estuaries has little meaning if there is no assessment of the impacts that such variation might have on the function of that ecosystem, such as community respiration, decomposition of organic matter, nutrient recycling and retention (Environment Agency 2002). Biological quality elements should include additional indicators that provide complementary information on the health of the ecosystem as a whole in relation to space, time and impact sources (Borja et al. 2003).

It is widely acknowledged that spatial and temporal variability affects most biological communities, especially those living in estuaries and coastal waters. This makes it difficult to establish for each biological element a single reference condition that is relevant to the whole water body, as the sensitivity and resilience of the ecosystems to human disturbance can vary greatly between locations. To make reference conditions more type-specific the Directive has further subdivided the water bodies in different typologies characterised by typical hydro-morphological and physico-chemical conditions. The Directive also implies that biological reference conditions of a water body can be predicted on the basis of the physical and chemical features characterising that type. This principle is not always applicable, as biological communities can show a high level of variation in areas sharing very similar physico-chemical characteristics. Also, the influence of these factors on the composition and abundance of biological communities might differ depending on the biological quality element considered. For example, nutrients and temperature are important factors for the prediction of changes in phytoplankton, whilst benthic infauna is largely influenced by the sediment granulometry. Multiple site-specific reference conditions should be defined in each water body, to help discerning between the natural variability of a biological element and the changes caused by human activities, thus avoiding the risk of ecological misclassification (Environment Agency 2002).

Once the type of parameters used to assess or predict the reference conditions of a water class is set, the next step is to define what the values should be if no changes, or minor changes occurred in the aquatic system as a consequence of human activities. Interpreting reference conditions as pristine conditions would be inappropriate under many aspects. Ecosystems are not static but evolve and adapt continuously to the environment reaching several potential good states that can be considered as reference conditions. Also, it would be unrealistic to imagine and define a state of a water body without any anthropogenic impacts, as since the medieval age catchments and coasts have been inhabited and modified by human activities (Chave 2001). Comparing the correct status of ecosystems with an historical landscape would inevitably cause the majority of water bodies to be classified at a lower ecological status, thus leading to enormous restoration costs.

It is recommended that the reference conditions must take into account both ecosystem and social and economic needs. This is also implied in the Directive, which promotes the concept of good ecological status alongside the sustainable use of water. It is therefore important to set reference conditions at a level which can accommodate a certain amount of water uses and services, so that water bodies can be restored and kept at good or high ecological status even in the presence of human pressures.

Water quality

Achieving a good chemical status of all surface waters, including transitional and coastal waters, is one of the environmental objectives of the Directive. The good chemical status is achieved when all the water quality parameters do not exceed the environmental quality standards listed in the Dangerous Substances Directive (76/464/EC). These indicate the maximum concentration of a particular compound that is allowed in water, sediment or biota. In the Water Framework Directive, however, a new series of objectives is added to those of the previous legislation, the Priority Substances.

Priority substances

Priority substances are 33 compounds represented mainly by organic compound. Half of these chemicals are defined as Priority Hazardous Substances, identified as particular toxic and persistent substances that can bioaccumulate in organisms. The Directive requires that objectives are set for these substances and that a programme of measures is established in order to gradually reduce pollution from priority substances and eliminate any emission or discharge of hazardous substances.

The criteria used for assessing the chemical status of waters do not take in consideration the ecology of the aquatic environment. Environmental quality standards are often based on annual mean values of a certain compound, while not taking into account maximum and minimal values. This can lead to errors in the evaluation of the potential effects of a certain pollutant in the organisms, as these are affected by the whole range of values over a certain period. Also, some compounds are present in the marine environment at dissolved concentrations that cannot be directly measured by analytical methods, but can be only be detected by its effects on the organisms. Methods for measuring these effects such as ecotoxicology tests and bio-essays should be therefore used alongside with the environmental quality standards.

The Directive also introduces a new combined approach for the control of compounds discharged in waters. This aims at limiting emission sources but also sets quality standards for compounds in the water bodies, so that pollution from both point and diffuse sources can be controlled. The approach will have important implications in terms of water management at national and international level, as the control, monitoring and evaluation of pollution sources will have to be co-ordinated across all the different compartments of the river basin district, from rivers to coastal waters crossing national borders (Chave 2001).

Bathing water quality

In coastal waters, the achievement of good chemical status of waters is essential not only for the ecology of aquatic ecosystems but represent also a guarantee for a safe use of waters for bathing (see also Georgiou chapter 4). In 1976 the Bathing Water Directive (Directive 76/1607EC) concerning the quality of waters used for bathing came into force. The Directive reflects the state of knowledge and experience of the early 1970's, in respect to its technical-scientific basis, the managerial approach and the involvement of the public (Chave 2001). In order to assess compliance with the Directive a range of physico-chemical, bacteriological and aesthetic criteria are specified. Recently changes in science and technology as well as in managerial experience have obliged the Commission to consider revision of EU environmental legislation where appropriate (see also Georgiou this volume). Further legislation has thus been proposed on more than one occasion by the EC in the form of revisions to the 1976 Directive, resulting in controversy among scientists, policy makers and public opinion. These discussions have centred on a number of dilemmas. There is a question mark over the level of protection to be afforded against minor illness acquisition by EC standards. At the same time, the costs of tightening these standards are considerably and the health gain associated with any tightening is likely to be measured in terms of self-limiting and minor illness, such that there is a question as to whether any expenditures on sewage cleanup represent effective and efficient use of resources. There is an expectation among the public that standards should be sufficient so as to prevent illness being acquired. Even minor levels of illness acquired through recreational bathing may be unlikely to be considered acceptable. Regulators and governments thus have to balance the public desire for better environmental quality with the economic impact of policy changes on both water bill payers and the financial health of water companies/boards. Furthermore, any new policy must be compatible with EU Water Policy, in particular the Water Framework Directive, which provides a coherent managerial framework for all water related EU Legislation.

Tools for the management of protected areas under the WFD

Protected areas are zones designated for special protection under the EU legislation. These include areas for abstraction of drinking waters, areas to preserve habitat and species of special interest, areas to protect economically important species and recreational zones. In coastal waters marine reserves have been created with differing objectives, such as preserving habitat and communities of ecological value, to protect fish and juvenile fish from overexploitation and scientific research (Cattaneo et al. 1984, Diviacco 1990, Diviacco et al 1992, see also Lindeboom and Bäck this volume). The creation of MPA's should also guarantee the sustainable management of regional fisheries (Agardy 1994) and increase the tourism value and attractiveness (Jones 1999). The environmental objectives for marine reserves and other protected areas set by the Water Framework Directive must be achieved not later than 15 years after the date of entry into force of the Directive (Directive 2000/60/EC). This is a challenging target, as MPAs are regulated by special rules and laws established locally; with the Directive coming into force, MPAs also become part of one or more water bodies thus their special regulations have to be integrated in the new Directive legislation. In many European countries MPAs are under the management of several different authorities and institutions that apply regulations based on the local need and features of the area. As protected areas will become part of a river basin, their policy and decision-making will have to conform to the common management of the river basin district. This implies that a high level of coordination must be established between the management of protected areas and the other components of the river basin district. This coordination, if successful, will contribute significantly to the success of marine reserves, as the restrictions applied to the protected areas cannot prevent and control impacts from land-based pollution, particularly during summer, when several coastal villages double their population due to tourism (see also Sarda chapter 16).

Difficulties in the implementation of the Water Framework Directive for MPAs are forecast also at local level. The designation and subsequently the management of MPAs is often subject of conflicts between scientists and conservation associations, who promote marine reserves, and fishermen, boating and tourist operators who strongly oppose the restrictions applied to the protected areas (Salmona and Verardi 2001). The Marine Protected Area of Portofino is a clear example of the difficulty in establishing the correct balance between the contrasting interests of the different parties involved (see box 1). Unless communication between the different social components improves, providing positive solutions to problems, conflicts will remain unresolved. Furthermore, larger public participation bringing more equity between the different parts of the society should be promoted, as regulations are often tailored around the needs of a few influential groups.

The technical implementation of the Directive will involve great changes in the management of the MPAs. As in the case of to non-protected coastal areas, MPAs lack the appropriate monitoring programmes. Monitoring and assessment of the status and evolution of a marine protected area are generally poor and not well coordinated by the responsible authorities. Current knowledge of the efficiency of the MPAs is often the results of single, short-term, ecological studies carried out by scientists within specific research projects. Furthermore most studies investigate the effect of MPAs spatially, between a protected and unprotected zone, whilst the temporal component is often neglected (Francour 2000). As a result, the information is often fragmented and not sufficient to forecast long-term effects of MPAs. Long-term monitoring is needed to allow a correct evaluation of changes in the flora and fauna as results of both natural variability and the protection effect.

For example, after 10 years monitoring in the Scandola Marine Reserve (Corsica, Mediterranean Sea) Francour (1994, 2000) observed that protected areas increased the resilience (buffer effects) of fish assemblages to environmental disturbance, despite the short-term fluctuations in fish diversity occurring in both protected and unprotected areas. This study highlights the importance of long-term surveys to assess the efficacy of MPAs.

Box 1. The Marine Protected Area in Portofino, Italy

The Portofino promontory is located in the Ligurian Sea and comprises 13 km of steep, rocky coastline. The marine area surrounding the promontory consists of a rocky bottom and marine caves, resulting in a high variety of habitats. This habitat diversity and the high water quality, characterised by low turbidity, high oxygen concentrations and hydrodynamics, favoured the development of a rich and diverse biocenosis (Cattaneo Vietti et al. 1988). In particular, the protected red coral and various types of madrepore are almost exclusive of this area. Another habitat of high ecological value is the seagrass (*Posidonia oceanica*) meadow, which hosts a diverse benthic and pelagic fauna.

At present, onshore and offshore tourism is the main activity on this coastline (Diviacco 1990). The small fishing ports have gradually been transformed in centres of recreation, and have become popular seaside resorts, often holiday target of VIP's. The increase in boating activities rapidly deteriorated the state of the marine ecosystems. Mechanical action of chains and anchors, sediment perturbation and resuspension caused by boat engines and jet-skies, rubbish and oil spills discharged into the sea by vessels, all seriously affected the benthic biocoenosis. Sport fishermen also contributed significantly to the degradation of coastal habitats, through collection of red corals, date mussel harvesting from the rocks and fishing with various gears (Cattaneo Vietti et al. 1984).

In 1998, the Ministry of the Environment created a marine reserve in Portofino to protect the ecosystem and regulate the recreational uses of the area. The reserve consists of: 1) a strict reserve zone, where only scientific research and controlled navigation for rescue and service are allowed; 2) a general reserve zone, where bathing and diving, small size boats, professional fishing for residents only and limited sport fishing are permitted; 3) a partial reserve zone, subject to similar restrictions as in zone 2.

Despite the fact that boundaries, zoning and temporary regulations were outlined for the protected areas by the Ministry, a proper legislation still does not exist and the above restrictions are often violated. The only regulation applied is the temporary bathing season ordnance that restricts boating, bathing and fishing. The creation of the MPA caused a series of debates that continued even after the official designation. Salmona and Verardi (2001) suggest that the conflicts represent a paradox: on one side there is a public awareness that local economy is mainly based on the ecological value of the area, which therefore needs to be protected, on the other side tourism industry would like to avoid any regulations.

Stakeholders opposing MPA are yachting associations, boating operators, local municipalities, tourist operators and tradesmen. All consider the MPA a serious threat to the local economy. Their opinion is that the protection of the ecological value of the area should have a lower priority than local economic needs.

Stakeholders promoting the MPA are scientists, environmental associations and a few tradesmen. They believe in the economic benefits that can derive from the implementation of the MPA. A preserved marine ecosystem, high water quality and clean beaches would, in long-term, guarantee a more sustainable economy. The perception of the ecological value is clearly different between the two parties. For the opposing stakeholders, the ecosystem does not provide any goods and services, thus there are no real economic benefits from preserving it. The supporting stakeholders appear to have a more complete concept of the ecological value, and consider the ecosystem as an important component of the local economy. Furthermore, the sectors against the MPA consider only short-term impacts from its implementation whilst those supporting it foresee also long-term effects. The duration and lack of conclusive solutions to this conflict has largely limited a correct implementation of the MPA. In addition, the success of the Portofino marine reserve has been seriously limited and delayed by the following problems:

1) limited social participation, as only few, influential economic sectors are involved; 2) minimal public environmental awareness, generally restricted to scientists and conservation associations; 3) lack of coordination between the MPA authority and other coastal and inland management institutions; 4) weakness in the enforcement of the MPA law with consequent violation of the regulations (Salmona and Verardi 2001).

The success of MPA can be achieved only through the public acknowledgement that an environmentally sound management will provide intermediate and long-term economic and ecological benefits.

Reference conditions and the implementation of ecological classification schemes

Translating the concept of reference conditions for an ecosystem into real values is one of the challenging requirements of the Directive. In an ideal world, reference conditions should be derived from current monitoring data of water bodies not impacted by any human activities. In most cases, however, this is not possible and alternative methods need to be adopted, such as deriving reference conditions from historical data on the abiotic and biotic features of the water body, or reference models using information from different sites. Modelling can be a powerful tool to hindcast reference conditions and assessing the ecological status of an ecosystem (Clarke et al. 2003, Nielsen et al. 2003). Several conceptual and numerical models are already available (see Herman et al. this volume, Windhorst et al. this volume). It is recommended that current models need to be improved and adapted to the Directive requirements and be transformed in effective management-oriented models. To be valid at a large scale, across various catchments or ecoregions, models require availability of appropriate monitoring programmes providing standardised sampling protocols and coordinated data management (Schmutz et al. 2000). Current models do not encompass the range of scales now required. It may be possible to extend their validation using for example historical records or paleological records, for example from sediments. However, even if not validated they can still be used to compare the effect of various scenarios.

One of the potential difficulties in using models for the implementation of the Directive is the lack of integration between river, estuarine and coastal water models. This reflects also the fragmented situation of monitoring schemes and methodologies applied so far to describe the status of the various water bodies. However, because of the limited time available, a unique model covering all the river basin area cannot be defined. Alternatively, the current models available can be linked to each other so that the output of one model represents the input of the following model. The problem of lack of harmonisation in monitoring procedures and models is evident also at international level. It will not be possible, for example, to transpose monitoring methods or models from one country to another, as the information provided is likely to be reliable only in the country where the system has been calibrated. Intense international collaboration and coordination work is therefore needed to overcome the differences in model approaches to reach a common scheme as required by the Directive (Chave 2001). The Directive defines the status of the aquatic environment using biological, hydro-morphological and physico chemical indicators. Similarly, an integrated approach should be used to link models on hydrology, climate and ecology at a catchment level. Dynamic models allow also prediction of past and future changes in the ecological conditions of ecosystem at different time scales. They therefore are essential tools for the management and preservation of ecosystems.

For monitoring and management purposes, models should be dynamic and deterministic, but still simple and effective. For rapid screening of the state of ecosystems, simplified, empirical models can be used to predict the worst outcome of given scenarios. This is not however always reliable. For example, in the deterministic assessment approach for the presence of compounds in the water bodies, the multiple effects resulting from interactions between compounds are not taken into account. For instance, the concentration of nutrient and metals has increased in many catchments all over the world. The effects described are eutrophication, caused by increased nutrient concentration. However, higher concentrations of metals might reduce algal growth and production. Also, only 33 compounds are assessed, so uncertainties over potential (latent) effects remain high. It is well established that the environmental effects can only be explained by a small percentage of these 33 compounds and that the effects are mainly caused by unknown compounds from the group of more than 100,000 anthropogenic compounds released in the aquatic environment. In vitro and in vivo bioassays may help to overcome this problem. By applying these techniques, the effect on the organisms is firstly studied and the compounds causing the effect can be isolated subsequently. In this way the risk of effects induced by unknown compounds is reduced dramatically.

In conclusion, each method assessing the reference conditions presents disadvantages and uncertainties (Schmutz et al. 2000). For a closer representation of reference conditions, an approach that integrates all the information provided by historical data, current field data from reference sites and models should be adopted. This approach will be relatively easy to adopt for many rivers and lakes, where a large amount of past and present field data is generally available. In contrast, ecological monitoring in coastal and estuarine systems is still scarce and focuses mainly on the assessment of the chemical status of waters. More rigorous, consistent and intense monitoring will be therefore necessary to develop a correct classification scheme.

Interactions between ecology, society, and economics

One of the innovative aspects of the Water Framework Directive is the concept of water pricing, which must guarantee a sustainable use of water and the protection of water resources. The Directive outlines the need for an economic analysis of water services that takes into account the recovery of environmental and resource costs due to negative impacts on the aquatic ecosystems (Directive 2000/60/EC).

The WFD supports an integrated management of water resources that involves both ecological and socio-economical realities, thus requiring active collaboration between ecologists, economists and public authorities. Despite the increasing efforts towards closer collaborations, the gap in communication between scientists, policy makers, coastal managers and economists is still considerable (Turner 2000). A different perception of the economic, ecological, cultural and aesthetic values of the coastal system is undoubtedly one of the difficulties in communication between the different groups. Public perceptions of these ecological values change with time, also as a consequence of growing environmental awareness through education and media. Given the generic goal of sustainable water resource management, there is an increasing focus on integrated frameworks in which water is an integral component of a catchment-wide ecosystem, a natural resource, and a social and economic good, whose quantity and quality determines the nature of its use (Turner et al. 2003). Such frameworks can make tractable the complexity of causes of coastal degradation, and the links to socio-economic activities across the relevant spatial and temporal scales. They can also provide the connection between coastal ecosystem change and the effects of that change (impacts) on people's economic and social well-being. For example, Bonn (2000) proposed the development of an integrated method for assessing river conservation value, based on the evaluation of a series of attributes of the river system such as naturalness, rarity, species richness and impacts. These attributes or criteria need however to be differentiated and ranked on the basis of their importance or value. Choosing the ecological attributes and ranking their value is often a subjective rather than objective judgement. Similarly to the river basin, the assessment of the ecological value in coastal systems needs attributes and criteria that are clearly defined through rigorous and consistent methods. A scoring system can be used to improve objectivity in the value ranking, but the derivation and interpretation of score should be easily understandable. This implies that the complex analysis of ecological processes and impacts be translated in a way that is understood and used by economists and coastal managers. Relevant indicators of environmental change can be derived so as to quantify and prioritise the ecosystem functions and requirements taking into account social and economic needs. In order to have a sustainable management of resources, ranking must be based on both "objective" ecological scores and subjective societal and economic value.

Ecosystem valuation can be a controversial task. Agencies in charge of protecting and managing natural resources must take difficult decisions in allocating resources. These decisions are based on society's values, which vary in time and can be different between countries and within countries, in one catchment area. Economic valuation can be a useful tool to protect and restore ecosystems functions and services. These are physical, chemical and biological processes that contribute to the maintenance of the system. Ecosystem services are benefits for nature or society resulting from these functions. The key to valuing a function is establishing the link between the function and some service flow valued by people. If that link can be established, then the concept of derived demand can be applied to assign monetary economic values. The marginal units of service flow are valued in terms of the willingness to pay for their provision, or willingness to accept compensation for their loss.

The valuation system sometimes seems not to represent equally all the members of the society. This is because the water management and decision making process often restrict stakeholder involvement and public participation (Van Ast and Boot 2003, Morrison 2000). Often, the public is not equally represented, and only a few influential, more powerful groups are considered. Under extreme distributional conflicts adjustments to the valuation system can be made by applying equity weights. Also there may be a need for harmonisation between scientific and social scales, for example by means of multi-criteria analysis.

7.12 Conclusions

Areas of potential difficulties in implementing the WFD were identified. These included (a) the use of ecological indicators and the establishment of a reference condition, (b) the use of models for integration and (c) the interaction between ecological and socio-economical needs. Current models used for the prediction of ecosystem behaviour of coastal waters generally are not geared towards the indicators specified in the WFD, or designed to match the spatial boundaries set-up by the river basin structure. Together, this will require a major update of these models and close scrutiny of integration across disciplinary boundaries. The interface between natural sciences and ecological and socio-economic perspectives on the coast is an area of interdisciplinarity that hitherto has only been explored on a project-based scale. Even here,, considerable progress has to be realised before widespread application across Europe is really feasible. Limited experience (cf. Ledoux et al. this volume) suggests that this is a feasible future goal.

References

Agardy MT (1994) Advances in marine conservation: the role of marine protected area. Trends Ecol Evol 9:267-270

Boon PJ (2000) The development of integrated methods for assessing river conservation value. Hydrobiologia 422/423:413-428

Borja A, Muxika I, Franco J (2003) The application of a Marine Biotic Index to different impact sources affecting soft-bottom benthic communities along European coasts. Mar Pollut Bull 46:835-845

Cattaneo Vietti R, Orsi Relini L, Wurtz M (1984) La pesca in Liguria. Centro Studi Un-ione Camere di Commercio Liguria, Genova

Cattaneo Vietti R, Sirigu AP, Tommei A (1988) Mare di Liguria. Centro Studi Unione Camere di Commercio Liguria, Genova

Chave PA (2001) EU Water Framework Directive on Water Management: An Introduc-tion. IWA Publishing (International Water Association), London

Clarke RT, Wright JF, Furse MT (2003) RIVPACS models for predicting the expected macroin-vertebrate fauna and assessing the ecological quality of rivers. Ecol Model 160:219-233

Diviacco G (1990) Indagine sulla situazione naturale ed antropica nell'area interessata all'Istitu-zione della riserva marina di Portofino. Instituto Centrale per la Rocerca Scien-tifica e Tecnologica Applicata alla Pesca, Roma

Diviacco G, Marini L, Tunesi L (1992) Parco marino di Portofino: criteri metodologici per la stesura della proposta di zonazione, Oebalia

EC (1976a) Dangerous Substances Directive 1976/464/EC of the European Parliament and of the Council

EC (1976b) Bathing Water Directive 1976/160/EC of the European Parliament and of the Council

EC (1998) Drinking Water Directive 1998/83/EC of the European Parliament and of the Council

EC (2000) Water Framework Directive 2000/60/EC of the European Parliament and of the Council. pp 1-70

EC (2002) The Water Framework Directive. Guiding principles on the technical require-ments. Environment Agency, Bristol, pp 1-84

EC (2003) Introduction to the new EU Water Framework Directive. European Commission

Francour P (1994) Pluriannual analysis of the reserve effect on ichthyofauna in the Scan-dola natural reserve (Corsica, northern-occidental Mediterranean). Oceanol Acta 17:309-317

Francour P (2000) Evolution spatio-temporelle à long terme des peuplements de poissons des herbiers à *Posidonia oceanica* de la réserve naturelle de scandola (Corse, méditer-ranée Nord-Occidentale). Cybium 24(3 suppl.):85-95

Georgiou S (2005) Bathing water quality. In: Vermaat JE, Bouwer LM, Salomons W, Turner RK (eds) Managing European coasts: past, present and future. Springer, Berlin, pp 75-101

Herman PMJ, Ysebaert T, Heip CHR (2005) Land-ocean fluxes and coastal ecosystems – a guided tour of ELOISE results. In: Vermaat JE, Bouwer LM, Salomons W, Turner RK (eds) Managing European coasts: past, present and future. Springer, Berlin, pp 21-58

Jones PJS (1999) Marine nature reserves in Britain: past lessons, current status and future issues. Mar Policy 23:375-396

Ledoux L, Vermaat JE, Bouwer LM, Salomons W, Turner RK (2005) ELOISE research and the implementation of EU policy in the coastal zone. In: Vermaat JE, Bouwer LM, Salomons W, Turner RK (eds) Managing European coasts: past, present and future. Springer, Berlin, pp 1-19

Morrison K (2000) Stakeholder involvement in water management: necessity or luxury? Water Sci Technol 47:43-51

Nielsen K, Somod B, Ellegaard C, Krause-Jensen D (2003) Assessing reference conditions ac-cording to the European Water Framework Directive using modelling and analysis of his-torical data: An example from Randers Fjord, Denmark. Ambio 32:287-294

Salmona P, Verardi D (2001) The marine protected area of Portofino, Italy: a difficult bal-ance. Ocean Coast Manage 44:39-60

Schmutz S, Kaufmann M, Vogel B, Jungwirth M, Muhar S (2000) A multi-level concept for fish based, ryver-type-specific assessment of ecological integrity. Hydrobiologia. 422/423:279-289

Turner, RK (2000) Integrating natural and socio-economic science in coastal management. J Mar Sci, 25:447-60

Turner RK, Georgiou S, Brouwer R, Bateman IJ, Langford IJ (2003). Towards an inte-grated environmental assessment for wetland and catchment management. Geogr J 169: 99-116

Van Ast JA, Boot SP (2003) Participation in European water policy. Phys Chem Earth 28:555-562

Windhorst W, Colijn F, Kabuta S, Laane RPWM Lenhart HJ (2005) Defining a good ecological status of coastal waters – a case study for the Elbe plume. In: Vermaat JE, Bouwer LM, Salomons W, Turner RK (eds) Managing European coasts: past, present and future. Sprin-ger, Berlin, pp 59-74

8. The EU Water Framework Directive: Challenges for institutional implementation

Erwin F.L.M. de Bruin[1], Frank G.W. Jaspers, and Joyeeta Gupta

Abstract

This chapter undertakes a limited analysis of the EU Water Framework Directive (WFD) and focuses in particular on some of the implementation challenges. It focuses on how the WFD aims to meet the goal of integrated water resource management. It then examines some of the challenges of implementing the Directive in a EU member state with advanced water policies (The Netherlands) and a potential EU member state (Turkey). It concludes that in the Netherlands the WFD essentially has led to more coordination among water management organisations. The establishment of an extra bureaucratic layer whose job is to ensure that integration takes place has facilitated this. In the case of Turkey, the difficulty is to find a way to actually differentiate responsibilities between different authorities and to have a more coherent water management system that challenges the current hierarchical power structure. This chapter then examines the impact of the WFD on coastal zones and concludes that the WFD has limited authority over the coastal zones. This is problematic because river flows do not simply end at an arbitrary distance from the coast and because there appears to be no real way of communicating with other instruments that deal with coasts and with the seas and oceans. On the other hand, the European Union's Coastal Zone Management initiatives will try and bridge the gap between the land, river systems, the coasts and the seas, and will try and link up with all the relevant EU regulations. The paper concludes that it would appear that the EU is moving steadily towards a democratic system of managing the waters and coasts of the region and that only time will tell how easy or difficult it is to harmonise policies in countries and regimes with vastly different histories and institutions.

[1] Correspondence to Erwin de Bruin: erwin.debruin@grontmij.nl

J.E. Vermaat et al. (Eds.): Managing European Coasts: Past, Present, and Future, pp. 153–171, 2005.

Introduction

The Water Framework Directive (WFD)[2], adopted in 2000 jointly by the European Parliament and the Council is a very ambitious legally binding document that aims to create a "good status" for all surface and ground waters throughout the European Union (EU) by 2015. The term 'good' reflects a new concept of ecological quality, which is based on biological, chemical and physical information (Chave 2001), but there remain questions about how this term will be interpreted (Lanz and Scheur 2001). To reach this aim, several steps will be followed, varying from assessing the pressures and impacts on a river basin to implementing a programme of specific measures. The Directive must be implemented internally by each EU member state. Accession countries that will join the EU in the near or more distant future will also have to implement the Directive. The purpose of this chapter is to analyse the challenges in the implementation of the goals of the Water Framework Directive within the EU, given that river basin and water management within the European Union is the result of a long historical process of fine-tuning the outcomes of complex negotiations between the riparians.

Institutional systems for water management can be traced back several hundreds of years to the Roman system of law and vary considerably across Europe.[3] There are 169 agreements that have historically been made with riparian states by various members of the European Union (Wolfe 2002). Very different national systems for water management have developed in the 15 EU member states and in relation to the major international river basins in this region. The ten new accession countries have also had quite different experiences in managing water as many of them have mostly been influenced by the Communist system of water management where water was mostly owned by the State. Some future potential member countries, such as Turkey, also have a completely different domestic water system influenced historically by Islamic precepts (Caponera 1992).

Since 1975, the EC has made an attempt at harmonising policies in different areas within the member countries by legislating some areas of water management including drinking water, the quality of various water bodies, urban waste water, nitrates, etc. Clearly, the maze of conflicting policies and laws within member countries was not very conducive to an integrated system of water resource management that also contributed to sustainable use of water resources. Citizens and environmental organizations were continuously demanding cleaner rivers, lakes, groundwater and coastal beaches.[4]

It is against this background, that the Water Framework Directive was adopted. The Water Framework Directive calls for a complete restructuring of water policy in the member states and in relation to the river basins. The key unit of focus is the river basin; the key goal good water status (which includes ecological and chemi-

---

[2] Directive 2000/60/EC of the European Parliament and of the Council, establishing a framework for Community action in the field of water policy.

[3] For some examples see for instance Alearts (1995), Correia (1998a) and Mostert (1999).

[4] Introduction to the New EU Water Framework Directive, http://europa.eu.int/comm/envronment/water/water-framework/ocerview.html; downloaded 13/8/2003.

cal protection for surface water, and chemical and quantitative status for ground water). The Directive combines controls on the source of pollution and on measures to promote qualitative objectives in water bodies. It calls on countries to establish river basin management plans based on active public participation and on establishing an effective pricing system. It repeals seven existing directives.

Authorities see it as the most significant legal instrument that provides a clear legal framework and institutional structure that can serve as the basis of catchment-based governance for the successful management of water quality and quantity. It is expected to have a major impact on water policy within the EU (Chave 2001, Holzwarth 2002). Certainly the timetable for implementation is impressive. By 2003, the Directive needs to be transposed in national legislation (Article 23) and river basin districts and authorities need to be created (Article 3). By 2006, a monitoring network (Art. 8) and the process of public consultation (Art. 14) must have been established. By 2009 river basin management plans must be finalised (Arts. 13 & 11); and by 2015 the environmental objectives must be met (Art. 4).

At the same time, critics argue that the provisions are ambiguous, the level of protection provided is very questionable and there are a number of opt-out clauses (Lanz and Scheur 2001). Quite noteworthy, the question - what is good status – is left open (Lanz and Scheur 2001).

The WFD is to be adopted throughout the Union, in the context of all the different national and river basin arrangements. Concepts such as river basin districts, cost recovery and integrated management at river basin level have to be implemented. The WFD will therefore have a profound effect on how European countries manage their water systems. It will create a much more integrated and precautionary approach to whole basin catchment management. It will lead to the reallocation of water abstraction and discharges. And it will encourage and stimulate a range of public participatory approaches to more inclusive water management generally.

The WFD poses not only serious challenges to the EU member states but also to the future members of the EU. Pilot projects, often financed by EU member states, are a common tool to get accession countries acquainted with the implications of the WFD for their own water management.

Comparing institutional systems for water management from the perspective of different countries or different river basins in order to find an ideal blueprint is not possible. These systems are largely based on historical and cultural factors, and one cannot assume that a successful system in one country or basin will be just as successful in another. This implies that also the way the WFD is implemented will be different for every country and water basin, as long as the aims of the Directive are reached, and the reporting requirements to the European Commission are met.

Against this background, this chapter addresses the question: What are some of the institutional and administrative challenges to the implementation of the WFD faced by (future) EU member states in order to achieve integrated water resource management? In order to address this question, this paper will first recapitulate the key principles of integrated water resources management and examine the WFD in terms of these principles. It will then examine two case studies – the case study of the Netherlands and that of Turkey. The Netherlands has been a EU member state since its establishment, while Turkey has still not formally entered political negotiations. Nevertheless, both countries are harmonising the domestic water management system with the WFD.

On the basis of the case studies, we will draw some conclusions on the types of implementation challenges that the WFD may face in the coming decades. Then finally this chapter examines how the WFD deals with the coastal zone in the context of integrated water resource management.

This paper is based on a literature analysis and on the practical experience of actually trying to implement the WFD, and this combination is reflected in the expertise of the authors. We take a country approach for our case studies, since the WFD is primarily to be implemented by national governments, and institutional systems are organised at country level.

A theoretical framework:
Integrated water resources management

The 2000 year historical evolution of water management has moved from local sectoral management to integrated water resources management. Although river basin management has long been an issue in the context of international treaty negotiations on rivers, the concept of integrated water resources management reached the international agenda, not via the treaties but via the general water policy making process at the 1992 Dublin International Conference on Water and the Environment. These principles were endorsed at the 1992 United Nations Conference on Environment and Development (UNCED, 1992) and subsequently at the three World Water Forums and the 2002 World Summit on Sustainable Development (WSSD 2002).

Integrated water resources management (IWRM) has been defined as follows: 'Integrated water resources management is the management of surface and subsurface water in qualitative, quantitative and ecological sense from a multi-disciplinary perspective and focused on the needs and requirements of society at large regarding water' (Van Hofwegen and Jaspers 1999). It has also been defined as 'a process that promotes the co-ordinated development and management of water, land and related resources, in order to maximize the resultant economic and social welfare in an equitable manner without compromising the sustainability of vital ecosystems' (GWP 2000).

The Global Water Partnership (2000) has recently published a book interpreting and explaining the Dublin principles, the source of integrated water resource management. The Dublin Principles recognize that fresh water is a finite and vulnerable resource, essential to sustain life, development and the environment. This Principle calls for a holistic approach to water (respect for the hydrological boundaries), a recognition that resource yields have limits, a need to constrain human activities, to manage upstream-downstream user relations and a holistic institutional approach. The second principle calls for a participatory approach which includes a recognition that participation is more than merely consultation, that in order to promote participation, decisions have to be taken at the lowest appropriate level and that participatory mechanisms need to be created and that there is a need to achieve consensus with the participants. The third principle focuses on the role of women in decision-making and the need for greater gender awareness. The fourth principle focuses on water as an economic good, where economic value includes the value to water users, the net benefits from return flows, the net benefits from indirect uses and an adjustment for societal objectives. Full supply costs is the operation and maintenance costs plus capital charges; the full economic costs includes in addition the opportunity costs and economic externalities; and the full costs include environmental externalities.

For IWRM to be implemented, it is important to keep the context in mind, to establish an enabling environment, where the government is enabler, regulator, controller and service provider. There is need for legislation and the political will to implement, for cross-sectoral and upstream and downstream dialogue, for clear mechanisms for promoting cooperation within international river basins, with clear responsibilities assigned to the different actors at different levels, for ways of financing policies through good pricing and for good conflict resolution mechanisms (GWP 2001, cf. Jaspers 2003).

Let us now see to what extent the Water Framework Directive has incorporated the features of IWRM. One of the most widely known characteristics of the WFD is that it advocates water management at the whole river basin level. In the Directive, a distinction is made between river (sub)basins and river basin districts. The river basin district is defined as the main unit for management of river basins. The river basin district consists of one or more river basins. A river basin is defined as 'the area of land from which all surface run-off flows through a sequence of stream, rivers and, possibly, lakes into the sea at a single river mouth, estuary or delta' (WFD Article 2). In other words, the WFD includes the total area of land and part of the sea that forms part of the basin (see analysis of the WFD on the coastal zone that follows). Furthermore, member states are requested to make river basin management plans. However, the WFD does not have jurisdiction over those parts of the river basins that fall outside the EU territory and does not cover the seas beyond 1km for biological purposes and 10 km for chemical purposes. Nevertheless, one could argue that in terms of meeting the objective of dealing holistically with the water resource, the WFD does remarkably well.

The preamble to the WFD mentions that 'there is a need for a greater integration of qualitative and quantitative aspects of both surface waters and groundwaters […]'. One of the purposes of the WFD is to contribute to mitigating the effects of floods and droughts. In the tools the Directive offers, there is a strong focus on water quality and ecology. The water quantity aspect mainly comes back in the allocation between consumptive and non -consumptive uses in order to protect aquatic biodiversity.

Water quantity issues and water quality issues cannot be seen separately, and as such should be dealt with by the River Basin Management Plans. As of yet, there is little guidance about how to deal with flooding or water sharing issues with regard to the implementation of the WFD.

Decision-making at the lowest appropriate level is an important aspect of IWRM, and as such also mentioned in the preamble. How this will be given effect in practice remains a key issue. The WFD is emphatic in its recognition of water primarily as a national heritage and not as an economic good (see preamble). Having said that cost recovery and stakeholder participation are issues that are dealt with by the WFD. This paper assesses two key challenges in the implementation of the WFD. The first is the huge administrative change required by the shift from existing water management systems to an administrative structure focused on catchment basins. The second is the challenge of undertaking integrated water resources management.

Case studies

The Netherlands

Dutch water management is handled by the public administration, although semi-private organisations can also play a role, e.g. the drinking water companies. Characteristic for the Dutch water management organisation is its high degree of decentralisation and a division between water management, environmental management and land use planning (Mostert 1999). The Ministry of Transport, Public Works and Water Management, the Ministry of Housing, Spatial Planning and the Environment and the Ministry of Agriculture, Nature Management and Food Quality are the main national-level actors. Under the Ministry of Transport, Public Works and Water Management one can find *Rijkswaterstaat*[5] (RWS), the directorate-general, which is carrying out the water management tasks for the Ministry. The provinces are responsible for groundwater quantity and quality management, while the Water Boards are responsible for surface water management (except national waterways). Municipalities have some direct tasks in water management, as well as in the field of spatial planning, by developing local land use plans (see also Table 1).

An important characteristic of the Dutch system is that the provinces, water boards and municipalities are autonomous regarding their specific tasks in their jurisdiction areas. This means that the planning documents issued by these organisations for their areas are binding, within the rules set by the national plans. They also have financial independence, as they are allowed to levy taxes to finance their work.

The Netherlands has a complex system of water laws (e.g. Correia 1998). The laws that structure the water management organisations are the Water Administration Act of 1900 and the Water board Act of 1992. Water management is regulated by the Water Management Act of 1989, the Groundwater Act (1981), the Pollution of Surface Waters Act of 1970, the Pollution of Seawater Act of 1975. Legislation on the management of Water Infrastructure includes the Flood Protection Act of (1995), the Delta Act of 1958, the Delta Act Major Rivers of 1995, the Reclamation Act of 1904, and the State Managed Infrastructure Act of 1996. In addition there are other laws such as the Drinking Water Supply Act 1958, the Soil Protection Act of 1987, the Spatial Planning Act of 1962, the Environment

---

[5] Rijkswaterstaat: Directorate-General for Public Works and Water Management, under the Ministry of Transport, Public Works and Water Management.

Protection Act of 1979, the Nature Conservation Act of 1998 and the Mineral Extraction Act 1965. Many of these laws have had several amendments over the years. Any consolidated effort at integrating water management in the Netherlands will also have significant implications for the implementation of these laws.

Table 1. Division of responsibilities in the Netherlands

Organisation	Main tasks and responsibilities
Ministry of Transport, Public Works and Water Management (V&W)	General water policy and legislation; flood management (primary river and sea dikes); management of national surface waters; navigation
Ministry of Agriculture, Fisheries and Food Quality (LNV)	General agricultural and nature policy; legislative policy regarding nature conservation with regard to species and areas; recreation
Ministry of Housing, Spatial Planning and the Environment (VROM)	General environmental policy; setting of water quality standards; legislation concerning among others soil, air and waste; drinking water and sewerage; land use planning
'Rijkswaterstaat' (RWS)	Carries out tasks delegated by the Ministry of Transport, Public Works and Water Management
Provincial Authorities	Coordination with other sector policies; construction and management of provincial waterworks; supervision of water boards and public waterworks in maintenance by third parties; groundwater resources management
Water Boards	(Surface) water quantity and/or quality management of regional waters; management of dikes, waterways, bridges and roads; Waste water treatment (building/operating treatment plants)
Drinking Water Companies	Drinking water abstraction, production and supply
Municipalities	Construction and maintenance of sewer systems; Some tasks on urban hydrology

Challenges of WFD implementation

The implementation of the WFD is a legally binding obligation in the Netherlands: the first deliverables are to be presented to Brussels in 2004. Since 1998, the potential for implementation has been reviewed by a National Project Group on the Implementation of the WFD. Under its leadership, several pilot studies were carried out to assess the implications of the WFD for the Dutch water management system (e.g. Bosma and Busch 2002). Thematical sub-groups have worked out various aspects in more detail, and a handbook has been prepared (Arcadis 2002). This preparatory work was carried out at the national level.

The responsible authority for the implementation of the WFD is the Ministry of Transport, Public Works and Water Management. Since the Netherlands is in the delta of four international river basin districts, i.e. those of Ems, Rhine, Meuse and

Scheldt, four River Basin Management Plans have to be prepared.[6] The creation of the plans will be coordinated by River Basin Coordination offices, which have been set up especially for this task in November 2002 (CRM 2003). The main work will be carried out jointly by *Rijkswaterstaat*-offices, Provincial authorities and the Water Boards. In order to achieve this aim, an elaborate project organisation has been set up, mainly within the existing institutional framework.

Fig. 1. River Basins and sub-basins in the Netherlands for WFD implementation (CRM 2003)

To illustrate the institutional complexity, the following gives some more information regarding the organisational structure of the management unit Rhine West (Broersen et al 2003). In this area there are four provinces, five *Rijkswaterstaat* regional directorates and 17 Water Boards. In addition to the social actors within the area of the plan, social actors in neighbouring plan areas may also be affected by the decisions taken in the plan area. Hence, the River Basin Coordination Office is preparing so-called 'Blue nodes', where these transboundary relations will be organised.

In Rhine-West itself, two platforms are being set up, a Regional Executive Platform, and a Regional Administrative Platform. The Executive Platform consists of

---

[6] For practical purposes, the Dutch Rhine basin is divided into four management units. Including a small piece of German territory, these form the (sub) river basin 'Rhine delta', which in its turn is part of the Rhine river basin district. One RBMP will be prepared for the Rhine delta (CRM 2003).

high-level officials such as the dike-reeves of Water Boards[7], provincial deputies and the directors of Rijkswaterstaat's regional directorates. This means that an official platform exists for all water-related government authorities in the region. This platform decides on the division of tasks and financial arrangements between the involved parties.

The Regional Administrative Platform prepares the Executive Platform meetings, takes care of coordination with the office of the River Basin Coordinator and other Rhine sub-basins, and coordinates the work of the so-called Product Teams. The Product Teams, which consist of employees of the involved organisations, work on the actual implementation of the Directive.

If one adds up the total expected input from the participating organisations in Rhine-West as mentioned in the project plan, one comes to a total of 12 person-years (annually), for coordination purposes and presence simply during the different meetings. This does not include the actual implementation work itself. According to an earlier study by the Inter-provincial platform and the Union of Waterboards, extra time input until 2005 for provinces and waterboards will be around 70 days per organisation per year[8] (IPO and UvW 2002). Next to this, also the regional directorates of Rijkswaterstaat and national level organisations need to reserve time for WFD implementation. All in all, quite a number of water managers will be working on implementation of the WFD during the coming years, either by performing analyses or through coordination activities.

One can see from this experience that the implementation of the WFD is a significant managerial task. This will absorb the attention and time of senior water managers in the public, private and voluntary sectors for some time to come. What is not yet known is how worthwhile all this preliminary effort will be. Framework directives of this sort usually require huge administrative and managerial preparation so it will be most interesting to appraise the outcomes of all this effort in comparison with existing institutional arrangements for managing water and whether these, in fact, will lead to a more integrated water resources management.

Turkey

In Turkey, water management is presently organised according to sectoral lines. Decision-making is strongly centralised. The main governmental actors at national level are the Ministry of Environment (MoE), the General Directorate of State Hydraulic Works (DSI), the General Directorate for Rural Services (GDRS), the Ministry of Agriculture (MoA), the Ministry of Health (MoH) and the State Planning Organisation (SPO). Other organisations, like the Ministry of Tourism

---

[7] All so-called Water Quantity Boards, and 2 representatives of the so-called Water Quality Boards are members of the Regional Executive Platform. The other Water Quality Boards are informed of the outcomes of the meetings, just like the Ministries of V&W, VROM and LNV (Broersen et al. 2003).

[8] According to estimates given by the organizations concerned. Low estimates speak of 35 to 70 days, high estimates of 70 to 140 days.

(MoT), the Ministry of Forestry (MoF) and the Bank of Provinces (BoP) have specific water management tasks (see Table 2).

In general, MoE is responsible for water pollution control, while DSI is responsible for the development and management of water resources. GDRS has water management tasks (among others water supply and treatment) in rural areas. MoA has some water management tasks related to agriculture (e.g. fishery), and MoH is responsible for drinking and bathing water quality. The SPO develops national Development Plans under the authority of the Prime Minister (OECD 1999).

Table 2. Water management organisations in Turkey (Grontmij 2003)

Organisation	Main tasks and responsibilities
Ministry of Environment (MoE)	water resource pollution prevention, environmental standards and inspection, EIA
State Hydraulic Works (DSI)	water resource investigations, river basin development, planning, construction and financing of water and wastewater treatment plants, water supply to municipalities with population above 100,000
Ministry of Health (MoH)	drafting drinking water legislation, setting drinking water standards, implementation and monitoring of these standards
Bank of Provinces (BoP)	planning, financing and constructing of water and wastewater treatment plants, water supply for populations between 3,000 and 100,000.
State Planning Organisation (SPO)	overall planning for investment for water resources (e.g. dams, reservoirs, water supply) and pollution control (e.g. sewerage and sewage treatment)
General Directorate Rural Services (GDRS)	drinking water and sewerage for villages (population <3,000)
Ministry of Agriculture (MoA)	Fisheries and Aquaculture legislation, pesticide control and monitoring
Ministry of Forestry (MoF)	Protection projects of water basins
Ministry of Tourism (MoT)	building wastewater infrastructure systems in tourist areas

The national-level organisations all have representations at a regional level, which may cover one or more provinces (OECD 1999). These regional offices carry out policies set out by the national level. Provincial offices also fall under the jurisdiction of the provincial authorities. DSI works with 26 regional offices, which more or less follow river basin boundaries.

According to the OECD (1999) there is 'limited co-ordination on environmental matters between sectoral ministries and different levels of government'. This view is shared by many Turkish water management organisations (Hermans and Muluk 2002). The fragmentation of tasks can be illustrated by the number of organisations that monitor surface waters (Grontmij 2003): For example, DSI monitors the quantity and quality of the water in the rivers; MoE has authority over domestic and industrial discharges; GDRS supervises the water quality for the purpose of drinking water production and irrigation; MoH has jurisdiction over the water quality of bathing water, and chemical and microbiological status for public health monitoring; MoT monitors the sea and lakes for the European Blue Flag campaign; BoP supervises the water quality on a project basis; and MoA monitors the pesticide and fertiliser run-off in water courses.

WFD implementation

In 2002, the Dutch government funded a project to assist Turkey in implementing the Water Framework Directive. The project is scheduled for completion in December 2003 (Grontmij 2002a). The project operates at two levels, national and regional. At the national level, among others a National Platform (NP) has been created, in order to reach agreement over river basin division, and task division in WFD implementation in general. In this National Platform, all major ministerial stakeholders are represented.

At the regional level, a pilot study is being carried out in the Büyük Menderes River Basin. The aim of the pilot is to make a River Basin Management Plan (RBMP) for the Büyük Menderes River Basin. Next to the RBMP itself, a practical Handbook is being prepared to facilitate replication of the process in other river basins in Turkey. A River Basin Working Group (RBWG) has been created, in which regional and provincial offices of the water management actors are represented. For practical reasons (the Büyük Menderes Basin runs through five provinces for instance), not all provincial stakeholders are represented in the working group itself. The RBWG was formed during a regional platform workshop, in which a large selection of stakeholders – both governmental agencies and user associations – chose representatives for the RBWG.

Within the project organisation, four project teams concentrate on respectively the national level implementation, capacity building, the regional pilot project, and communication aspects. Under the National Platform and River Basin Working Group, specialised working groups are set up to work on specific aspects of the Water Framework Directive. These include tasks such as the division of river basin districts, access to information (national level), ecology, and measures (regional level).

Current status

The National Platform has been working on the division of Turkey into a number of river basin districts. At the moment (summer 2003), the status of the platform after the project's completion is being discussed. All members have expressed the wish to continue the platform, and to shape it into a discussion platform in which also non-WFD related water management aspects can be discussed. In this way, the platform can be used to improve inter-sectoral cooperation between ministries on water related issues.

At the regional level, work on the River Basin Management Plan is under way. From the start of the project it was clear that major traditional water pollution problems still need to be addressed. Industrial and domestic wastewater (only three municipalities in the whole basins have sewerage systems, one has a working waste water treatment plant), as well as boron pollution from a thermal power station are the main problems. During a stakeholder analysis (Hermans and Muluk

2002), it became clear that the responsible organisations also regarded the difficulties of establishing co-operation amongst all the relevant agencies and ministries as a major bottleneck. Political influence on functional decision making, as well as lack of staff, budget and other resources, were also often mentioned. The actors largely agree on the problems, and in general also have a clear idea of what to do about them.

Although the major problems are easily identifiable, the implementation of the WFD requires more analysis. Especially at the start of the process, considerable attention has to be put on gathering information. The characterisation of the river basin, as well as the pressures and impacts analysis, are 'data-focussed' activities. Local staff involved sometimes wondered why all these detailed analyses were necessary, as the causes of the problems are clear. However, a co-operative and practical spirit facilitated considerable progress.

At both the national and regional level it can be noticed that the increased contacts between people from different organisations have increased cooperation. Especially at the regional level this cooperation is working out very well. At the national level 'political' considerations still play an important role, but even here communication has improved considerably in recent times.

The implementation of the WFD in Turkey remains at an embryonic level. Much has to be done to establish a pilot project or projects that can show the way for integrated catchment management. To improve institutional co-ordination and to establish clear political leadership are also vital elements in the early stages of the Turkish experience. Monitoring of effective progress by the EU may help because Turkey's accession will, in part, depend on good intent over the implementation of EU directives generally, and not just in the environmental arena.

Comparative aspects

This section compares the case studies in relation to three issues – institutional integration, integrated water resource management and the challenge of subsidiarity.

The two case studies show some of the challenges that countries face in developing water management systems that are based on the river basin approach. In many countries, water management is essentially carried out by government agencies whose jurisdictions follow administrative boundaries (state, province, region etc). In the Netherlands, regional water tasks are essentially divided between provinces and water boards. Although water boards are essentially functional administrations purely for water management, they do not always follow hydrological boundaries. The Netherlands has just completed the process of developing catchments basin districts as outlined above. Credit for this change is due to the WFD, as this new territorial division has been made in response to the requirements in the WFD. However, as far as structure is concerned, the existing organisations did not change much. Instead, a new organisational layer has been inserted, namely the River Basin Coordination office. This office only has a coordinating task. The actual work will be done within the existing system, which has caused a large coordination effort, as described above. It also has provided an official platform for discussion between the water management organisations, which did not exist earlier.

In Turkey, one of the actors involved (DSI) has regional offices, which are more or less based on hydrological boundaries. The other actors are organised along provincial boundaries. River basin districts are presently being formed.

The implementation of the WFD is an important step towards more integrated water management in general. Neither in the Netherlands nor in Turkey is this integration perfect. In the Netherlands, two implementation processes are actually going on: that of the Water Framework Directive and the implementation of 'Water Management in the 21st Century' (RWS 2000). In this national water management policy, issues like climate change, land subsidence and safety against flooding are dealt with. Interestingly, there are many common grounds between the WFD and Water management policy for the 21st Century, such as division in river basins, managing protected areas, economic analysis of water use, communication etc. (Grontmij 2002). Attempts are being made in the Netherlands to integrate these two implementation processes from 2004 onwards. This would be a very important step towards IWRM, partially due to the WFD. In Turkey, different organisations are responsible for overlapping tasks. Integration is a difficult process, which has only just started.

Finally, we turn to the issue of subsidiarity, or decision-making at the lowest appropriate level. This is an important aspect of IWRM, and is also mentioned in the WFD. However, there are few indicators as to how one determines what the most appropriate level for water management is. This eventually depends on the scale of the problem to be addressed.

In the Netherlands, which is part of four international river basins, water management is organised partly at national, partly at regional level. International issues are dealt with through various transboundary river commissions.

In Turkey, decision-making is strongly centralised at national level. One could reason that a more appropriate level would be that of the individual river basin, and that therefore more decision-making power should be given to regional-level organisations. This is only possible if there is sufficient capacity at lower levels, and at the moment this is lacking in Turkey. Capacity building and institutional development are therefore important conditions for the decentralisation processes. At the same time, it is unclear to what extent the central government will be willing to transfer power to the regional levels. How subsidiarity is likely to be handled remains a matter of conjecture.

The implementation of the WFD requires considerable coordination of and management capacity within the water management organisations, as the cases illustrate. The implementation of the Water Framework Directive has only just started. The longer term administrative and management implications of this directive are unsure. These depend on whether the member states will indeed consider the WFD as a radically new approach to water management in Europe, or as just another piece of EU legislation. All this in turn appears to require political sensitising, and ultimately political will.

Extrapolation to the coast

The coastal zone and river basin districts

Thus far we have examined the implementation challenges of the WFD within the national context. Another dimension of the implementation challenge of the WFD is the way the Directive deals with coastal waters. The following section attempts to elaborate on some of these challenges.

The Water Framework Directive acts on all waters, including transitional and coastal waters. It is therefore of high relevance to coastal zone managers. The implementation of the Directive should result in a better water quality and actions to make this happen are to be taken throughout the river basin. Coastal zone managers should therefore participate in the structures being set up for the implementation of the WFD.

Under the WFD, the coastal zone is divided into river basin districts. This could imply that coastal zone management of a specific coastal area becomes fragmented, but in view of the authors this risk is small. The WFD provides an opportunity for coastal zone managers to influence the behaviour of water managers in the upstream part of the watershed.

The land-sea continuum of the coastal zone

The fundamental assumption lying behind the analysis that follows is that the hydrological systems do not end at a certain distance from the coasts and that the hydrological and ecological systems of the open sea, the exclusive economic zones and the coastal waters are closely related.

In the literature the coastal zone normally refers to the land area and adjacent ocean space "in which land ecology and use directly affect ocean space ecology, and vice versa" (Ketchum 1972). Historically different countries have defined the inland boundary options and the ocean-ward boundary options differently. Thus the Netherlands and Sweden defined the coastal zone to go up to the outward boundary of the exclusive economic zone. Spain, on the other hand, went only as far as the territorial boundaries – the 12 mile nautical zone (Sorensen and Mc Creary 1990). These boundaries have continued to expand as knowledge about how the intensity of cross-coastal zone interactions has increased and as technology has modernised. For example, fisheries in the open sea can affect the ecological status of the coastal waters (Young 2003).

The centre of gravity of the WFD is the river. Although Article 1 of the WFD defines surface waters to include coastal waters, there is a limit to the extent of control (Farmer 2001). This is possibly necessary in order to make the WFD practical. There are, however, some practical challenges to the fact that the WFD only extends to one nautical mile into the sea, except in relation to the chemical status of the waters, in which case the boundary extends to 12 nautical miles.

The question then is, are there not other institutional frameworks that deal with these issues, and if yes, why should the WFD deal with these? In fact, there are a large number of other institutional arrangements that deal with many of these aspects. However, the lack of coherence between these arrangements and the WFD may create new challenges. This lack of coherence has two dimensions. The first is the physical dimension. The flow of rivers does not stop at the boundary laid down by the WFD. The Rhine can have a considerable influence on the North Sea, and not just within the coastal areas (Admiraal et al. 1998) Klein and Buuren (1990). Hence, eutrophication is considered problematic in the North and Baltic Seas. River loads are thus the source of this problem, whilst nutrient loading may not be considered a prime issue by river management authorities. Here is thus a mismatch between the managerial and system boundaries of cause and effect. There is also the problem of salt water intrusion because of over-exploitation of coastal aquifers especially in the Mediterranean. Unless there is a mechanism that allows for communication between the other regimes and the WFD such problems will not be addressed.

Institutional coherence

The second problem is that of institutional coherence between the various arrangements designed to deal with coastal regions. To the extent that coasts mark the boundary of a country, coastal areas and the international seas are generally subject to a number of other legal systems. In other words, any discussion of the management of Dutch coasts, inevitably brings us a to discussion of the management system of the North Sea. The North Sea is not only subject to all international laws that apply to seas in general, but also subject to all agreements specifically made in relation to the management of the North Sea. Let us elaborate. For example, the UN Convention on the Law of the Sea 1982, the International Convention on the Prevention of Pollution from Ships 1982, the Convention on the Prevention of Marine Pollution by Dumping of Wastes and Other Matter 1972, the Agreement for Cooperation in Dealing with Pollution of the North-Sea by Oil and Other Harmful Substances 1983, the Convention for the Protection of the Marine Environment of the North-East Atlantic 1992, the Paris Memorandum of Understanding on Port State Control, the Convention on the Conservation of European Wildlife and Natural Habitats, all have an influence on the seas. International and Regional policies such as those in the 1972 Stockholm Declaration on the Human Environment, Agenda 21 of 1992 and the North Sea Conferences also influence decision making on the North-Sea. Finally many organizations have some sort of jurisdiction on the seas including the IMO, the WMO, the OECD and UNECE. As such coordinating among the different regimes is in itself a very challenging task.

In particular there is need for institutional coherence within the EU. The EC has developed an Integrated Coastal Zone Management Strategy (COM/2000/547). This strategy recognizes a number of problems in coastal areas including coastal erosion, habitat destruction, loss of biodiversity, contamination of soil and water resources and problems of water quality and quantity. The Strategy aims to provide a link between the various EU policies namely Article 6 of the Habitats Directive, the Strategic Environmental Assessment of the EU transport policy, the proposed Strategic Environmental Impacts Directive, the Common Fisheries Policy, the Rural Development Policy, the marine regimes, the Council Directive on

Pollution Caused by Certain Dangerous Substances Discharges to the Aquatic Environment and of course the Water Framework Directive.

The CZM strategic document specifies that "In view of the fact that many of the driving forces that create pressures on the coastal zones are actually located upstream in the river basin, the proposed Water Framework Directive should particularly yield results in the coastal water and beach area. It will be important to ensure that implementation of the proposed Water Framework Directive includes consideration of the impact of water management activities on sediment regimes". According to this document there is a strong expectation that the WFD will indeed ensure that the coastal areas will not suffer from complications arising by fresh water flows through the river system. This strategy may indeed help to bridge the gap between the river regimes and the coastal and open seas regime. The CZM Recommendation prepared by the CEC (2000) calls on nations to undertake a national stocktaking, develop national strategies on the basis of the stock-taking that develop in particular "the means of bridging the land/ sea gap in national legislation, policies and programmes". This Recommendation (Council 2002) was adopted by the Commission in a modified form in 2002, and calls for, inter alia, improved coordination of the actions taken by all the authorities concerned both at sea and on land, in managing the sea-land interaction. This recommendation does not have the same status as a directive. However, it provides a good basis for member states to shape their coastal zone management. In pre-accession projects (SENTER 2002, 2003) it can be noticed that WFD implementation in coastal areas takes this recommendation into account as well. The requirements of the WFD, combined with the code of practice of the Recommendation, can provide a good basis for WFD implementation in the coastal zone.

Conclusions

We believe that the adoption of the European Water Framework Directive marks a turning point in water management in Europe. It calls for a complete restructuring of water policy by emphasising the need to undertake water management in terms of hydrological boundaries or catchment areas. This not only calls for a major change in administrative set-ups, but it also calls for the administrative system to be flexible and to adapt as and when catchment areas change in the future. The Water Framework Directive also calls, inter alia, for the development of integrated water resources management and this too poses serious scientific and management challenges to countries.

The two case studies in this paper show firstly that not only are EU member countries preoccupied with developing institutional responses to this challenge, but also aspirant countries are seriously trying to adapt their domestic water policies.

The case studies illustrate that in a developed water management system as is the case in the Netherlands, the WFD leads to more cooperation between the different water management organisations. In the Netherlands, this involved creating a new bureaucratic level whose aim is to coordinate the implementation effort. In countries with even more fragmentation of tasks, such as Turkey, it is difficult to achieve this coordination. Also here, a platform has been created which could be seen as a first step towards integrated water management. The question here remains whether endogenously induced institutional change is possible. If so, the WFD could be a strong support to achieving integrated water resources management, as the legal requirements for at least the current EU member states are concrete in this respect.

The goal of the WFD to achieve integrated water resources management through river basin management is a laudable one. However, by drawing the hydrological boundaries artificially in the coastal areas, integration may suffer in order to achieve practicality. However, such integration may be achievable in the long-term through closer coordination with the regimes that deal with the seas. And this task could perhaps be left to the current initiatives on coastal zone management being undertaken by the European Commission.

Acknowledgements

Erwin de Bruin and Frank Jaspers acknowledge the Matra project: Implementation of the Water Framework Directive in Turkey (MAT01/TR/9/3). Joyeeta Gupta worked on this paper as part of EUROCAT (EU Project number EVK12000 00510) and the NWO project on 'Intergovernmental and Private International Regimes and their compatibility with good governance, the rule of law and sustainable development.'

References

Admiraal W, De Ruyter-van Steveninck ED, De Kruijf HAM (1988) Environmental strategy in five aquatic ecosystems in the floodplains of the river Rhine. Sci Tot al Environ 78:58-75
Alaerts GJRF (1995) Institutional arrangements in the water and sanitation sector - European country studies. IHE Working Paper EE-4, Delft
Arcadis (2002) Handbook Water Framework Directive (in Dutch).
    http://www.kaderrichtlijnwater.nl
Bosma J, Busch S (2002) Pilots Water Framework Directive Eems & Midden Holland - final report (in Dutch). Lelystad
Broersen K, Ten Brinke M, Zigterman E, Van der Does J, Van Alphen J, Soppe G, Looijen H, Jansen S, Hoozemans R (2003) Plan of approach for sub-basin Rhine-West. Haarlem
Burak S, Gönenç IE, Erol A, (1994) Institutional and legislative structure of water quality management in Turkey. Wat Sci Tech 30:261-268
Caponera D (1992) Principles of water law and administration. Balkema, Rotterdam
Bro C (2001) Analysis of environmental legislation for Turkey
Chave P (2001) The EU Water Framework Directive: an introduction. IWA Publ

Commission of the European Communities (2000) Communication from the Commission to the Council and the European Parliament on integrated coastal zone management: a strategy for Europe. COM (2000) 547 final

Commission of the European Communities (2000) Proposal for a European parliament and Council recommendation concerning the implementation of integrated coastal zone management in Europe

Correia FN (1998a) Institutions for water resources management in Europe. Balkema, Rotterdam, The Netherlands

Correia FN (1998b) Selected issues in water resources management in Europe Balkema, Rotterdam, The Netherlands

CRM (2003) Project plan implementation European Water Framework Directive in Rhine and Meuse (in Dutch). Coordination Bureau Rhine and Meuse, Arnhem

EEB (2001) Making the EU Water Framework Directive work: ten actions for implementing a better European water policy. EEB, Brussels

European Council (2002) Recommendation of the European Parliament and of the Council of 30 May 2002 concerning the implementation of Integrated Coastal Zone Management in Europe (2002/413/EC), Off J Eur Community L 148/24, 6-6-2002

Grontmij (2002a) Implementation of the Water Framework Directive in Turkey - inception report MAT01/TR/9/3. Houten

Grontmij (2002b) Common grounds between the Water Framework Directive and Water Management 21st Century. RIZA, Lelystad

Grontmij (2003) Institutional and legal strengthening in the field of water management in Turkey (draft). De Bilt

Hermans LM, Muluk ÇB (2002) Actor analysis for the Büyük Menderes River Basin Management Plan (draft). IHE, Delft

Holzwarth F (2002) The EU Water Framework Directive - A key to catchment-based governance. Water Sci Technol 45:105-112

IPO, UvW (2002) Consequences European Water Framework Directive for provinces and water boards (in Dutch). Inter-provincial Platform and Union of Waterboards, The Hague

Jaspers FGW (2003) Institutional arrangements for integrated river basin management. Water Policy 5:77-90.

Ketchum B (1972) The Water's edge: critical problems of the coastal zone. MIT Press, Cambridge, MA

Klein AWO, Van Buuren JT (1992) Eutrophication of the North Sea in the Dutch Coastal Zone, 1976-1990. Tidal Waters Division, Directorate General for Public Works and Water Management, The Hague

Lanz K, Scheur S (2001). EEB Handbook on EU water policy under the Water Framework Directive, European Environmental Bureau, Brussels

Mostert E (1999) River Basin Management and planning - Institutional structures, approaches and results in five European countries and six international basins. RBA Series on River Basin Administration, Research Report nr. 10, RBA Centre, Delft

OECD (1999) Environmental performance reviews – Turkey. OECD, Paris

RWS (2000) A different approach to water - water management policy in the 21st Century. Ministry of Transport, Public Works and Water Management, Directorate-General for Public Works and Water Management, The Hague

SENTER (2002) Terms of reference 2002 - implementation of the Water Framework Directive and integrated coastal zone management in transitional and coastal water in Romania. The Hague

SENTER (2003) Terms of Reference 2003 - Development of a framework for formulation of regional water management plans in the water districts of the coastal zone of Croatia. The Hague

Sorensen JC, McCreary ST (1990) Coasts: institutional arrangements for managing coastal resources and environments, USDIA and US Agency for International Development. Coastal Publications No. 1

UNCED (1992) Agenda 21, Rio Declaration and Agenda 21. Report on the UN Conference on Environment and Development. Rio de Janeiro, Brasil, 3-14 June 1992. A/CONF.151/26

Van Hofwegen PJM, Jaspers FGW (1999) Analytical framework for integrated water resources management - guidelines for assessment of institutional frameworks. Balkema, Rotterdam

WSSD (2002) Johannesburg Declaration on Sustainable Development and the Plan of Implementation. Report of the World Summit on Sustainable Development, A/CONF./199/20

Young OR (2003) Environmental governance: the role of institutions in causing and confronting environmental problems. Int Environ Agreements: Polit Law Econ 3:377-393

9. Inclusive and community participation in the coastal zone: Opportunities and dangers

Tim O'Riordan[1]

Abstract

Inclusive and community participation applies to negotiating procedures that are designed to encompass a wide and representative range of interested parties to guide environmental management. For such a democratic procedure to prove effective, the participatory procedures need to be accepted by policymakers and those responsible for delivery. These procedures must also be fully representative so as to be accepted to all stakeholders. Finally the process must be pragmatic and timely. This chapter examines both the theory and practice of inclusionary procedures for coastal management for long-term and uncertain coastal futures. It concludes that such procedures cannot easily be put in place unless there is a change in the design and management of coastal governance. Some suggested proposals are enhanced as part of long-term research evaluation of changing approaches to governance for sustainable development.

Introduction

There is growing expectation and requirement for inclusive community involvement in coastal management. This is evident in the Marine Site guidelines for establishing Natura 2000 habitats on the coast throughout Europe, and in the evolution of coastal habitat management plans and shoreline management plans or their equivalents in continental Europe. In general, EU directives in the offing, notably following up the Strategic Environmental Assessment and Water Framework Directives, require an increasing element of articulated involvement and social ac-

---

[1] Correspondence to Tim O'Riordan: t.oriordan@uea.ac.uk

J.E. Vermaat et al. (Eds.): Managing European Coasts: Past, Present, and Future, pp. 173–184, 2005.
© Springer-Verlag Berlin Heidelberg 2005.

knowledgement in coastal planning. A critical appraisal of the value of, and dangers associated with, inclusive participatory involvement is therefore timely.

Such an approach is seen as legitimate, in that it obtains community consent and benefits for specialised local knowledge. It is also regarded as effective in that broad support is likely to lead to less contested outcomes, and a basis for strategic acceptance for long term planning across large sectors of coastline. Thus the very nature of integrated coastal management would seem to require and benefit from stakeholder involvement and acquiescence for emerging policy.

Buckeley and Mol (2003) have helpfully summarised other reasons why greater participation is becoming the norm. These are:

1. The state apparatus is becoming more "democratic" and less hierarchical;
2. Scientific prediction is framed by considerable uncertainty, much of which requires some indication of policy response to shape outcomes;
3. The precautionary principle requires explicit incorporation of public values;
4. Complexity of outcome involves step-by-step understanding and acceptance of the options that have to follow from each decision stage to the next;
5. Voluntary agreements in planning and business practice, including regulatory compliance more generally, involve a greater extent of participation by interested publics, for legitimacy of non-formal practices;
6. Citizens generally are becoming more critical of governance, and more demanding of their say.

This may all seem dramatically plausible. But there are dangers to pursuing and relying on an inclusionary approach as this paper articulates. The points listed below have Europe-wide significance.

Long term strategic redesign for the coast may not be regarded as acceptable to shorter-term policy-designed institutions and financing arrangements. So there may be no local recognition of the "longsight" and no powers for guaranteeing land-sea management measures and coastal redesign sustained over, say, two generations. Hence, even if participation was "good", it may not be capable of handling the complexities of science, management, time and space that integrated coastal redesign will demand.

Planning powers for strategic intervention in coastal processes and development are not strong enough to ensure adequate safeguard of coastal protective systems and reconstruction of existing development. This kind of intervention would only be possible with openness and imagination and 'give-and-take' attitudes amongst citizen participants and official agencies that are not commonly associated with community participation at the local level. Planning is a means: powers, financing and political will backed by community support to get things done are also vital ingredients.

Precautionary science, leaving room for manoeuvre and creating opportunities to adapt to changing circumstances, requires a degree of vision and continued involvement that may not be easy to engineer into long term coastal design. In essence, there may be a disjunction between the "mood of the people" and the application of precautionary science. This may require a more interactive process between science-based management and stakeholders, possibly facilitated by training. In addition, coastal reconfiguration involving naturally functioning "soft" defences is not yet scientifically guaranteed to work in every instance. So agreement could be reached over an outcome that is functionally a failure. Such an outcome could lead to disenchantment with both the scientific and participatory procedures.

Structures for decision making for the coast currently preclude long term planning and coastal reconfiguration. Also stakeholder participation techniques are still ill-designed for the imagination and innovation of creative integrated coastal management over many generations. Arguably we do not have a democracy for this "style" of intergenerational management.

If these observations hold, then the scope for evolving inclusive community participation may be limited. This restriction can be increased by inappropriate statutory limitations on organisational structures, the restricted pattern of financing and evaluation of opportunities, and the outlook of participants, who may be locked more into the "do-able in the present" than the "possible in the future". The very act of inclusion may carry with it the constraining, rather than the enabling and compatible framework that delimits innovative opportunity. Only when decision structures are designed to be more accommodative and holistic for incorporating ecosystem functioning and adaptive management, combined with fresh approaches to public-private-civil partnerships as experimented with around the coast, and introduced in other countries (O'Riordan et al. 2000), will it be possible to move on with legitimate and effective community involvement in integrated coastal management.

It is also necessary to observe that coastal futures will involve a creative mix of "hard" and "soft" defences. Where there is substantial commercial and residential property, there will always be "hard" defences, unless the soft defences are extraordinarily robust. The scope for soft defences is as much a matter of sediment and ecological dynamics, as it is of economics and social acceptance.

The pros and cons of deliberative inclusion

It has long been a human dream that people (demos) rule (kratos) their destiny. The notion of inclusion carries with it expectations of being heard, of obtaining a favourable outcome, of involving everyone who has a stake and of sharing power with those who must rule.

In essence, there are two purposes of inclusionary participation (Owens 2000). One is to inform, and the other is to enable citizens to be partners in a management programme. The enablement model forms the basis of this paper. Its purpose is to examine how to improve the participatory process for coastal redesign beyond its present pattern of innovation. Informing is not a neutral activity. Effective participation depends on the structure of power, the legal basis of the management strategy, and the particular framing of the advocates in management teams. Even the apparent necessary and innocuous practice of informing, shapes biases and disturbs interpretation. For example, it is usually said that it is not possible to safeguard land for 50 years hence in coastal UK without purchase by a management body. But it is possible to create covenants and lease arrangements that could do this job. It is just that such an arrangement might be messy and unpopular. So the simple aim of informing can be influenced by pragmatic awkwardness of an option, not its lack of appropriateness. Similarly, the function of informing new scientific interpretations of ecological thresholds, or resilience tolerances, is also influenced by trust, authority and the management message. For this to happen, the scientific "voice" must be deemed to be credible. The indicators of tolerance must be seen to be intelligible and recognisable in everyday life. If the possible management prescription should allow scope for economic manoeuvre, then it will be accepted via participatory buy in. For example, a model of fish stock depletion could create an indicator of the size of young fish in a given catch. This is easily measurable by local fisherman (and may well be already done). If the limits to fishing become recognised by such a measure, and if the science community works with the civic community to find alternative livelihoods of the fisherman, then the act of "informing" becomes a sustainable experience for all concerned (see Roberts 2000).

The trick is to see integrated assessment not as a simple informing process, but as caring, interacting and negotiating pattern of science society relationships. This is the key to the discussion that follows.

Enablement is set in a civic model of deliberation. Here is where players help to shape an outcome based on regulation of their values with others. In essence, enablement is a creative act of reaching an initially unknown outcome, one that is shaped by the procedures themselves.

Bohman (2000, pp 237-248) provides a strong philosophical basis for a deliberative democracy:

1. It allows for the diversity of values of a modern multi-cultural society to be discovered, expressed and encountered;
2. It generates a public use of reason, which allows the civic state to argue its way into a common understanding, if not shared agreement;
3. It allows for different capabilities of political entry so that those with smaller resources and lower capabilities for debate and negotiation can be ensured a hearing and follow-through delivery;
4. It creates a deliberative majority for technical and bureaucratic expression of analysis and justification. Without the former, the latter will always remain illegitimate;
5. It establishes a mechanism for constitutional refurbishment through which engaged publics challenge the institutions on the grounds of their political authority and capacity for integrated and responsive management. In essence, deliberation is a necessary precursor to institutional reform;
6. It enables a deeper discussion of the very purpose and structure of democracy. The deliberation platform provides an invaluable basis for repositioning democracy in a changing constitution and age.

Bohman argues a fine theoretical justification for a fresh approach to deliberative democracies. But the world is not so neatly packaged. Susan Owens (2000, pp 1144-1147) suggests five reasons why the enablement model for deliberation may fail:

1. Not all interests declare themselves at the outset, and some may shift position as the process evolves. The other case, inclusivity cannot be guaranteed and objection or conflict may emerge, as outcomes are determined;
2. An excessive zeal for reaching consensus may mean that real conflict or contradiction is ruled out or diffused even when it may seriously impede final acceptance;
3. The deep unwillingness of those in power genuinely to share their power when the results of participation are announced, either because their powers are constrained, or because of a shortfall in budget;
4. The very mechanisms for deliberation and inclusion set the guiding framework, the language of discourse and the style of involvement. This means that differently conducted exercises could result in different outcomes, without the same players;
5. The particular policy frameworks within which inclusive processes are conducted will influence how information is transmitted and processed, what signals are regarded as most significant for guiding discussions, and how and when funding can be put in place.

In essence, there is no example of a deliberative and inclusionary process that is not subject to considerable bias and distortion of both procedure and effectiveness. In the case of the coastal zone there are a number of issues that severely stretch the effectiveness of inclusionary participation. These are examined in the section that follows. It is important at this stage to tease out further why deliberative and inclusionary procedures may not result in sound management outcomes:

1. All the relevant information may not be available, so actors may be working with imperfect understanding or misleading outcomes;
2. Patterns of power, limited budgets, and predetermined agency responsibility may make it difficult or impossible to deliver what is requested by stakeholders. Even when these limitations are spelt and broadened, it is still possible that they will be ignored;
3. Stakeholders may arrive at outcomes that are ecologically impossible to deliver. Saltmarshes may not survive when recreated and shifting muds and dunes may be undermined by poor substrate conditions.

These may appear self evident difficulties, yet they arise on many occasions. For example, on the North Norfolk coast near the village of Cley, the natural shingle sea defence is no longer capable of holding back a rising and stormy sea (see O'Riordan and Ward 1997 for details). A negotiated solution involving 37 stakeholders in a deliberative process reached a consensus in favour of a clay wall running across the existing nature reserve. This would safeguard part of a Natura 2000 habitat, and protect the coast for 40 years or more. But the cost of the wall grew with both the preparatory documentation (nowadays a major expenditure of coastal planning) and with the environmental requirements of its construction. Consequently, it became too expensive for the justification of the protection of the Natura 2000 site. This was the case even in the face of the UK interpretation of the Habitats Directive. This interpretation requires equivalent compensation of a similar nature reserve for any Natura 2000 site lost. The final solution was not negotiated. It is most likely to be a salt marsh reconstitution with a low wall right next to the coastal road. This was not the preferred solution in 1997, but circumstances have created an outcome that by-passed deliberation, and which has benefited from the loosening of the legal straight jacket of the UK interpretation of the EU Habitats Regulations. Nowadays it is possible to establish new nature reserves as part of a biodiversity action plan. There is no requirement to set a replacement reserve nearby.

Integrated coastal management and the challenge to inclusive participation

If coasts are to be redesigned over a century to withstand sea level rise, coastal reconfiguration and changing political and governance frameworks, then the task for managing inclusive participation is immense.

There is no political mechanism for examining the coast as an integrated system of erosion, transport, deposition and coastal defence that extends across many hundreds of kilometres. Stakeholders are too spaced out, they do not naturally take to making decisions 100 years hence when the options are highly uncertain, and where new policy that might enable longsight to be incorporated into the coast is not yet imaginable.

The notion of sustainable coastal management would provide for robust biodiversity and coastal geomorphology, as well as effective use of the coastal zone in the face of uncertain climate change, sea level rise, carbon and transport policy and reconstitution of key Euro-habitats. Thus there is no mechanism to create an inclusive negotiating process for sustainable management of future coasts, because there are so many interacting uncertainties as to what will happen to coastal integrity over the next 50 years.

The introduction of the EU Habitats Directive places fresh administrative and policy biases over the removal and relocation of key habitats. Under the terms of the Directive the integrity of biodiversity should be retained, even if a key habitat will be lost to sea level rise. This arrangement is based on a range of procedures and administrative requirements that absorb huge amounts of time and formal analysis, much of which precludes effective stakeholder intervention at the local level (O'Riordan 2002).

Current patterns of planning and financing coastal defence measures, certainly in the UK, make it very difficult to design coasts for 25 or more years hence. The coastal planning authorities have to take sea level rise into account in their strategic planning function, but their advice in zoning limitation can still be overridden by politicians anxious to generate income for coastal development. Financing coastal management remains a three year pattern with emphasis on formal cost benefit analysis that impedes creative ecological evaluation, and inhibits long term funding of coastal areas that should be designed for sea level rise and coastal change.

Compensation for land loss to flooding and flood hazard is not automatically available in the UK, nor indeed elsewhere. This severely inhibits stakeholder dialogue as the lack of any guarantee of compensation influences the willingness to participate let alone negotiate. Admittedly the UK is designing measures to provide funding for areas that may materially contribute to coastal management. But this approach is hit and miss, it relies on coastal science, which is imperfect, and it may still leave landlords or tenant farmers in the cold. Add to this the possibility of lack of insurance cover for future flood damage and one can see how strategic land may be blighted by unsuitable planning measures. Coastal networks of cooperating landowners are not always ready to respond to calls for effective participation when they feel they are being unfairly treated.

Decision structures are inevitably influenced by vociferous protest rather than inclusive democracy. Vigorous protest, when carefully targeted, causes decision bodies to pause and re-examine. The options under consideration are often promoted by the protest rather than by the precautionary shoreline management principles. The very procedures for discussion often require tortuous regulation across a wide range of governmental agencies and non governmental organisations. The outcome is often the result of a protracted and costly process of lobbying and reaction. The net result is either impasse, or a very expensive proposal, which cannot be funded from the monies available or by cost-benefit rules. Getting a "new" coastline is costly, paperwork demanding, exasperating for negotiators, and often unsatisfactory (see O'Riordan 2002, 41).

Current consultative procedures for integrated shoreline management generally fail to deliver long term sustainable strategies that are environmentally robust, cost effective and socially understood and acknowledged, even if not fully endorsed. Even elaborate forms of inclusionary democracy fail completely to overcome all of the pitfalls noted above.

The workshops produced a chart of possible relationships between forms of participation, especially the primary drivers as indicated at the outset of this paper, and the implications for both management and institutional design.

The science community needs to clarify ecological tolerances and thresholds to resilience in a manner that is not only intelligible to lay people. This process must also be communicated in such a manner that civil society in its various forms are actually empathetic to the issues raised and indicators used. This means a science-society dialogue as to how tolerances can be placed in the lifeworlds of local people. Control of nutrients for farmers to recognise and respond to (with appropriate incentives where necessary) may require a more inclusionary process of farm by farm management, relating to both visions and tolerances.

Visionary futures should be set in discussions and narratives that enable stakeholders to understand an implication for management choices of "longsight" that unite the aspirations of individuals to the wider concerns of all other stakeholder interests. This is a process of dialogue that also involves a science-civil understanding.

There seems, as yet, to be no entirely suitable approach to sustainable and integrated coastal management as the current pattern of powers, responsibilities and funding arrangements operate. It may be necessary to try an alternative design of the institutional patterns of coastal management to see if this could provide a basis for more in depth, legitimate and informed deliberation and regulation.

Redesigning the management of coastal futures

This section puts forward one possibility for the redesign of coastal management in the UK (Fig. 1). It is possible that some aspects of this approach may be followed in mainland Europe. One value of this volume is to compare notes and experiences.

The central management structure would be a coastal management partnership (CMP). This would run along a stretch of coast to match the coastal cells of the current shoreline management plan. The CMPs would be an amalgam of county and district authorities and statutory agencies, plus representatives for landownership, nature conservation, public recreation, fishing, parish councils and local traders and farmers. A mechanism to establish this joint body is available under the UK Local Government Act of 1972. It would be a non statutory association of organisations joined by a common benefit and a statutory shoreline management plan. This would provide the management structure with powers to deliver as well as powers to listen and to respond. It would also have its own budget and full political representation.

The actual mechanism for delivery of integrated coastal management might be in some form of a public private partnership (PPP). Such an arrangement is being tried out in the UK in the Norfolk and Suffolk Broads and in the Pevensey Levels in Sussex (see Ayling and Rowntree 2002). The point about the PPP, which is an offshoot of the private finance interactive generally, is that would provide a reliable basis for funding over a prolonged period. So long as the overall cost benefit analysis was favourable for the total PPP, then at any stage a particular piece of coastal management would be allowed to go ahead on the basis of sustainability principles alone. It would provide for much greater flexibility at the detailed management level, and a mechanism for more creative stakeholder input as to final design detail. Where the suite of projects can be kept within budget, and this would be for the CMP to determine through its annual reporting, then the PPP provides a basis for unusual proactive and interactive shoreline management.

Right now the PPP, as currently designed, would not be ideally suitable for the kinds of management tasks outlined in the figure, and summarised in the text above. This is because it is legally bound in formal contractual timetables and deliverables. But it can offer the inflexibility of tactic and management style that enables it to cope with the fluidity and freedom offered by a creative and inclusionary process. Nevertheless, the PPP is essentially a vehicle for ensuring private profit, so it is not especially a public service device, despite is name. This means that costs are challenged, routeways to delivery are short-circuited to save costly time, and there is little scope for imaginative cooperative agreements to connect flood alleviation to other public interest needs. The latter includes navigation, recreation, public amenity and heritage protection, and the maintenance of a local economy through creative enterprise based on flood protection and exuberant amenity. In principle all of these linked values should be connected to a viable public private partnership. But the cold practice of profit maintenance and expedient provision of the deliverables, such niceties are not fully appreciated.

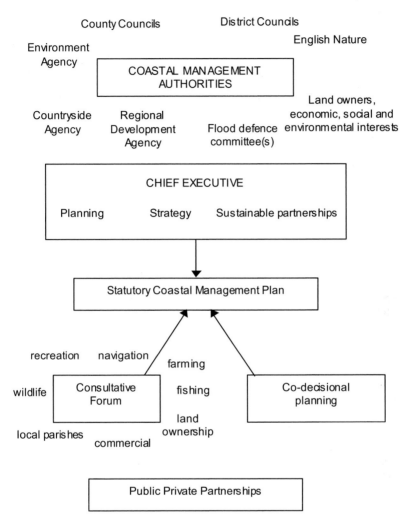

County Councils   District Councils

Environment   English Nature
Agency

COASTAL MANAGEMENT
AUTHORITIES

Countryside   Regional   Land owners,
Agency   Development   Flood defence   economic, social and
Agency   committee(s)   environmental interests

CHIEF EXECUTIVE

Planning   Strategy   Sustainable partnerships

Statutory Coastal Management Plan

recreation   navigation
farming
wildlife   Consultative   fishing   Co-decisional
Forum   planning
land
local parishes   ownership
commercial

Public Private Partnerships

Fig. 1. Coastal management authorities and public private partnerships. Source: O'Riordan 2002

Nevertheless, the PPP does form a potentially viable vehicle for this new approach to coastal management. This test is to reform it, not to abandon it. The CMP would embrace a range of economic, recreational and fisheries issues, so would attract a wide variety of established stakeholders. The aim is to be inclusive yet workable. One model is that of a small "political" representation on the partnership but a larger incorporation of stakeholder interests via a single, or set of, participatory forums. It would be helpful if such forums were designed to be both deliberative and inclusive, so that their judgements and community actually carried weight into the partnership.

The statutory shoreline management plan would be a format outcome of the work for the partnership. It would provide the framework for economic development and environmental protection set in the context of sustainable development. This would mean a partnership in planning, economic wellbeing and social improvement aimed at evolving a social advantage out of robust, environmentally valuable, coastal management. The role of sustainability partnerships is outlined at the end of this paper.

For the kind of integrated shoreline plan outlined in this workshop, the PPP will require some modification. This would include greater flexibility to the legal-contracted framework subject to an agreed audit of performance. There would also have to be more flexible legal capacity to establish a wide range of enabling partnerships for imaginative coastal management. And there would have to be scope for creative networks for coordinating community forums into the process of evolving management as part of the civic dialogue introduced earlier.

The main advantages of public private partnerships for integrated and sustainable shoreline management are:

1. There is an overall budget that provides a guaranteed programme of work for 25 or more years. So the long term time being is built in with supportive funding that would not require annually to be accounted for;
2. The overall budget is designed to be cost effective, but its parameters would include ecological accounting so as to place value on ecosystem-based management of the shoreline, as well as the aesthetic and psychological benefits of retaining and creating natural shorelines through managed realignment;
3. Each project of reconstruction would be designed by the same firm of engineers, biologists and social scientists and citizens groups so that would be continuing in the design process and in the sequence of implementation of flood management;
4. Stakeholder forums could be formed at various geographical parts of the CMP area. These would be cooperative partnerships of a range of stakeholders networked to a consultative panel and linked by Internet. Every year they would meet in person across the CMP to endorse the next sequence of management. And each management scheme would be inclusively processed taking into account the issues raised in the previous section;
5. Replacement habitat, compensation for creative shoreline management, long term purchase or leasing of management agreements would be more possible in the new structure of financing and management. Links to agri-environment schemes and effective cooperating partnerships with other agencies and voluntary organisations would allow for fresh opportunities for creative additional design of sustainable options for local enterprise, educational arrangements and food production, to add to the enhancement of heritage and wildlife. All these would be promoted by sustainability partnerships;
6. The planning function would be subsumed within the statutory shoreline management plan. This would allow for much better coordination of the planning formulation with prolonged shoreline management, thereby overcoming a period of friction in current arrangements.

It is worthy of note that various partnerships, many with overlapping membership, are already in place. In the Humber sub-region in the UK, for example, there is a Humber Action Plan, a ports and estuaries strategy, a Humber Economic Development Action Plan, and a Humber Industry Nature Conservation Association. All of these involve consultative forums, and such a convoluted pattern of discussion is common in major estuaries. So the CMP would need to involve a rationalisation and restructuring of such partnerships into a common endeavour. This will probably be the case elsewhere in Europe. It also raises an important question as to how all this rationalisation is effectively to take place. Irrespective of the CMP idea, this process of streamlining participatory networks is long overdue on the coastal scene.

The scheme outlined in Fig. 1 has yet to be tested. The critical test is whether it can be truly produced a sustainability outcome compared with present arrangements. And the equally important issue is whether inclusive negotiation would work better in a more enterprising responsive framework.

The notion of sustainability partnerships requires more attention. Sustainability begins with a robust and functioning ecosystem service. Reliable and robust coastal protection by tide, sediment, substrate, vegetation and integrated management is a first base condition. Water stewardship, biodiversity enhancement and tourist-economic related activity on the redesigned coast would be a second base condition. Local communities designed to be inclusive and capable of developing enterprise on the basis of a robust coastline, would be a third base condition. The sustainability partnership would locate ecologically sound, socially responsive and economically reliable patterns of coastal activity.

What still needs to be done is to examine how far it is possible to devise a PPP that is sufficiently encompassing and yet financially and contractually flexible to accommodate the uncertainties and precautionary management aspects of long term shoreline management. This will be a difficult contractual task and will involve more flexibility in legal analysis that is commonly the case.

In essence, the very design of the PPP will have to involve deliberative and inclusionary stakeholder involvement. This is not normally the procedure for UK Treasury rules but it may prove vital if there is to be stakeholder forum "ownership" of the PPP, and a better understanding of the civic discourse that will guide all future negotiations.

Stakeholders themselves will need to exert an element of a sustainability democracy that spans generations and much shoreline space. This will require an element of training, visualisation procedures so that future coastal images can be creatively and effectively portrayed and mechanisms to enable land protection and leasing to be put in place well before land is needed, but without causing planning blight (see Gill et al. 2002).

The sustainability partnerships will require a degree of ecological system functioning and evaluation to justify the ecological-social benefits of natural shoreline protection. This is not yet a basis for cost benefit analysis and will have to be negotiated in at the start of the proceedings. Funding for such partnerships may come from health, crime prevention and social care budgets if the schemes promoted by such partnerships generate better health, more security and local enterprise for vulnerable people close to crime.

Reporting styles between civic sphere and the implementary agencies will need a fair amount of common language and understanding. Again there is a need for some sort of training preparatory phase to all of this to ensure that there really is a common approach.

All of this is creative, innovative and evolutionary. The challenge of interactive and sustainable shoreline management is enormously demanding on current patterns of administration, financing and consultation. Research of the kind proposed here can do wonders to map out the possibilities and the blockages. Right now there is a huge amount of ferment within official worlds over new management and consultative structures on the coastline. So this is an opportune time to promote this particular kind of institutional review.

References

Ayling B, Rowntree J (2002) The Broads flood alleviation strategy: a review of the public private partnership programme. In Gill J, O'Riordan T, Watkinson A (ed) Redesigning the Coast. CSERGE Working Paper PA 02-01, UEA, Norwich, pp 49-54

Bohman J (2000) Public deliberation: pluralism, complexity and democracy. MIT Press, Cambridge MA

Buckeley H, Mol A (2003) Participation and environmental governance: consensus, ambivalence and debate. Environ Val 12:143-154

Gill J, O'Riordan T, Watkinson A (2002) Redesigning the coast. Tyndall Centre for Climate Change Research, University of East Anglia, Norwich

O'Riordan T (2002) Redesigning the coast: results of a workshop. CSERGE Working Paper PA 2002-01, University of East Anglia, Norwich

O'Riordan T, Ward R (1997) Building trust in shoreline management: creating participatory consultation in shoreline management plans. Land Use Policy 14:257-276

O'Riordan T, Andrews J, Samways G,. Clayton K (2000) Coastal processes and management. In O'Riordan T (ed) Environmental science for environmental management. Prentice Hall, Harlow, pp 243-266

Owens S (2000) Engaging the public: information and deliberation in environmental policy. Environ Plann A 32:1141-1149

Roberts CM (2000) Why does fisheries management so often fail? In Huxham M, Sumner D (eds) Science and environmental decision making. Prentice Hall, Harlow, pp 170-192

10. Group report: Institutional and capacity requirements for implementation of the Water Framework Directory

Wietze Lise[1], Jos Timmerman, Jan E. Vermaat, Tim O'Riordan, Tony Edwards, Erwin F.L.M. de Bruin, Areti D. Kontogianni, Kevin Barrett, Ton H.M. Bresser, and Emma Rochelle-Newall

Abstract

Different aspects of institutional and capacity requirements need to be considered to effectively put the Water Framework Directive (WFD) in place. This chapter tries to find the most effective way of managing the river coast continuum, to ensure an appropriate role for public participation, EU ('Brussels') policymakers and catchment managers. We find that coordinated action is needed to oversee the river-coast continuum. Cultural differences, varying from one EU member state to another, can influence the style and role of implementation of the WFD. There is a clear role for formal public participation in implementing the WFD, as the process of a higher level of organised public participation is unstoppable. A dialogue is needed that brings together community intelligence and scientific systems understanding for the sustainable management of flood-prone rivers and coasts.

Introduction

Achieving a 'Good Ecological Status' for all watercourses and catchments, as required by the WFD, has large implications for the future of integrated water resources management, especially in the coastal zone. The ecological situation in the coastal zone is influenced on the one side from land with out-flowing rivers and shoreline activities including maintaining shoreline morphology, and on the other side, by activities in the wider sea including fisheries, transportation and mineral extraction. The transboundary aspects of this, where activities in one country have influence on the coastal zone of another country, are obvious. Implementation of the WFD, even more strongly than previously, influences institutional arrangements and the capacities needed to put the stewardship principles of the WFD into practice.

This chapter reports on the possible institutional and capacity requirements for implementation of the WFD. It builds on earlier chapters by De Bruin et al (this

---

[1] Correspondence to Wietze Lise: wietze.lise@ivm.falw.vu.nl

J.E. Vermaat et al. (Eds.): Managing European Coasts: Past, Present, and Future, pp. 185–198, 2005.

volume) and O'Riordan (this volume). The chapter deals with a delineation of the water body and management structures, both in spatial and in legal terms, the means to achieve the goals of the WFD, with emphasis on the subject of public participation, and the institutional changes that are anticipated to reach such goals. This chapter follows the structure of the discussion and is divided into the following four themes:

1. Boundaries of the water body and management structures;
2. Relation of WFD to other legislation;
3. The role of public participation in implementing WFD;
4. Institutional change for implementing the WFD and cost consequences.

The *first theme* compares the natural boundaries of water bodies with their actual management structures. The following questions are discussed: How to define the watershed boundaries of the coast in relation to the catchment, as seawater also interacts with inland water via groundwater, tidal rivers and estuaries? Do we need a managing body for the open sea and the catchment and what should such a managing body look like?

The relation between the WFD and other legislation is discussed in the *second theme*, namely trying to find answers to the following questions: What is the role of legal issues in implementing the WFD? How does the WFD relate to other legislation like the EU Habitat Directive and broader legislation on the control of toxic substances? How will "incorporation in the WFD" be actually specified and implemented?

Theme three discusses the role of public participation in implementing the WFD. The following questions drive the discussion: What is the scope for public participation in the WFD? What are the current trends with respect to public participation? What are the criteria for successful public participation in the WFD? What are the drivers for public participation?

The fourth theme deals with the required institutional changes to bring about the implementation of the WFD including the cost consequences. The following questions drive the discussion: To what extent do countries have flexibility in implementing the WFD? Do we need to change institutional arrangements for implementation of the WFD in different countries and, if so, how do we redesign the institutional structure? What are the economic consequences and how should costs and benefits, be traded off; and cost recovery and how to achieve cost efficiency achieved?

In sum, the four themes in this chapter each address different aspects of the institutional and capacity requirements of the WFD. All four have to be considered to effectively put the WFD in place. These themes are linked by a single overarching question: What is the most effective way of managing the river coast continuum and what should be the appropriate new roles for public participation, EU (Brussels) and catchment managers?

Boundaries of the water body and management structures

Coastal water in the WFD is defined as follows: " 'Coastal water' means surface water on the landward side of a line, every point of which is at a distance of one nautical mile on the seaward side from the nearest point of the baseline from which the breadth of territorial waters is measured, extending where appropriate up to the outer limit of transitional waters" (European Commission 2000, art. 2.7).

The question is whether the first nautical mile into the sea is an appropriate boundary for the identification and implementation of the WFD. It can be answered positively, if the hypothesis –a good ecological status in the first mile implies a good ecological status in the rest of the sea– holds. However, since river plumes continue far beyond the one-mile zone, persistent substances and nutrients and, consequent algal blooms can be found in the wider sea, even if they do not occur in the coastal zone (OSPAR 2002). Furthermore, the wider sea can have an impact on the one-mile coastal zone through, for example, shipping accidents and malpractices during fishing. Besides these *anthropogenic* causes, environmental problems in the coastal zone can be induced by *natural* causes (e.g. the natural flow of water, possibly induced by storms). Finally, although the coastal zone is identified as a conceptual and operational entity, it is generally characterised as a transition zone or interface and not as a distinctly defined system bound of a one-mile distance (Von Bodungen and Turner 2001). Consequently, the one-mile boundary of the WFD is too limited to support integrated coastal zone management and to achieve a good ecological status. The European Commission acknowledges this and a strategy to protect and conserve the marine environment is currently under development (European Commission 2002).

One third of the regional seas worldwide have regional conventions. Ledoux et al (this volume) provide an overview of the conventions covering European regional seas. It is difficult to gauge sense the contribution of these conventions in the achieved reduction in nutrient emissions and other polluting substances (European Commission 2002). Nevertheless, such agreements can be very useful as they bring the problems of larger scales to the public attention, and put pressure on governments to get polluters to reduce emissions. Emission reduction may be possible in a voluntary manner when less pollution goes hand-in-hand with bigger profits. However, more stringent reduction measures tend to meet opposition. Besides, the socio-economic conditions in the region play a role too. For example, some agreements have worked better in the Baltic Sea, than in the Mediterranean Sea. These differences may be related to the comparative wealth of riparian countries, as well as the nature of the pollution in relation to economic (including recreational) activity.

While the border of the water body as pointed out in the WFD makes sense in the context of water management, there is a need to cope with varying interests of riparian countries. Joint bodies[2] exist for transboundary catchments like the Rhine, Elbe and Danube, but such commissions are not yet formed for all relevant European catch-

---

[2] Joint body in the definition of the convention on the protection and use of transboundary watercourses and international lakes means "any bilateral or multilateral commission or other appropriate institutional arrangements for cooperation between the Riparian Parties" (UNECE 1992).

ments. Hard law (i.e. statute law which is enforced by criminal proceedings or economic sanctions) does not always provide full support to these commissions, but most have strong political support. A complicating factor is that not all countries sharing a catchment are EU member states (i.e. Switzerland in the Rhine catchment). Nevertheless, the WFD has provisions for these kinds of situations. However, these river commissions are not always in direct connection with the regional seas commissions.

Therefore, a distinct need was identified for a joint body that incorporates the full catchments, which can settle trans-boundary issues and manage the catchment-coast continuum. The objective of such a joint catchment and open sea management body is to translate the joint responsibility of riparian countries in the catchment into coordinated action. A stronger political and legal foundation of regional seas commissions, without attempting to redefine the WFD, could be pursued. The point here is that strong, co-ordinated and scientifically supported analysis of coastal-river management for water stewardship and sustainable development generally should be the articulated purpose.

Relation of WFD to other legislation

As discussed in Ledoux et al (this volume), a range of EU-directives and other legal arrangements exists that steer water management. In this section, we will deal with how the WFD relates to other legislation. In this context, it is important to distinguish between hard law and soft law (implemented by voluntary agreements and codes of practice)[3]. Conventions such as discussed briefly in the previous section can be labelled as 'soft law'.

Legal issues play an important role in managing rivers at the catchment scale. The further away from the source the impact is, the harder it is to prove causality, and thus hold someone legally liable beyond reasonable doubt. Hence, one first has to prove that there is an impact and, second, there has to be a proven causal chain. Compliance with current international and national standards would be a defence against prosecution or civil action. From a legal point of view, it is also important to make a distinction between natural and human causes, as already touched upon in the previous section. For instance, an event where dunes are washed away because of a storm can be caused solely by natural conditions, but also by poor maintenance. In such instances it may be difficult to get a clear-cut proof. From this, we can draw the conclusion that hard law may lead to slow legal procedures.

Yet, soft law can be more effective than hard law, because soft law can work as a catalyst to achieve targets faster. Moreover, it is desirable to seek means to strengthen soft law, for instance through publicity. This is even more important, as soft law often complements hard law and much hard law is implemented in a soft way.

Let us now turn to the question of how the WFD relates to other legislation. We observe that a range of previous water related EU directives are incorporated into

---

[3]   For a discussion on the dividing line between hard law and soft law see Tanza (2002).

the water *framework* directive. Besides, a number of directives become redundant, as they are taken care of within the WFD. For instance, the Shellfish Directive becomes obsolete under the WFD as it aims for the same objective, namely to obtain a good ecological status. However, other agreements, like the Ramsar agreement are not included in the WFD. Furthermore, biodiversity is not dealt with in the WFD. The Bathing Water Directive, which is currently under revision, is mentioned in the WFD, but will not be replaced by it.

Hence, it is not straightforward to specify and implement the actual 'incorporation into the WFD'. On the one hand, in the case of the Habitats Directive, the creation or strict protection of habitats may conflict with the interests of local communities, for example when land may have to be surrendered to the sea. Also, various stakeholders may fear a reduction in their freedom to engage in possibly polluting economic development (e.g. port authorities, land owners, fisheries and aquaculture entrepreneurs). From an international perspective, river basins and the Habitats Directive, are based on natural geographical boundaries rather than administrative ones. Watersheds will thus cut across local, regional and international borders. On the other hand, the Habitat Directive calls for sanctuaries in the sea. This may facilitate the achievement of a good ecological status, the principal objective of the WFD.

Also, there has been much discussion in the EU legislation on the control of toxic substances, leading to a list of priority substances in 2001 being added to the WFD (European Commission 2002), but this list is not comprehensive. Within the REACH programme the European Commission tries to further regulate and control the production and release of toxic substances.

Over the last two decades we have seen a reduction of inflow of dangerous substances into the coastal zone. The implementation of the Priority Substances Directive will lead to a further lowering of the toxic load from the catchment, contributing to the achievement of a better ecological status. In addition, it is important to harmonise the monitoring and reporting regimes both between directives and throughout Europe. Initiatives to resolve this issue are on their way elsewhere, but mainly for land and less for the sea (e.g. Harmoni-CA (http://www.harmoni-ca.info/); and Monitoring Tailor-Made (Timmerman et al. 2001), and calibration exercises across transboundary catchments, which is part of the Common Implementation Strategy of the WFD). In conclusion, we see good opportunities that different directives will mutually enforce one another. However, real-world implementation of the "incorporation into the WFD" remains a challenge to be awaited. All we do here is offer clear guidelines to assist this strategy.

Flexibility in implementing the WFD and public participation

In relation to public participation, the WFD states that "Member States shall encourage the *active* involvement of *all* interested parties in the implementation of this Directive, in particular in the production, review and updating of the river basin management plans" (European Commission 2000, art. 14.1) (italics added).

This phrase is discretionary, allowing for a variety of interpretations. The key point here is that some sort of participatory and deliberation processes will be required, if implementation of the WFD is to meet its political and statutory aims. The term "interested parties" implies active involvement of at least the most important stakeholders; this involvement can range from inclusionary processes, where they are consulted, through participation of selected stakeholders, to deliberative participatory processes, where they are part of the decision making process as equal partners (see e.g. Turner 2004).

We can see from the citation that the WFD also gives ample opportunity for a broader and direct participation of the general public. An increasing involvement of the public in decision-making, next to traditional democratic representation, furthers the need to fully engage in deliberative participation. Current trends indicate that three processes are ongoing, which have caused the growing importance of public participation:

1. *Negativism*. There is a general feeling of democratic deficit. On the one had, some governments seem no longer to deliver the results as desired by the public and public trust in political decisions has fallen dramatically and is not recovering, causing a crisis of legitimacy. On the other hand, governments are constrained by global economic interests and by multi-lateral obligations, and lack the capacity to meet the many and frequently conflicting local population needs.
2. *Pragmatism/efficacy*. There is a growing awareness among governments that decisions are often no longer acceptable without participation of the public in the decision-making process. Without this public consensus, decisions may fail.
3. *Citizenship/sustainability*. Nowadays people want to be able to shape their own futures. We have a self-evolving society, which is a recent trend. ICT has a role in this, in that it opens up information to an ever-wider audience and enables people to coordinate and direct actions. It is now possible to visualise images of future flooding, or landscapes or coastal patterns to allow stakeholders to see for themselves how future patterns of landscape and policy may evolve. Such images are critical in the participatory process.

Norris (1999) studied 28 countries and one of his conclusions is that Scandinavian countries still have powerful coalition governments, who cooperate creatively to establish wide acceptance for their decisions.[4] For example, in setting the UK carbon tax, the government worked with industry via a consensus approach (DETR 2000). However, this did not stop complaints about increased costs once the carbon tax was implemented, particularly as competitors in some other countries were not subject to such progressive environmental taxes. In the specific context of Norway, the implementation of a carbon tax did not ask for rigorous changes of the mainstream economy; the story may be quite different elsewhere. When the coalitions are less powerful or countries have single party governments, the qualifying negativism plays a much larger role.

This brings us to the question as to what makes public participation successful? Indicators of community trust are necessary to provide the answer. Public participation is an interactive process, being much more than either top-down or bottom-

---

[4] See also CEEP (2002) on governance.

up. In this interactive process there is a need for both larger structures of guidance that also set the boundaries of the process, and smaller structures of self-evolvement in which the decisions grow. Participatory methods are certainly not the panacea of future coastal management, as there are dangers in putting institutions based on deliberative processes in place (see O'Riordan this volume):

- We do not always know the stakeholders, as information and understanding of processes is limited;
- Openness is not necessarily the best strategy, as an open dialogue between stakeholders with different powers is difficult to achieve. As a result, a solution, which is optimal for everyone, may not exist, because not all partners strive for optimal solutions;
- There is a danger of bias, when only a small number of voices are heard. In such a situation, power differences may increase, as the public participation process may strengthen involvement of certain groups over others;
- The existing mandatory frame (existing legislation, but also financial constraints) for the deliberative process may lead to disharmonious rules. It may lead, for instance, to infeasible budget requirements;
- Perverse outcomes may emerge, because of excessive demands by particular individuals or groups. This is also known as the "squeaky wheel" syndrome. Such outcomes may be rigid, inappropriate, inconsistent (compared with national or international requirements), short-term, poor compromises, or contrary to people's well being and good water management.

These difficulties may be overcome by reshaping decisions. The greatest danger may be when society and its decision-makers do not engage in participation at all. Moreover, the process of a higher level of public participation is evolving, irrespective of whether it is good or bad. Consequently it is necessary to anticipate on participatory processes. For this, the following can be suggested as rules-of-thumb guidelines for improved public participation:

1. Establish an open dialogue between scientists and practitioners where community intelligence is valued at the same level as scientific intelligence;

2. Provide for a genuine ability to share outcomes. If certain outcomes are mandatory beforehand, an open dialogue will never be possible;

3. Devise visioning procedures (Turner this volume). Have a catchment forum, and sub-catchment ones for large areas, which enable various stakeholder interest and catchment managers to meet face to face (Janssen et al. 2003). ICT offers many possibilities for visualising possible design outcomes for rivers and coasts. Such a visualisation helps to rule out certain measures and choose among alternative viable measures. But visualisation on its own is no panacea. It is a tool for more effective civic engagement.

Based on this, we distinguish among five important factors that drive inclusive participation in Table 1. First, the publicly perceived democratic deficit implies that the general public no longer easily accepts "top-down" political decisions: there is a need for informed public participation throughout the decision process. Second, the legal/regulatory mandate needs to be adjusted to account for changes in society towards public participation. Third, society is changing into a self-organising citizenship. Fourth, besides scientific intelligence, there is a need for inclusion of local knowledge and vernacular guidance in decision-making. Finally, as coastal management deals with long time horizons, there is a need for visioning futures through scenarios and storylines. These 'drive' the demand in the society and require to be attended when participation is to be incorporated.

Note that participatory procedures always take place in the context of political power, procedural legitimacy, and the statutory framework of agency commitments. So the framework of policies and power relations, which may be hidden from the public, shapes participation. Just because there is participation, does not mean to say that it is legitimate or well executed. It is vital that the wider institutional biases are properly understood before any participation programme is evaluated.

Based on these societal drivers, an attempt was made to derive a number of traits, required for a successful management style for the coast and the institutions associated (Table 1). Integrated coastal zone management is complex and interdisciplinary and dwells therefore on integrated assessment (Turner, this volume). We identify how integrated assessment is to be deployed to meet the requirements of our five drivers, and what indicators would be useful to assess their successful deployment. We included a third column here, labelled thresholds, since we felt that to assign such thresholds would be important indicators of societal transition towards an institutional incorporation of deliberate participation.

Table 1 summarises the drivers and the implications for managing integrated coastal futures. It is aimed to be in parallel with the scenarios table issued elsewhere in this book. The purpose is to summarise the key drivers as outlined in the text, and to run these against management and institutional arrangements that may have to be modified for effective public participation to be put in place.

The table emphasises that public participation is driven by a loss of trust in conventional politics and political decision-making. It is also promoted by legal rules set in directives and regulations, as is the case for the WFD. It is further promoted by citizens who now feel they have a responsibility to shape their own lives and catchments. And there is a technology and a decision format available via visioning and participatory geographical information systems. The vertical columns apply to six measurements of comprehensive integration. Much of the right hand side of the table reveals the need for capacity building and skills training in the more adventurous aspects of participation.

Furthermore, this table points out (in the last row) that interdisciplinary problems need integrated assessment, which can be performed by undertaking Driver-Impact-Response and scenario analyses. As an outcome, existing institutions need to be examined and in some cases redesigned, which can be achieved through a comparative analysis of institutional drivers. This will be discussed in more detail in the next section.

Institutional change and cost consequences

The WFD is often considered rather prescriptive in its implementation (Ledoux et al. this volume). The directive nevertheless leaves much room for its implementation to the countries themselves, as long as they achieve the targets of a good ecological status. Thus, the WFD largely sets the playing field with issues like river basin approach and transboundary cooperation, on which countries can make their own match. For example, managed realignment[5] can also be used as a flexible instrument in implementing the WFD and is a very important tool for the Habitats Directive. Managed realignment of coastal defences can also be an alternative to solid dykes (Rupp and Nichols 2002). As institutional arrangements differ from country to country, it is interesting to compare WFD implementation in various countries.

Public tasks, such as the management, monitoring, enforcement, as well as implementation of amelioration measures, are carried out by institutionalised organisations, often with delegated powers from the government. Sufficient legal and financial support is a prime condition for their appropriate functioning, but public recognition, as well as a mechanism for public engagement (O'Riordan this volume) is equally significant. These organisations vary from country to country and a variety of mechanisms are used for public engagement. Often these are for single issues, but they can develop into multi-issue groups.

One important element of ICZM is that often measures, such as the ones required in the WFD, have long-term effects. Handling complex issues with a long-term perspective of say 70 years requires an adaptive design approach and institutional management. Important elements here are:

1. A participatory deliberative culture that embraces the precautionary principle (e.g. EEA 2002);
2. The notion of long-sightedness is difficult to introduce: how to get people to think two generations or more ahead? Long-sighted democracy needs self-adaptive community networks that are based on sustainability;
3. Insight into planning and decision-making is needed as, for instance, planning a marina will freeze land for 70 years, while a better environmental solution for this land could have been to turn it into salt-marshes;
4. A precautionary society that enables decisions, before certainty in cause-effect mechanisms is confirmed and thus allows for mistakes;
5. Coastal partnerships designed as open and adaptive structures. Such mechanisms are needed to bring people together.

---

[5] Managed realignment 'involves setting back the line of actively maintained defences to a new line inland of the original – or preferably to rising ground – and promoting the creation of inter-tidal habitat between the old and new defences' (Rupp and Nichols 2002).

Table 10.1. Requirements for management and institutions from drivers for inclusive participation

Drivers	Management			Institutions		
	Integrated assessment	Indicators	Thresholds	Comprehensiveness of function	Responsiveness – capacity - skills	Pro-active and precaution
1. Democratic deficit	Governmental	Politically intelligible and socially derived	Opportunity not fear; Taking responsibility forward	Test for comprehensiveness	Training/ Capacity building	Shared understanding and responsibility
2. Legal/regulatory mandate	Due process following established procedures	Legal requirement as laid down in regulations	Justification through participative science	Proving legitimacy	Testing for skills training needs	Designing for worst case
3. Self-organising citizenship	Network forums at local scale	Locally grounded	Locally grounded	Stakeholder inclusion	Creating community leadership	Locally grounded
4. Local knowledge and vernacular guidance	Dialogue of shared knowledge	Shared knowledge	Trust and confidence	Dialogue and Media	Using local training and schools to image futures	Shared knowledge
5. Visioning futures	Scenarios Storylines	Socially intelligible	Mapping impact trails	Reasoning the Remits	Training workshops	Testing for opportunity and compensation for new livelihoods
Comparative analysis of institutional drivers	Interdisciplinary→integrated assessment→ Driver-Impact-Responses and scenarios			Institutional examination and redesign based on critical evaluation of coastal management arrangements		

In deliberative processes, it is important to have clear and shared objectives. Re-design for longer-term coastal decision-making of organisations will only take place when it is viable and if real improvement can be achieved. The setting of objectives should therefore also be done in a deliberative process. Although this can be a laborious task, the result is often worth the effort (e.g. Gregory 2000).

Distinct cultural/political differences across Europe should be taken into consideration in long-term planning for sustainability. In pre-accession countries, for instance, political problems (distrust in institutions, uncertainty about the present and the future, anxiety about day-to-day living) discourage people from long-term planning. In pre-accession countries such as Turkey, in Greece, and also elsewhere, substantial fractions of the human population connect significance in their daily lives to a kind of predetermination or *kismet* overruling the future. This may be an important cultural driver, which may discourage long-term planning. Engagement of all societal strata in a deliberate, long-term participatory decision-making process may meet unexpected opposition here.

Elsewhere, public participation is only invoked after planning conflicts have magnified and stakeholder positions are falsified. Cultures, with a traditional working class-elite, conflict with massive strikes and property looting and will not easily develop consensus platforms and negotiations.

Turkey, as pointed out in De Bruin et al. (this volume), has a water management structure, but this does not comply with all the principles of integrated river basin management (De Bruin et al. this volume). Turkey is implementing the WFD in the context of her pre-accession status and is very interested in utilising the WFD methodology. Clear objectives and implementation strategies could improve catchment management in Turkey.

Greece does not perceive the sea level rise as a problem, whereas water scarcity is considered the main problem. Such a perception can lead to overreacting. In extracting water in certain parts of Greece for example, farmers over-extract in June in their fear of facing scarcity in July and August when water is needed for the rice crop. While the total amount of water should be sufficient, through this behaviour, the farmers themselves create problems. Through a participatory process, in which the problem situation was discussed, the situation could be improved.

Institutional change also has financial consequences. Is it possible, for instance, to charge a German farmer for not reducing nutrient loads that lead to negative impacts in the Dutch coast? This example shows that the principle of full and fair cost recovery is difficult to achieve. In the case that costs are recovered, it may still be necessary to decide whether revenues should go to nature conservation, flood protection or the general revenue. Is it possible to compensate people when they are financially disadvantaged in the provision of an ecological benefit? This is possible by financing ecosystem functioning via a trust fund or stewardship fund. Alternatively, environmental bonds may be issued. For example, in the case of coastal water pollution, bonds may be auctioned on reaching a desired reduction in the load to the sea. The bonds, which should be tax-free and financially attractive, will only be paid out once the target is reached. In this way, cost-effectiveness can be reached without government planning (see also Horesh (2003), for a more general discussion, or Lise and Van der Veeren (2002) who calculated a possible cost optimal solution of the eutrophication problem in the North Sea from the Rhine basin).

In the previous section (Table 1) institutional change was called for, coupled to changing needs. A number of priority questions arise, which call for a careful analysis by also taking local priorities into account. Important questions would be: What should institutions dealing with coastal issues look like? What is one looking for in institutional design and capacity building? What expert system or decision-making tools do we need?

One answer is that we need community forums for validation that are built on new institutional design. .We need an institutional design that facilitates the dialogue between community intelligence and scientific intelligence. This is also a modelling challenge: how to include local intelligence into models? Community intelligence may be important for setting minimum conditions for viable ecosystem functioning, but the participatory process should also be used to explain and gain acceptance of national and international requirements and constraints on action. Knowledge should be gathered from different sources: local – national – international. A process is needed that integrates from the small scale to the large scale. In this process, it is also important to show the environmental benefits in order to get public support, as without it, it is becoming more and more difficult to take decisions.

Uncertainty has to be kept in mind when presenting scientific results, especially when it concerns results that look into the future, because scientific outcomes have errors and policy objectives change. How should we deal with this? A self-evolving process, which is flexible and adaptable, may be a valid alternative when top-down solutions are not possible. There is also a role for the media, namely by informing citizens so that uncertainty about public behaviour will reduce and public confidence in the carefulness of the decision process will increase.

Box 1. Code of practice for sustainable coastal management:

legitimacy
speaking free    fairness and joint responsibility respect
shared understanding

- visioning
- independence of jurisdiction
- acceptance of a fair and trustful process
- openness and transparency
- accountability and responsiveness
- media friendly
- network of stakeholder partnerships and forums
- interactions based on trust and believe
- independent interaction

A code of practice would assist institutional review to include such elements for sustainable long-term coastal management (Box 1). O'Riordan (this volume) elaborates on these issues. It is vital that the procedures be independently evaluated and validated, and presented to all parties for their understanding, before any deliberative process is concluded.

Conclusions and recommendations

In this chapter we treated four themes, namely the boundaries of the water body and management structures, the relation of the WFD to other legislation, the role of public participation in implementing the WFD, and the required institutional changes for implementing the WFD and cost consequences. Based on these themes we now try to address the main question: What is the most effective way of managing the river coast continuum and what should be the role of public participation, EU ('Brussels') and catchment managers?

From the boundaries of the water body and management structures, it was concluded that coordinated action is needed to oversee the river-coast continuum. The use of soft-law should be encouraged here, as hard law will follow suit much later at such a large scale.

The relation of the WFD to other legislation indicates that there should be clear objectives as to the process of achieving public involvement in the WFD programmes and projects. In addition there should be codes of practice as to how to ensure effective active involvement. These codes should define how to ensure legitimacy and representativeness, as well as trust and responsibility in the deliberative experience. Sensitivity to the cultural, geographical, and project-based circumstances of countries and localities is useful and would require 'open' decision-making.

Hence, there is a clear role for public participation in implementing the WFD. While there are some dangers of putting public participation in place, the greatest danger may be not to participate at all. Moreover, the process of a higher level of public participation is unstoppable, irrespective of whether it is good or bad. Consequently it is necessary to anticipate and design participatory processes.

Related to the required institutional changes for implementing the WFD and cost consequences, we need clear objectives about which areas require public involvement. This may be possible by showing the benefits and knowing the problem. We also need to account for cultural differences, which can influence the implementation of the WFD. For that, a dialogue is needed which brings together community intelligence and scientific systems understanding.

Waiting in the wings is the possibility of new institutional forms for co-operative, integrated and long-range river and coastal management. O'Riordan (this volume) offers one model. This is primarily based on UK experience and opportunities. There is scope for an EU wide discussion of new institutional forms for coastal and river management under conditions of climate change and sustainability planning.

References

CEEP (2002) White paper on "Governance". European centre of enterprises with public participation and the enterprises of general economic interest (CEEP), Brussels.
http://europa.eu.int/comm/governance/contrib_ceep_en.pdf

De Bruin EFLM, Jaspers FGW, Gupta J (2005) The EU Water Framework Directive: challenges for institutional implementation. In: Vermaat JE, Bouwer LM, Salomons W, Turner RK (eds) Managing European coasts: past, present and future. Springer, Berlin, pp 153-171

DETR (2003) The UK's climate change strategy. Department of the Environment Transport and the Regions (DETR). London

EEA (2002) Late lessons from early warnings: the precautionary principle 1896-2000. Office for official publications of the European communities (OPOCE), European Environmental Agency (EEA), Copenhagen

European Commission (2000) Directive 2000/06/EC of the European Parliament and of the Council of 23 October 2000 establishing a framework for Community action in the field of water policy. Off J Eur Community. L 327/1-L327/72

European Commission (2002) Communication from the commission to the council and the European parliament, towards a strategy to protect and conserve the marine environment. Commission of the European communities, Brussels

Gregory R (2000) Using stakeholder values to make smarter environmental decisions. Environment 42: 34-44

Horesh R (2003) Investing for the future: environmental policy bonds.
    http://www.geocities.com/socialpbonds/epbs.html

Janssen MA, Goosen H, Omtzigt N (2003) Simple mediation and negotiation support tools for water management in the Netherlands, personal communication

Ledoux L, Vermaat JE, Bouwer LM, Salomons W, Turner RK (2005) ELOISE research and the implementation of EU policy in the coastal zone. In: Vermaat JE, Bouwer LM, Salomons W, Turner RK (eds) Managing European coasts: past, present and future. Springer, Berlin, pp 1-19

Lise W, Van der Veeren RJHM (2002) Cost-effective nutrient emission reductions in the Rhine River basin, Integr Assess 3: 321–342

Norris P (1999) Critical citizens. Global support for democratic governance. Oxford University Press, Oxford

O'Riordan T (2002) Redesigning the coast: results of a workshop. CSERGE Working Paper PA 2002-01, University of East Anglia, Norwich

O'Riordan T (2005) Inclusive and community participation in the coastal zone: opportunities and dangers. In: Vermaat JE, Bouwer LM, Salomons W, Turner RK (eds) Managing European coasts: past, present and future. Springer, Berlin, pp 173-184

OSPAR (2002) Progress report, 5th international conference on the protection of the North Sea, 20–21 March, Bergen

Rupp S, Nicholls RJ (2002) Managed realignment of coastal flood defences: a comparison between England and Germany. http://www.survas.mdx.ac.uk/pdfs/delft_pa.pdf

Tanza A. (2003) Achievements and prospects of the water law process in the UNECE region. In: Bernardini F, Landsberg-Uczciwek M,Haunia S, Adriaanse M, Enderlein RE (eds). Proceedings of the International Conference on Sustainable Management of Transboundary Waters in Europe,Szczecin, pp 263-278

Timmerman JG, Cofino WP, Turner RK (2001) Introduction to Monitoring Tailor-Made III special issue, Reg Environ Change 2: 55–56

Turner, R.K. (2004) Environmental information for sustainability science and management. Pages 153-167 in: Timmerman, J.G. and S. Langaas (eds.) Environmental information in European transboundary water management. IWA Publishing, London

Turner, RK (2005) Integrated environmental assessment and coastal futures. In: Vermaat JE, Bouwer LM, Salomons W, Turner RK (eds) Managing European coasts: past, pre-sent and future. Springer, Berlin, pp 255-270

UNECE (1992) Convention on the protection and use of transboundary watercourses and international lakes. Helsinki

Von Bodungen B, Turner RK (2001) Science and integrated coastal management: Dahlem University Press, Berlin

11. Climate change and coastal management on Europe's coast

Robert J. Nicholls[1] and Richard J.T. Klein

Abstract

Climate change and sea-level rise due to human emissions of greenhouse gases is expected to accelerate through the 21st Century. Even given substantial reductions in these emissions, sea-level rise will probably be significant through the 21st Century and beyond. This poses a major challenge to long-term coastal management. While Europe has a high adaptive capacity, climate change will produce problems that have not been faced previously, and solutions need to be reconciled with the wider goals of coastal management. A recent European survey of the current response to sea-level rise and climate change shows a few countries engaged in proactive planning, while most are ignoring the issue, or only beginning to recognise its significance. While a proactive response should minimise the actual impacts and need for reactive responses, ignoring sea-level rise and climate change will almost certainly increase vulnerability.

A common theme that emerges is the need for more impact and vulnerability assessment that is relevant to coastal management needs. This should include the consequences of sea-level rise and climate change on coastal areas from the local to the European scale. This will require continued development of broad-scale assessment methods for coastal management. It is also important to assess coastal adaptation and management as a process rather than just focus on the implementation of technical measures. Lastly, the uncertainties of climate change suggest that management should have explicit goals, so that the success or failure of their achievement should be regularly monitored and the management approach adjusted as appropriate.

[1] Correspondence to Robert Nicholls: rjn@soton.ac.uk

J.E. Vermaat et al. (Eds.): Managing European Coasts: Past, Present, and Future, pp. 199–225, 2005.

Introduction

Climate change is one of the main challenges for environmental management through the 21st Century. Even in areas such as Europe, which has a high adaptive capacity due to relative wealth, access to a strong science base and well-developed management institutions, climate change may produce conditions not previously experienced, and management will need to evolve to cope with this in a variety of ways.

This chapter explores the implications of climate change for coastal management around Europe through the 21st Century and beyond. It builds on a number of earlier reviews and assessments of climate change and Europe's coasts such as Tooley and Jelgersma (1992), Jeftić et al. (1992, 1996), Nicholls and Hoozemans (1996), Nicholls (2000), de la Vega-Leinert et al. (2000), de Groot and Orford (2000), Kundzewicz et al. (2001), Brochier and Ramieri (2001) and Nicholls and de la Vega-Leinert (2004). When assessing long-term coastal management needs, it is fundamental to consider the changing balance of pressures at the coast (e.g. Turner et al. 1998a, 1998b; Turner, this volume). Here the emphasis is on pressures due to climate change and sea-level rise, but this is placed in the broader context of the changing European coast.

The chapter is structured as follows. First, it reviews the uses and trends within the coastal zone of Europe. Then it considers climate change and sea-level rise scenarios for the 21st Century. The potential impacts of these changes in Europe are considered, including both the natural system and socio-economic system changes. Possible responses to these impacts are then considered and placed into the broader context of coastal management. Lastly, some key issues for further investigation are identified and linked to the opportunities and threats for coastal management in Europe. Rochelle-Newall et al. (this volume) explores these key issues in more detail.

The coastal zone in Europe

The coastal zone in Europe is varied with a range of distinct environments in terms of geomorphology and wave/tidal conditions. Five distinct areas are recognised: the Black Sea, the Mediterranean, the Baltic, the North Sea and the Atlantic seaboard (Figure 1). These areas can be subdivided based on natural characteristics into the physical units that will respond to climate change and sea-level rise: coastal cells and sub-cells, estuaries, deltas, etc. These 'natural' divisions are further fragmented by intensive and varying human use, as the coastal zone is a focus for important population and economic centres. Human activities within the coastal zone include industry, urban and residential, tourism and recreation, transport, fisheries/aquaculture and agriculture (Rigg et al. 1997). One third of the European Union (EU) population is estimated to live within 50 km of the coast, with the proportion being 100% in Denmark and 75% in the United Kingdom and the Netherlands. Coastal urban agglomerations are important with a collective population of 120 million people in the EU alone (Papathanassiou et al. 1998).

Even though Europe is already highly urbanised, coastal urbanisation continues due to coastward migration and tourism development, particularly around the Mediterranean. In addition to direct human uses, the coast is an important habitat of international significance with freshwater, brackish and saline marshes and intertidal and shallow subtidal habitats and it supports important fishery resources. Lastly, Europe's coast is culturally and archaeologically significant as exemplified by Istanbul, Athens, Venice, London, Amsterdam and St. Petersburg, to name a few historic coastal cities.

As populations have grown and economic activity has intensified so a range of pressures have emerged in the coastal zone, including a legacy of significant land claims around estuaries and lagoons (e.g. French 1997, 2001, Papathanassiou et al. 1998). Significant assets and populations are located in floodprone coastal plains subject to erosion, and large lengths of coast are defended (Quelennec et al. 1998). Hard defences generally reduce sediment availability to the coastal system, intensifying erosional pressures and hence increase defence needs. Hard defences also lock the coastal position and hence contribute to a coastal squeeze of intertidal habitats on retreating shorelines (French 1997, Nicholls 2000). Human changes outside the immediate coast have also had adverse consequences on coastal areas, such as deltaic areas that have become threatened because they have been sediment-starved due to changing catchment management, particularly dam construction (e.g. Sanchez-Arcilla et al. 1998).

Given that Europe has a reasonably stable and ageing population, it might be thought that future problems will be minimised. However, present trends suggest coastal pressures will continue and intensify. The different possible pathways of development within Europe will lead to different sets of coastal problems and hence management needs. Turner (this volume) and the group report of Theme 5 (Nunneri et al. this volume) discuss three possible scenarios for Europe's coasts.

The widespread coastal impacts of human interventions were not foreseen, and only now are their full implications being appreciated. This is driving important changes to more flexible and strategic approaches to coastal management, including more soft engineering, sediment recycling and managed realignment (e.g. Hamm and Stive 2002, Rupp and Nicholls 2003) and long-term analysis of future changes (e.g. DEFRA 2001; the Eurosion Project, http://www.eurosion.org). Environmental designations also protect many coastal areas, and compensation for habitat destruction is now often required, although the long-term success of these policies remains to be assessed.

Climate change and the European coast

Climate change is already a pressure with rising sea levels evident around most of Europe's coasts, excluding parts of Scandinavia (Figure 2). In the 21st Century this rise is expected to continue and accelerate due to global warming. There are also observed inter-annual and inter-decadal fluctuations in the characteristics of storms during the 20th Century, but with no evidence of long-term trends (e.g. WASA Group 1998). This means that long-term climatic observations are re-

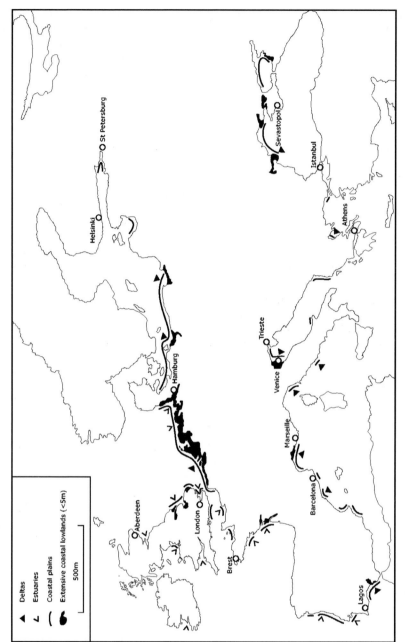

Fig. 1. Major estuaries and deltas, coastal plains and extensive coastal lowlands (<5-m elevation) in Europe, as well as the stations in Figure 2 and selected locations threatened by sea-level rise

quired to accurately estimate storm statistics. A decline in the formation of seasonal sea ice in parts of the Baltic due to rising sea temperatures has been observed, which is now allowing winter storms to cause significant erosion when before the coast was frozen and protected (e.g. Kont et al. 2004). Note that the Caspian Sea has also been impacted by significant sea-level rise (ca 1 m) over the late 20th Century. However, as this is an enclosed sea not linked to the global ocean, these changes are not considered further.

Human-induced climate change is caused by the emission of so-called "greenhouse" gases, which trap long-wave radiation in the upper atmosphere and thus raise atmospheric temperatures. Carbon dioxide is the most important of these gases and its atmospheric concentration has exponentially increased since the beginning of the industrial revolution due to fossil fuel combustion and land-use change. In 1800, the atmospheric concentration of carbon dioxide was about 280 parts per million (ppm); today it is about 350 ppm and rising. Similar increases have been observed for other greenhouse gases such as methane and nitrous oxide (Houghton et al. 2001).

By 2100, carbon cycle models project atmospheric carbon dioxide concentrations of 540 to 970 ppm, with a range of uncertainty of 490 to 1260 ppm (Houghton et al. 2001). Based on these projections and those of other greenhouse gases, the Intergovernmental Panel on Climate Change Third Assessment Report projects an increase in globally averaged surface temperature by 1.4 to 5.8°C over the period 1990 to 2100. It is very likely that nearly all land areas will warm more rapidly than the global average, particularly those at high latitudes in the cold season, including much of Europe (Houghton et al. 2001).

These simulations of global warming have led to a predicted global-mean sea level rise of 9 to 88 cm between 1990 and 2100, due largely to thermal expansion and melting of land-based ice, especially small glaciers. The central estimate of a 48-cm rise represents an average rate of global-mean sea-level rise of 2.2 to 4.4 times the estimated rate of rise over the 20th century. Importantly, even with drastic reductions in greenhouse gas emissions, sea level will continue to rise for centuries beyond 2100 because of the long response time of the deep ocean to reach equilibrium to a surface warming (Wigley and Raper 1993, Church et al. 2001). Thus an ultimate sea-level rise of 2 to 4 metres is possible for atmospheric carbon dioxide concentrations that are twice and four times pre-industrial levels, respectively (Church et al. 2001). Melting of the Greenland ice sheet and instability to the West Antarctic Ice Sheet could contribute significant additional sea-level rise over the coming centuries (Vaughan and Spouge 2002, Woodworth et al. 2004).

For coastal areas, it is not the global-mean sea level that matters but the locally observed, relative sea level, which takes into account regional sea-level variations and vertical movements of the land (Figure 2). A major uncertainty is how sea-level rise will manifest itself at regional scales, such as in the North Atlantic. All the models analysed by Church et al. (2001) and Gregory et al. (2001) show a strongly non-uniform spatial distribution of sea-level rise across the globe. However, the patterns produced by the different models are not similar in detail. This lack of similarity means that confidence in projections of regional sea-level changes is low, and it is possible that sea-level rise on Europe's coast could be ±50% of the global-mean changes already described (Hulme et al. 2002). This uncertainty needs to be taken into account in impact analysis.

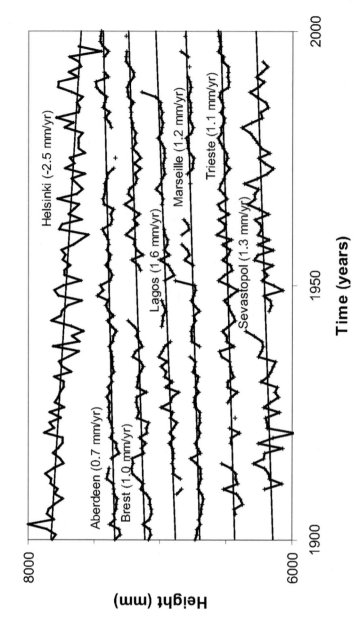

Fig. 2. Sample long-term relative mean sea-level curves from around Europe through the 20th Century, including the linear trend. While sea levels are falling at Helsinki (and some other Scandinavian stations), elsewhere there is a rising trend. Height is arbitrary to allow each record to be seen. Location of stations is shown in Figure 1 (data from the Permanent Service for Sea Level: http://www.pol.ac.uk/psmsl/)

Land uplift and subsidence can also be significant. Parts of Scandinavia experience land uplift due to global-isostatic adjustment at a sufficient rate that projected global-mean sea-level rise may be completely offset and relative sea level may continue to fall, albeit at a lower rate (Johansson et al. 2004). Other areas, such as deltas and coastal lowlands, are characterised by a strong downward movement of the land, which will add to global-mean sea-level rise (Emery and Aubrey 1991, Bird 1993, Suanez and Provansel 1996). This subsidence is often greatly enhanced by land claim and/or sub-surface fluid withdrawals as happened widely in Europe's coastal lowlands, such as around the North Sea. As another example, widespread 20th Century human-induced subsidence in the North Italian coastal plain produced over 2,300 km^2 of land below sea level, which is now protected from inundation by dikes (Bondesan et al. 1995). Thus, many of the areas threatened by sea-level rise are also prone to human-induced subsidence: coastal management needs to recognise this link to avoid this problem being exacerbated during the 21st Century.

Table 1. Possible implications of climate change on coastal zones in Europe, excluding sea-level rise (see Table 2)

Climate factor	Direction of change	Biogeophysical effects	Socio-economic impacts
Air and sea temperature	Increasing	Northerly migration of coastal species	Changes to fisheries, nature conservation implications
		Decreased incidence of sea ice at higher latitudes	Improved navigation, but increased coastal erosion during winter months
			Increased coastal tourism (Mediterranean, Southern North Sea and Baltic)
Water Resources/ Run-off	Drying in south, wetter further north	Changed fluvial sediment supply	Increased erosion (or accretion)
		Changed peak flows	Changed flood risk in coastal lowlands
Coastal storms	Increase in westerlies in northwest Europe (?)	Changed occurrence of storm damage and flooding	Increased risk of flood and storm damage
Atmospheric CO_2	Increasing	Increased productivity in coastal ecosystems	Uncertain

As already noted, sea-level rise is not the only climate-related effect relevant to coastal zones (Table 1). However, confidence in model projections of other manifestations of climate change is generally still low. The North Atlantic Oscillation

(NAO) Index[2] is expected to increase under global warming leading to warmer, wetter and windier winters in northwest Europe (Hulme et al. 2002). This in turn will change the frequency, intensity and spatial patterns of coastal storms, but it is hard to quantify the significance of these changes. A rise in mean sea level will lead to a decrease in the return period of storm surges without any other change, but it remains unclear if changing storms will additionally change the variability of storm surges themselves (e.g. Flather and Williams 2000, Lowe et al. 2001).

Moreover, there could be a weakening of the Gulf Stream due to global warming (Hulme et al. 2002). This would lead to significant cooling of the Northern European landmass. However, given that global warming is also occurring, the net effect is less certain. A cooling event would have such widespread impacts across northern and western Europe, so it is not meaningful to focus on the coast alone for this issue, and it is not shown in Table 1.

Climate change impacts around Europe's coasts

Framework for analysis

Following the uncertainties about other climate change factors, the main focus of most assessments has been the impacts and responses to sea-level rise. A common framework as illustrated in Figure 3 provides a useful basis for interpretation and comparison between studies. In particular, it highlights the varying assumptions and simplifications that are made within all the available studies and hence helps to establish common issues and make limitations more explicit.

Relative sea-level rise, due to whatever cause, has a number of biogeophysical impacts such as erosion and increased flood potential. In turn, these can have direct and indirect socio-economic impacts depending on the human exposure to these changes. There are also important feedbacks as the impacted systems adapt to these changes, including the human exploitation of beneficial changes and adaptation to adverse changes. Hence, the coastal system is best defined in terms of interacting natural and socio-economic systems. Figure 3 has been modified from the original in Klein and Nicholls (1999) to reflect the terms used by Smit et al. (2001), but the basic meanings remain the same. Both systems may be characterised by their *exposure, sensitivity* and *adaptive capacity* to change, both from sea-level rise and related climate change, and these factors may all be modified by other *non-climate stresses*. Sensitivity simply reflects each system's potential to be affected by changes such as sea-level rise, exposure defines the nature and amount to which a system is exposed to climate change, while adaptive capacity describes each system's stability in the face of change. Collectively, sensitivity, exposure and adaptive capacity determine each system's *vulnerability* to sea-level rise and other drivers of change.

---

[2] The NAO Index measures the difference in barometric pressure between the Azores and Iceland. It indicates the direction and strength of atmospheric flow across northwest Europe, especially in winter.

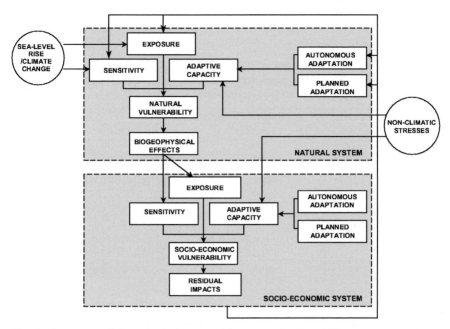

Fig. 3. A conceptual framework for coastal impact and vulnerability assessment of sea-level rise (adapted from Klein and Nicholls 1999)

Both systems are dynamic and adapt to change (e.g. Smit et al. 2001). *Autonomous adaptation* represents the spontaneous adaptive response to sea-level rise (e.g. increased vertical accretion of coastal wetlands within the natural system, or market price adjustments within the socio-economic system). Autonomous processes are often ignored by coastal management, and yet have a significant influence on the magnitude of many impacts. Further, natural autonomous processes may be reduced by human-induced, non-climatic stresses (Bijlsma et al. 1996). *Planned adaptation*, which must emerge from the socio-economic system, can serve to reduce vulnerability by a range of measures, which are discussed in more detail later in this chapter.

Dynamic interaction occurs between the natural and socio-economic systems in the coastal zone, including the natural system impacts on the socio-economic system and planned adaptation by the socio-economic system influencing the natural system. This results in the natural and socio-economic systems interacting in a complex and poorly understood manner, which can only be fully understood via integrated assessment. Importantly, adaptation normally acts to reduce the magnitude of the potential impacts that would occur in its absence[3]. Hence, actual impacts are normally much less than the potential impacts that are estimated in the absence of adaptation. Hence, impact assessments that do not take adaptation into account will generally overestimate impacts (determining *potential* rather than *actual* impacts).

---

[3] Adaptation that increases potential impacts and vulnerability is termed maladaptation (Smit et al. 2001).

Impacts of sea-level rise

The most significant biogeophysical effects of sea-level rise are summarised in Table 2, including relevant interacting climate and non-climate stresses.

Table 2. The main effects of relative sea-level rise, including relevant interacting factors (adapted from Nicholls (2002). Some factors (e.g. sediment supply) appear twice as they may be influenced both by climate and non-climate factors

Biogeophysical effect		Other relevant factors	
		Climate	Non-climate
Inunda-tion, flood and storm damage	Surge (open coast)	Wave and storm climate, morphological change, sediment supply	Sediment supply, flood management, morphological change, land claim
	Backwater effect (river)	Run-off	Catchment management and land use
Wetland loss (and change)		CO_2 fertilisation Sediment supply	Sediment supply, migration space, direct destruction
Erosion	Direct effect (open coast) Indirect effect (near inlets)	Sediment supply, wave and storm climate	Sediment supply
Saltwater Intrusion	Surface Waters	Run-off	Catchment management and land use
	Ground-water	Rainfall	Land use, aquifer use
Rising water tables/ impeded drainage		Rainfall	Land use, aquifer use

Most of these impacts are broadly linear functions of sea-level rise, although some effects such as wetland loss show a threshold response and are more sensitive to the rate of sea-level rise, rather than the absolute change. Some responses are instantaneous, such as an increase in risk of flooding, while others such as erosion lag behind sea-level rise. Most existing studies have focused on one or more of the first three factors in Table 2: (1) inundation, flood and storm damage, (2) erosion and (3) wetland loss (Nicholls, 1995). Hence, most assessments of the biophysical impacts of sea-level rise are incomplete. These studies often make simple assumptions, such as wetlands being submerged as sea levels rise with no consideration of their possible responses (see Viles and Spencer 1995). Hence the studies are assessing exposure and aspects of sensitivity, and largely ignoring adaptive capacity. In addition, the non-climate stresses identified in Figure 3 and Table 2 are often ignored.

The natural-system effects of sea-level rise in Table 2 have a range of potential socio-economic impacts, including the following identified by McLean et al. (2001):

- Increased loss of property and coastal habitats;
- Increased flood risk and potential loss of life;
- Damage to coastal protection works and other infrastructure;

- Loss of renewable and subsistence resources;
- Loss of tourism, recreation, and transportation functions;
- Loss of non-monetary cultural resources and values;
- Impacts on agriculture and aquaculture through decline in soil and water quality.

The indirect impacts of sea-level rise are more difficult to analyse, but they have the potential to be important in many sectors, such as human health, fisheries, and nature conservation. Europe supports internationally significant numbers of shore-birds, especially in the winter. Shorebird numbers depend on intertidal areas, so sea-level rise could reduce the carrying capacity for these shorebirds. Human migration is another possible consequence of sea-level rise if coastal areas are abandoned or degraded. Forced migration is unlikely in Europe, but sea-level rise could generate migrants to Europe from other parts of the world. Hence, sea-level rise could produce a cascade of impacts through the coastal system, although analysis to date has focussed mainly on the direct impacts.

The impacts of sea-level rise have been investigated in a range of policy-driven sub-national, national and regional/global case studies (e.g. Nicholls and Mimura 1998, de la Vega-Leinert et al. 2000), as well as in more science-orientated studies which examine the biophysical processes of sea-level rise and their linkages (e.g. Cahoon et al. 1999, Capobianco et al. 1999). A range of socio-economic analyses have also been undertaken (e.g. Fankhauser 1995a, Tol 2002a 2002b). While these studies are policy relevant as they discuss issues such as the costs of sea-level rise in monetary terms, they are also experimental in terms of exploring the coupled natural and socio-economic dynamics of the coastal zone (see Figure 3). Most of the policy-driven studies have been relevant to national to international issues, especially reducing greenhouse gas emissions. They are less useful when adaptation and coastal management are considered, due to their broad scale.

Possible impacts of climate change on Europe's coastal areas

In global terms, Europe appears much less threatened by sea-level rise than many developing country regions (e.g. Nicholls 2003). However, coastal ecosystems do appear to be threatened, especially those on the Baltic, Mediterranean and Black Seas. In the worst case, these habitats could be severely reduced or eliminated during the 21st Century. This is due to the low tidal range in these areas, and the limited scope for onshore migration due to the intense human use of the coastal zone.

Most national-scale assessments in Europe comprise semi-quantitative analyses and/or inventories of the potential impacts of sea-level rise, with limited consideration of adaptation (Nicholls and Mimura 1998, Nicholls and de la Vega-Leinert 2004). As one might expect, low-lying coastal areas are most sensitive to sea-level rise, such as the large coastal lowlands bordering the North Sea. Figure 1 indicates the coastal plains and lowlands, estuaries and deltas that are threatened around Europe, as well as selected low-lying cities and areas of historical/cultural significance. The most common scenario has been a 1 m rise imposed on the present socio-economic situation (e.g. Table 3), so results over-emphasise the impacts of sea-level rise over other change factors. However, the results do confirm what has already been stated about the importance of the coastal zone in Europe. Table 3 suggests that >13 million people could be affected by flooding given a rise in sea level, just considering five countries. However, the national results vary between countries, with the Netherlands having the highest potential human impacts, and Poland and Estonia the lowest (Table 3).

Coastal wetlands and intertidal habitats also appear highly threatened in national and sub-national studies (Table 3), although their capacity to respond to sea-level rise requires more assessment as already discussed. Given that increased protection of human activities in coastal areas is a likely response to climate change and variability, these potential impacts combined with coastal squeeze are an important long-term challenge to coastal management in Europe. As discussed later, managed realignment of flood defences and 'depolderisation' is being seriously evaluated, including trials, in parts of Europe (Goeldner 1999, Goeldner-Gianella 2001, Rupp and Nicholls 2003). There are often sufficient sites of land claim to maintain the current stock of saltmarsh and other intertidal habitat, but at the cost of extensive areas of freshwater coastal grazing marsh of significance to nature conservation which have developed on the land claim areas (e.g. Watkinson et al. 2003). Thus, there are again trade-offs to consider with this policy.

In terms of adaptation, these studies have usually made simple assumptions that are consistent with the inventory approach, such as application of a uniform national response (Table 3). These studies show that the poorer countries in Europe face the largest relative burden of adaptation costs: Poland has higher relative adaptation costs than The Netherlands, despite the potential impacts being at least an order of magnitude lower. However, the adaptation process and the capacity of the coastal communities to adapt have not been evaluated. Thus, while we can be confident that Europe can afford significant levels of adaptation, we are much less clear what would be most appropriate. In the UK, an integrated assessment of future flooding concluded that coastal areas will become relatively more threatened by flooding relative to inland areas (Evans, 2003). This reflects the effect of sea-level rise and suggests that in national terms, coastal adaptation will be essential and this will require more resources relative to present management costs. This result will likely be relevant in neighbouring European countries.

One important result of significance to coastal management is the importance of the scale of assessment. Sterr (2004) has investigated the vulnerability of Germany to sea-level rise, at national, state (Schleswig-Holstein) and case studies (within Schleswig-Holstein) levels. As the scale of study increases, so the size of the hazard zones declined due to the use of higher resolution data. However, the potential impacts do not change significantly as the human values remain concentrated in the (smaller) hazard zones. Turner et al. (1995) examined the optimum response to sea-level rise in East Anglia, UK, using cost-benefit analysis. At the regional scale, it was worth protecting the entire coast. In contrast, at the scale of individual flood compartments, 20% of flood compartments should be abandoned, even for the present rates of sea-level rise. This conclusion is consistent with current trends in coastal management policy for this region. This shows that realistic assessment of adaptation options requires quite detailed analysis to capture the potential variation in responses within a region, rather than assuming a uniform adaptation response.

Responding to climate change

Given the potential impacts in Europe identified in the previous sections, some response to climate change is prudent. To date, the European Union's response to climate change has stressed policies to reduce greenhouse gas emissions (usually termed mitigation within the climate change debate). While the authors support this policy, it has become increasingly apparent that it needs to be augmented by adaptation to the inevitable climate change (Parry et al. 1998, Metz et al. 2002). Given the strong commitment to sea-level rise, in coastal areas this need for adaptation is greatest and will continue for centuries (Lowe and Nicholls 2004). This commitment to coastal adaptation needs to be built into long-term coastal management policy. However, while there is consensus on the need for mitigation across the European Union, proactive adaptation across Europe's coasts is much more patchy and variable. The SURVAS Project found that concern about sea-level rise, including the level of preparation varied greatly around Europe, a few northern countries are already preparing for accelerated sea-level rise, while many southern countries are ignoring observed 20th Century rise, let alone preparing for projected accelerated rise (Tol et al. 2004).

Proactive adaptation to climate change is aimed at reducing a system's vulnerability by either minimising risk or maximising adaptive capacity. Five generic objectives of proactive adaptation can be identified (Klein and Tol 1997, Klein 2001), which are relevant to coastal zones:

- *Increasing robustness of infrastructural designs and long-term investments*—for example by extending the range of relevant climatic factors (e.g. still water level) that a system can withstand without failure and/or changing a system's tolerance of loss or failure (e.g. by increasing economic reserves or insurance);
- *Increasing flexibility of vulnerable managed systems*—for example by following adaptive management approaches, which explicitly allow adjustments and learning, and/or reducing economic lifetimes (including increasing depreciation);
- *Enhancing adaptability of vulnerable natural systems*—for example by reducing non-climatic stresses (e.g. reactivating natural sediment supplies) and/or removing barriers to migration (e.g. promoting managed realignment);
- *Reversing maladaptive trends*— for example by introducing zoning regulation in vulnerable areas prone to repeated flood events that prohibits redevelopment after major damage;
- *Improving societal awareness and preparedness*—for example by informing the public of the risks and possible consequences of climate change and/or setting up disaster response and early-warning systems.

For coastal zones another classification of three basic adaptation strategies is often used (IPCC CZMS 1990, Klein et al. 2001):

- *Protect*—to reduce the risk of the event by decreasing the probability of its occurrence;
- *Accommodate*—to increase society's ability to cope with the effects of the event;
- *Retreat*—to reduce the risk of the event by limiting its potential effects.

Table 3. Potential impacts of a 1-m sea-level rise in selected European countries, assuming the 1990s situation and no adaptation, plus adaptation costs to protect the human population (taken from Nicholls and de la Vega-Leinert 2004)

Country	Coastal floodplain population		Population flooded per year		Capital value loss		Land loss		Wetland loss	Adaptation costs	
	no. 10^3	%	no. 10^3	%	10^9 US$	% GNP	Km^2	%	km^2	10^9 US$	% GNP
Netherlands	10,000	67	3,600	24	186	69	2,165	6.7	642	12.3	5.5
Germany	3,120	4	257	0.3	410	30	n.a.	n.a.	2,400	30	2.2
Poland	235	0.6	196	0.5	22.0	24	1,700	0.5	n.a.	4.8	14.5
										+0.4/yr	+1.2/yr
Estonia	47	3	n.a.	n.a.	0.22	3	>580	>1.3	225	n.a.	n.a.
Ireland	<250	<5	<100	<1.8	0.17	0.2	<250	<0.3	<150	<0.42/yr	<0.6/yr

While each of these strategies are designed to protect human use of the coastal zone, if applied appropriately, they have different consequences for coastal ecosystems. Retreat and accommodation avoid coastal squeeze as onshore migration of coastal ecosystems is not hindered. In contrast, protection will lead to a coastal squeeze, although this can be minimised using soft approaches to defence such as beach nourishment and sediment recycling. In terms of timing, accommodate and retreat are best implemented in a proactive manner, while protection can be both proactive or reactive.

The recent shoreline management guidelines used in England and Wales and adapted for use in the Eurosion Project are also useful to consider as they are being applied at a national level (DEFRA 2001, Cooper et al. 2002), and potentially more widely across Europe. They comprise a set of proactive strategies for shoreline management[4]. The original strategies were entirely geometric (MAFF et al. 1995), but based on experience with the first generation of shoreline management plans, five strategies will be considered in the second generation plans, which are just commencing:

- Hold the Line;
- Advance the Line;
- Managed Realignment;
- Limited Intervention;
- No Intervention.

Figure 4 shows the linkages between these three sets of responses. The three coastal adaptation strategies roughly coincide with the first three of the five proactive adaptation objectives.

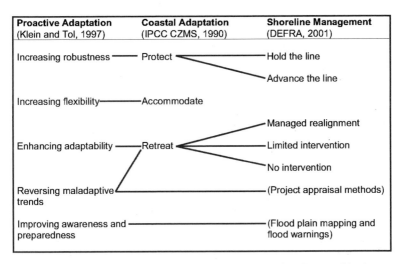

Fig. 4. Linkages between the different adaptation approaches discussed in the text

---

[4] Strategic management that addresses long-term responses to coastal flood and erosion hazards.

Protecting coastal zones against sea-level rise and other climatic changes would involve increasing the robustness of infrastructural designs and long-term investments such as seawalls and other coastal infrastructure. A strategy to accommodate sea-level rise could include increasing the flexibility of managed systems such as agriculture, tourism and human settlements in coastal zones. A retreat strategy would serve to enhance the adaptability of coastal wetlands, by allowing them space to migrate to higher land as sea level rises. The shoreline management options can also be mapped to these two approaches. Reversing maladaptive trends and improving societal awareness and preparedness are not explicitly addressed in the DEFRA (2001) guidelines. Managed realignment could be seen as reversing maladaptive trends for areas of land claim. It can also be argued that both approaches are addressed in other elements of shoreline management policy: project appraisal is now based on cost-benefit analysis, while flood warning and disaster preparedness are now central elements of flood management in England and Wales (see http://www.defra.gov.uk/environ/fcd/). However, it is striking that while the retreat strategy is now considered, the accommodation option is not considered, except in regard to warning systems. The authors are unaware of accommodation being applied anywhere in Europe, except locally. This contrasts with flood management in the USA, which use accommodation extensively, such as raising coastal buildings above surges and waves on deeply embedded pilings. This suggests that the feasibility of accommodation strategies should be evaluated more explicitly, especially within some of the more innovative approaches being advocated for future coastal development (e.g. Waterman et al. 1998).

Climate change and coastal management

Integrated coastal zone management (ICZM) has been widely recognised and promoted as the most appropriate process to deal with these current and long-term coastal challenges, including climate change and sea-level rise. It is a proactive policy process to address resource-use conflicts, as well as find the balance between short-term economic and longer-term environmental interests. By considering short, medium and long-term interests, ICZM aims to achieve sustainable development by stimulating economic development of coastal areas and resources, whilst reducing the degradation of their natural systems (Cicin-Sain 1993, WCC'93 1994, Ehler et al. 1997, Cicin-Sain and Knecht 1998). Demonstrating that coastal management plans are sustainable is difficult. However, any sustainable plan must address the issue of climate change. Hence the effectiveness of how climate change and sea-level rise are considered in coastal management plans is one useful measure of commitment to integration and sustainability. ICZM has been endorsed as a integrated response to climate change by the Intergovernmental Panel on Climate Change (IPCC CZMS 1992, Bijlsma et al. 1996, McLean et al. 2001). The European Union are exploring ICZM via a demonstration programme (European Commission, 1999a, 1999b) and an EU ICZM Recommendation was ratified in 2002 (see their website:

http://europa.eu.int/comm/environment/iczm/home.htm). National stocktaking is presently being conducted and this may ultimately result in an EU directive on ICZM.

The responses to sea-level rise and climate change discussed in the previous section need to be implemented in this broader context of coastal management, and the responses need to be consistent with the wider objectives of coastal management (Klein et al. 1999, Klein 2003). Current pressures may have adversely affected the coastal ecosystem's integrity and thereby its ability to cope with additional pressures such as climate change and sea-level rise. In Europe given its high level of development, large coastal populations, and high levels of interference with coastal systems this is a particularly significant factor. It can be argued that natural coastal buffers such as dunes and wetlands should be preserved and enhanced, as climate change indicates the value of this buffering capacity. Equally, improving shoreline management for non-climate change reasons will also have benefits in terms of responding to sea-level rise and climate change. This illustrates that when current coastal pressures are not adequately dealt with in the short term, coastal zones will become increasingly more vulnerable to the consequences of climate change and sea-level rise (WCC'93 1994).

Adaptation to climate change in integrated coastal zone management

In an integrated coastal policy that aims to address both climate and non-climate issues, the potential for conflict between development objectives and adaptation needs should be minimised. In view of the fact that coastal zones are usually host to a number of, often competing, sectoral activities, coastal zone management to date has been designed primarily to satisfy sectoral needs. Given the additional challenge of climate change in coastal zones, the purpose and design of coastal management will have to be revisited. In order to do so, it is important that all stakeholders—governments, universities and government-sponsored laboratories, the private sector, non-governmental organisations and local communities—are aware of the need to reduce coastal vulnerability to climate change. In addition, successful coastal management requires that the planning, design and implementation of adaptation technologies be based on the best available information as well as on the regular monitoring and evaluation of their performance.

Accordingly, Klein et al. (1999, 2001) showed that effective and efficient coastal adaptation to climate change is not just a set of technical options. Rather, it can be conceptualised as a multi-stage and iterative process, involving four basic steps, within the wider frame of coastal management (Figure 5):

1. Information development and awareness raising;
2. Planning and design;
3. Implementation;
4. Monitoring and evaluation.

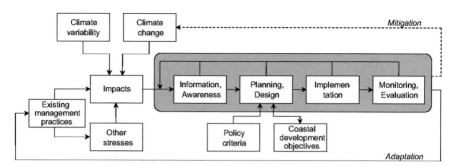

Fig. 5. Conceptual framework showing in the shaded area the iterative steps involved in coastal adaptation to climate variability and change (from Klein et al. 2001)

Climate variability and/or climate change – together with other stresses on the coastal environment brought about by existing management practices – produce actual or potential impacts. These impacts trigger efforts of mitigation to remove the cause of the impacts or adaptation to modify the impacts. The process of adaptation is conditioned by policy criteria and coastal development objectives and interacts with existing management practices.

Figure 5 is a schematic framework based on the long-term coastal management experiences in The Netherlands, the United Kingdom and Japan, with an emphasis on coastal protection. In each of these countries, management approaches have been adjusted over the past decades to reflect new insights and priorities, including concerns about climate variability and, more recently, climate change. It is important to note that Figure 5 represents an idealised decision framework, which does not capture the multitude of actors involved in decision-making, the uncertainty with which these actors are faced, the other interests they have or the institutional and political environments in which they operate.

Given the uncertainty of climate change and the ability of our models to predict its impacts, monitoring and evaluation is fundamental. This will allow an adaptive management approach as we learn more about the problems of climate change, the tools and methods that we use to analyse them, and hence how to effectively management these issues (see National Research Council 1995, Willows and Connell 2003, Tol et al. 2004, Mee this volume).

Improving information on climate change for coastal zone management

Coastal zone management requires more information on the possible impacts of climate change and the choices that are faced in responding to these threats. The variable response to the threats of climate change around Europe's coasts identified by Tol et al. (2004) partly reflects this lack of information, as well as variation in the historical and 20th Century experiences of coastal hazards. This will require further vulnerability assessments with the specific purpose of improving coastal zone management and policy. Earlier assessments were often more focussed on broader needs than coastal management, so their results are of limited value to coastal policy makers. The detailed assessments that have been conducted in the Netherlands (Hekstra 1986, Stive et al. 1990, De Ronde 1993, Baarse et al. 1994, Peerbolte 1994, De Ruig 1998, Jacobs et al. 2000) and shoreline management in the UK (DEFRA 2001, Foresight Project on Flood and Coastal Defence, http://www.foresight.gov.uk) illustrate studies that are helping to improve coastal policy for climate and other long-term changes. Needs include understanding the impacts of sea-level rise and climate change in the context of the non-climate stresses (i.e. an analysis of multiple stresses), defining the options to respond to climate change and identifying when they might be best implemented. Positive benefits of climate change may also occur for some sectors such as coastal tourism (Maddison 2001), and these should be acknowledged and fully exploited.

Some specific issues that might be addressed in future studies were identified at the SURVAS Overview Workshop (Nicholls and de la Vega-Leinert 2001). These reinforce some of the earlier remarks in this chapter. The main recommendations that are pertinent to the future direction of climate change and coastal management in Europe concern the conduct of future vulnerability assessments of coastal areas and are as follows:

- Guided sensitivity analysis of coastal areas to the full plausible range of climate change can usefully proceed in parallel with developments in climate and related science. Scenarios of regional sea-level change and changes in storminess should be developed as promptly as possible (see Beersma et al. 2000);
- It is important to place the impacts/adaptation needs of sea-level rise in a broader context of change and today's coastal problems including consideration of:
  - Other climate change, including extreme events such as storms;
  - Non-climate environmental, land use and socio-economic changes.
- Evaluate the full range of possible impacts, including the natural system and the socio-economic system, and the direct and the indirect impacts (which have often been ignored);
- Consider impacts on the entire coastal zone, including the sub-tidal and inter-tidal areas (mainly impacts on fisheries and ecosystems);
- Identify 'flagship' impacts on cultural or natural sites (e.g. Venice or the Camargue, France) that are likely to attract widespread public concern and attention;
- Consider adaptation as a process rather than just a set of technical measures or fixes, as there may be important constraints on this process, and once a measure is implemented it requires monitoring;
- Consider the timing of different adaptation measures, particularly identifying those that would be most effective if implemented in the near future;
- Identify the constraints and barriers to adaptation, and how the capacity to adapt could be enhanced.

Conclusions and further work

Some important conclusions concerning climate change and the long-term management of the European coastal zone are as follows:

1. *Climate change and variability are already an issue.* During the 20th Century, sea levels have risen 10 to 20 cm around much of Europe's coast, while storm frequency and track have shown significant interdecadal variability. These climatic factors are already contributing to a range of problems, including increasing flood risk, coastal erosion and coastal squeeze. However, the relative importance of these historic influences on current problems could be better quantified;

2. *Future climate change is expected to be greater than historic experience.* By 2100, global-mean sea levels could be 9 to 88 cm higher than in 1990, while air and sea temperature will have risen significantly. Storm frequency and intensity may increase across northwest Europe and the large interdecadal variability is almost certain to continue;

3. *Potential impacts on human systems are significant.* Of particular concern is increased flood risk and storm damage in low-lying coastal areas. While the North Sea coast presently has the largest exposure, the risk of flooding will increase more around the Mediterranean, Black Sea and the southern Baltic, assuming no human adaptation. Storminess must also be considered;

4. *Intertidal habitats and ecosystems are also threatened.* The Baltic, Mediterranean and Black Sea coasts are most vulnerable to sea-level rise due to their low tidal range and in the worst case, intertidal ecosystems could be largely eliminated in these areas by the 2080s. This is a major challenge for coastal management;

5. *Coastal zones face many other pressures over the 21st Century.* Profound socio-economic and other changes will continue, although future trends are highly uncertain. This will interact with climate change, and exacerbate or ameliorate vulnerability to climate change;

6. *Actual impacts of climate change are highly uncertain.* They will depend on the magnitude of climate and other change and the success of human adaptation to that change. Many of the impacts of sea-level rise and climate change could be avoided or managed effectively given appropriate proactive measures;

7. *There is conflict between sustaining coastal ecosystems and maintaining human coastal activity.* The natural ecosystem response to rising sea levels is onshore migration, but this is stopped by fixed sea defences, producing coastal squeeze. Thus, there is a fundamental conflict between protecting socio-economic activity and sustaining the ecological functioning of the coastal zone in Europe under rising sea levels. This conflict needs to be explicitly acknowledged and resolved by coastal management policy. It suggests a need for more soft protection, managed retreat, and possibly accommodation strategies;

8. *Global sea levels are likely to continue to rise for many centuries irrespective of future greenhouse gas emissions.* Therefore, sea-level rise will remain an important issue for coastal management beyond the 21st Century. Coastal management and land use planning should prepare for these changes, recognising that there is a long-term 'commitment to adapt to sea-level rise'.

Thus, climate change is a major challenge for coastal management through the 21st Century and beyond. The scale of the challenge is significant as most available tools and methods are designed for more immediate local to sub-national problems. Strengthening our capacity for long-term coastal management is fundamental to our response to climate change and sea-level rise. Three issues require further debate and investigation.

First of all, protecting human use and sustaining the natural functioning of the coast: how can we most effectively marry these two often conflicting goals (especially given the long-term commitment to sea-level rise)? While near-universal protection of human activities could probably be provided, it is unlikely that this would be the preferred option. Finding better solutions needs to take account of coastal dynamics, which are often ignored in coastal management. For example, the EU Habitat and Birds Directives (SPAs and SACs) take a rather static view of existing coastal habitats, rather than encouraging a dynamic view of an evolving coastal landscape. Rather than preserving existing coastal sites, the focus could be on preserving and enhancing stocks of habitats, but accepting that their location is not fixed. This suggests the need for European scale assessment of coastal ecosystems.

Secondly, considering the available tools and methods for coastal management, what useful tools and methods exist for climate change issues and what new tools and methods are required to effectively manage the challenges of the 21st Century? While many useful tools and methods already exist, given the long-term implications of climate change, coastal policy and management requires new broad-scale integrated assessment and management tools across a range of scales: local, sub-national (or regional), national and Europe. Assessments at each of these scales will provide useful information to coastal zone management, and if the studies are consistent across the scales, they will allow nesting of the results, maximising their use for policy purposes (e.g. Hall et al. 2003, Polsky et al. 2003). Long-term coastal morphological evolution is an issue that requires particular attention (e.g. Capobianco et al. 1999, De Vriend and Hulscher 2003), as our predictive capability remains quite low, and morphological change influences all other impacts. Dynamic approaches to coastal management also require long-term morphological predictions. It is noteworthy that studies at the European scale are quite limited and often semi-quantitative or qualitative. Possible approaches include combining European-scale assessments such as Eurosion (http://www.eurosion.org) with global-scale integrated assessment models such as DINAS-COAST (http://www.dinas-coast.net), or the more detailed Tyndall Centre Regional Coastal Simulator (http://www.tyndall.ac.uk). However, there are a number of constraints on such integrated model development, including defining the integrated frameworks and appropriate constituent models, as well as identifying the necessary data and scenarios to implement them. Data at broad scales is often rather limited and of variable quality: data on socio-economic and institutional aspects is particularly weak.

Thirdly, what is the appropriate role of proactive versus reactive (wait and see) adaptation policies in long-term management of the coastal zone, and how can proactive approaches be best facilitated? Changing climate and socio-economic conditions presents both opportunities and threats to future coastal developments –

we could develop strategic proactive policies that effectively manage the threats and also fully exploit the opportunities, or conversely, we could ignore the issue which will maximise our vulnerability to climate change and sea-level rise. A recent European survey suggests the latter approach is presently the norm (Tol et al. 2004). The European scale assessments discussed in the previous points could play a role in facilitating debate about coastal adaptation policy, and encourage further investigation via more detailed studies.

Acknowledgements

This chapter has benefited from the authors' work on the EU-funded projects SURVAS (Contract No. ENV4-CT98-0775) and DINAS-COAST (Contract No. EVK2-2000-22024). Richard Tol, Hamburg University, is thanked for his comments on an earlier draft of this chapter. Steve Chilton, Middlesex University drew Figure 1, while Nassos Vafeidis, Middlesex University conducted the analysis that identified areas <5-m elevation.

References

Baarse G, Peerbolte EB, Bijlsma L (1994) Assessment of the vulnerability of the Netherlands to sea-level rise. In: O'Callahan J (ed) Global climate change and the rising challenge of the sea. Proc 3d IPCC CZMS workshop, NOAA, Silver Spring, 211-236

Beersma J, Agnew, M, Viner D, Hulme, M (2000) Climate scenarios for water-related and coastal impacts. ECLAT-2 Workshop Report No. 3, KNMI, De Bilt

Bijlsma L, Ehler CN, Klein RJT, Kulshrestha SM, McLean RF, Mimura N, Nicholls RJ, Nurse LA, Pérez Nieto H, Stakhiv EZ, Turner RK, Warrick RA (1996) Coastal zones and small islands. In: Watson RT, Zinyowera MC, Moss RH (eds) Climate change 1995- impacts, adaptations and mitigation of climate change: Scientific-technical analyses Contr of Working Group II to the Second Assessment Report of the Intergovernmental Panel on Climate Change, Cambridge University Press, Cambridge, 289–324

Bird ECF (1993) Submerging coasts, Wiley, Chichester

Bondesan M, Castiglioni GB, Elmi C, Gabbianelli G, Marocco R, Pirazzoli PA, Tomasin A (1995) Coastal areas at risk from storm surges and sea-level rise in Northeastern Italy. J Coast Res 11: 1354-1379

Brochier F, Ramieri E (2001) Climate change impacts on the Mediterranean coastal zones. Nota di Lavoro 27.2001, Fondazione Eni Enrico Mattei, Milano, Milano

Cahoon DR, Day JW Jr, Reed DJ (1999) The influence of surface and shallow sub-surface soil processes on wetland elevation: A synthesis. Curr Topics Wetland Biogeochem 3: 72-88

Capobianco M, DeVriend HJ, Nicholls RJ, Stive MJF (1999) Coastal area impact and vulnerability assessment: A morphodynamic modeller's point of view. J Coast Res 15: 701-716

Church JA, Gregory JM, Huybrechts P, Kuhn M, Lambeck K, Nhuan MT, Qin D, Woodworth PL (2001) Changes in sea level. In: Houghton JT et al (eds), Climate change 2001: the scientific basis, Contr Working Group I to the Third Assessment Report of the Intergovernmental Panel on Climate Change, Cambridge University Press, Cambridge, pp 639–693

Cicin-Sain B (1993) Integrated coastal management. Ocean Coast Manage 21: 1–352

Cicin-Sain B, Knecht RW, (1998) Integrated coastal and ocean management: Concepts and practices. Island Press, Washington

Cooper NJ, PC Barber, MC Bray, Carter DJ (2002) Shoreline management plans: a national review and an engineering perspective. Proc Inst Civil Eng, Water Mar Eng 154: 221-228

DEFRA (2001) Shoreline management plans: A guide for coastal defence authorities, Department for Environment, Food and Rural Affairs, (downloadable at http://www.defra.gov.uk/environ/fcd/pubs/)

De Groot TM, Orford JD (2000) Implications for coastal zone management. In: Smith D, Raper SB, Zerbini S, Sanchez-Arcilla A (eds) Sea level change and coastal processes: Implications for Europe, European Commission, Brussels, pp 214-242

De la Vega-Leinert AC, Nicholls RJ, Tol RSJ (2000) Proceedings of the SURVAS expert workshop on European vulnerability and adaptation to accelerated sea-level rise. Flood Hazard Research Centre, Middlesex University, Enfield, (downloadable at http://www.survas.mdx.ac.uk)

De Ronde JG (1993) What will happen to The Netherlands if sea level rise accelerates? In: Warrick, RA, Barrow, EM, Wigley, TML (eds) Climate and sea level change: observation, projections and implications. Cambridge University Press, Cambridge, pp 322-335

De Ruig JHM (1998) Coastline management in the Netherlands: human use versus natural dynamics. J Coast Conserv 4: 127-134

De Vriend HJ, Hulscher SJMH (2004) Predicting aggregated coastal evolution, J Coast Res, in press

Ehler CN, Cicin-Sain B, Knecht R, South R, Weiher R (1997) Guidelines to assist policy makers and managers of coastal areas in the integration of coastal management programs and national climate-change action plans. Ocean Coast Manage 37: 7–27

Emery KO, Aubrey DG (1991) Sea levels, land levels, and tide gauges. Springer Verlag, New York

European Commission (1999a) Towards a European integrated coastal zone management (ICZM) strategy: general principles and policy options. Office for Official Publications of the European Union, Luxembourg

European Commission (1999b) Lessons from the European Commission's demonstration programme on Integrated Coastal Zone Management (ICZM). .Office for Official Publications of the European Union, Luxembourg

Evans E (2003) Assessing future risks of flooding and coastal erosion: synthesis of results. Foresight Flood and Coastal Defence Project preliminary paper, Office of Science and Technology, Department of Trade and Industry, London (downloadable at http://www.foresight.gov.uk/)

Fankhauser S (1995) Protection versus retreat: estimating the costs of sea-level rise. Environ Plann A 27: 299-319

Flather R, Williams J (2000) Climate change effects on storm surges: Methodologies and Results. In: Beersma J, Agnew, M, Viner D, Hulme, M (eds) Climate scenarios for water-related and coastal impacts. ECLAT-2 Workshop Report No. 3, KNMI, De Bilt, pp 66-72

French PW (1997) Coastal and estuarine management, Routledge, London

French PW (2001) Coastal defences. Routledge, London

Goeldner L (1999) The German Wadden Sea coast: reclamation and environmental protection. J Coast Conserv 5: 23-30.

Goeldner-Gianella, L (2001) Depolderisation schemes on the western coast of Europe. In de la Vega-Leinert AC, Nicholls RJ (eds) Proceedings of the SURVAS Overview Workshop on "The Future of Vulnerability and Adaptation Studies", Flood Hazard Research Centre, Middlesex University, London, pp 50-53 (downloadable at http://www.survas.mdx.ac.uk)

Gregory JM, Church JA, Dixon KW, Flato GM, Jackett DR, Lowe JA, Oberhuber JM, O'Farrell SP, Stouffer RJ (2001) Comparison of results from several AOGCMs on global and regional sea level change 1900-2100. Clim Dyn 18: 225-240.

Hall JW, Dawson RJ, Sayers PB, Rosu C, Chatterton JB, Deakin R (2003) A methodology for national-scale flood risk assessment. Proc Inst Civil Eng, Water Marit Eng 156: 235-247

Hamm L, Stive MJF (2002) Shore nourishment in Europe. Coast Eng 47: 79-264

Hekstra GP (1986) Will climatic change flood the Netherlands? Effects on agriculture, land-use and well-being. Ambio 15: 316-326

Houghton JT, Ding Y, Griggs DJ, Noguer M, van der Linden PJ, Dai X, Maskell K, Johnson CA (eds) (2001) Climate change 2001: the scientific basis. Contribution of Working Group I to the Third Assessment Report of the Intergovernmental Panel on Climate Change, Cambridge University Press, Cambridge

Hulme M, Jenkins G, Lu X, Turnpenny JR, Mitchell TD, Jones RG, Lowe J, Murphy JM, Hassell D, Boorman P, McDonald R, Hill S (2002) Climate change scenarios for the United Kingdom: The UKCIP02 Scientific Report. Tyndall Centre for Climate Change Research, School of Environmental Sciences, University of East Anglia (downloadable at http://www.ukcip.org.uk)

IPCC CZMS (1990) Strategies for adaptation to sea level rise. Report of the coastal zone management subgroup, Ministry of Transport, Public Works and Water Management, The Hague

IPCC CZMS (1992) Global climate change and the rising challenge of the sea. Report of the Coastal Zone Management Subgroup, Ministry of Transport, Public Works and Water Management, The Hague

Jacobs P, Blom G, Van der Linden T (2000) Climatological changes in storm surges and river discharges: The impact of flood protection and salt intrusion in the Rhine-Meuse delta. In: Beersma J, Agnew M, Viner D, Hulme M (eds) Climate scenarios for water-related and coastal impacts. ECLAT-2 Workshop Report No. 3, KNMI, De Bilt pp 35-48

Jeftić L, Milliman J, Sestini G (1992) Climate change and the Mediterranean—environmental and societal impacts of climatic change and sea-level rise in the Mediterranean Region. Edward Arnold, London

Jeftić L, Kečkeš S, Pernetta JC (1996) Climate change and the Mediterranean—environmental and societal impacts of climate change and sea level rise in the Mediterranean region. Volume 2, Arnold, London

Johansson MM, Kahma KK, Boman H, Launiainen J (2004) Scenarios for sea level on the Finnish coast. Boreal Environ Res, accepted.

Klein RJT (2001) Adaptation to climate change in German official development assistance—An inventory of activities and opportunities, with a special focus on Africa. Deutsche Gesellschaft für Technische Zusammenarbeit, Eschborn

Klein RJT (2003) Coastal vulnerability, resilience and adaptation to climate change: an interdisciplinary perspective. PhD thesis, Christian-Albrechts-Universität zu Kiel

Klein RJT, Nicholls RJ (1999) Assessment of coastal vulnerability to climate change. Ambio 28: 182–187

Klein RJT, Nicholls RJ, Mimura N (1999) Coastal adaptation to climate change: can the IPCC Technical Guidelines be applied? Mitigation Adapt Strat Global Change 4: 51–64

Klein RJT et al (2001) Technological options for adaptation to climate change in coastal zones. J Coast Res 17: 531–543

Klein RJT, Tol RSJ (1997) Adaptation to climate change: options and technologies—an overview paper. Technical Paper FCCC/TP/1997/3, United Nations Framework Convention on Climate Change Secretariat, Bonn

Kont A, Jaagus J, Aunapb R, Ratasa U, Rivisa R (2004) Implications of sea-level rise for Estonia, J Coast Res, accepted.

Kundzewicz ZW et al. (2001) Europe. In: McCarthy JJ, Canziani OF, Leary NA, Dokken DJ, White KS (eds) Climate change 2001: Impacts, adaptation and vulnerability, Cambridge University Press, Cambridge pp. 641-692

Lowe JA, Gregory JM, Flather RA (2001) Changes in the occurrence of storm surges around the United Kingdom under a future climate scenario using a dynamic storm surge model driven by the Hadley Centre climate models. Clim Dyn 18: 179-188

Maddison D (2001) In Ssarch of warmer climates? The impact of climate change on flows of British tourists. Clim Change 49: 193-208.

Ministry of Agriculture, Fisheries & Food (MAFF), Welsh Office, Association of District Councils, English Nature and National Rivers Authority (1995) Shoreline management plans: A guide for coastal defence authorities. Ministry of Agriculture Fisheries & Food, London

McLean R, Tsyban A, Burkett V, Codignotto JO, Forbes DL, Mimura N, Beamish RJ, Ittekkot V (2001) Coastal zone and marine ecosystems. In: McCarthy JJ, Canziani OF, Leary NA, Dokken DJ, White KS (eds) Climate change 2001: impacts, adaptation and vulnerability, Cambridge University Press, Cambridge, pp 343-380

Mee LD (2005) Assessment and monitoring requirements for the adaptive management of Europe's regional seas. In: Vermaat JE, Bouwer LM, Salomons W, Turner RK (eds) Managing European coasts: past, present and future. Springer, Berlin, pp 227-237

Metz B, Berk M, Den Elzen M, De Vries B, Van Vuuven D (2002) Towards an equitable global climate change regime: compatibility with Article 2 of the Climate Change Convention and the link with sustainable development. Clim Policy 2: 211-230

National Research Council (1995). Science, policy and the coast: improving decisionmaking. National Academy Press, Washington

Nicholls RJ (1995) Synthesis of vulnerability analysis studies. Proc WORLD COAST 1993, Ministry of Transport, Public Works and Water Management, the Hague, pp 181-216 (downloadable at http://www.survas.mdx.ac.uk)

Nicholls RJ (2000) Coastal zones. In Parry ML (ed). Assessment of the potential effects of climate change in Europe. Jackson Environment Institute, University of East Anglia, pp 243-259

Nicholls RJ (2002) Rising sea levels: potential impacts and responses. In: Hester R, Harrison RM (eds) Global environmental change. Issues in Environmental Science and Technology, 17, Royal Society of Chemistry, Cambridge, 83-107

Nicholls RJ (2003) Coastal flooding and wetland loss in the 21st Century: changes under the SRES climate and socio-economic scenarios. Global Environ Change, in press.

Nicholls RJ, De la Vega-Leinert AC (2001) Introduction/recommendations. Proc SURVAS overview workshop on the future of vulnerability and adaptation studies, Flood Hazard Research Centre, Middlesex University v-vi, (will be (downloadable at http://www.survas.mdx.ac.uk)

Nicholls RJ, De La Vega-Leinert, AC (2004) Implications of sea-level rise for Europe's coasts J Coast Res, accepted.

Nicholls RJ, Hoozemans FMJ (1996) The Mediterranean: vulnerability to coastal implications of climate change. Ocean Coast Manage 31: 105-132

Nicholls RJ, Lowe J (in rev) Benefits of climate mitigation for coastal areas Global Environ Change

Nicholls RJ, Mimura N (1998) Regional issues raised by sea-level rise and their policy implications. Clim Res 11: 5–18

Nunneri C, Turner RK, Cieslak A, Kannen A, Klein RJT, Ledoux L, Marquenie JM, Mee LD, Moncheva S, Nicholls RJ, Salomons W, Sardá R, Stive MJF, Vellinga T (2005) Group report: integrated assessment and future scenarios for the coast. In: Vermaat JE, Bouwer LM, Salomons W, Turner RK (eds) Managing European coasts: past, present and future. Springer, Berlin, pp 271-290

Quelennec R, Oliveros C, Uhel R, Devos W (1998) CORINE coastal erosion, Office for the Official Publications of the European Communities, Luxembourg.

Parry M, Arnell N, Hulme M, Nicholls R, Livermore M (1998) Adapting to the inevitable. Nature 395: 741.

Papathanassiou E, Bokn T, Skjolodal H, Skei J, Green N, Bakke T, Severinsen G (1998) Marine and coastal environment. In: Europe's environment: the second assessment. Elsevier, Kidlington, Oxford

Peerbolte EB (1994) Hazard appraisal: modelling sea-level rise and safety standards. In Penning-Rowsell EC, Fordham M (eds) Floods across Europe: flood hazard assessment, modelling and management, Middlesex University Press, London, pp 107-134

Polsky C, Schroter D, Patt A, Gaffin S, Martello ML, Neff R, Pulsipher A,, Selin H, (2003) Assessing vulnerabilities to the effects of global change: An eight-step approach. Belfer Center for Science and International Affairs Working Paper, Environment and Natural Resources Program, John F. Kennedy School of Government, Harvard University, Cambridge (downloadable at http://ksgnotes1.harvard.edu/bcsia/sust.nsf/publications)

Rigg K, Salman A, Zanen D, Taal M, Kuperus J, Lourens J (1997) Threats and opportunities in the coastal areas of the European Union: A Scoping Study. National Spatial Planning Agency, Ministry of Housing, Spatial Planning and the Environment, the Hague

Rochelle-Newall E, Klein RJT, Nicholls RJ, Barrett K, Behrendt H, Bresser THM, Cieslak A, De Bruin EFLM, Edwards T, Herman PMJ, Laane RPWM, Ledoux L, Lindeboom H, Lise W, Moncheva S, Moschella P, Stive MJF, Vermaat JE (2005) Group report: global change and the European coast – climate change and economic development. In: Vermaat JE, Bouwer LM, Salomons W, Turner RK (eds) Managing European coasts: past, present and future. Springer, Berlin, pp 239-254

Rupp S, Nicholls RJ (2003) The application of managed retreat: A comparison of England and Germany. In: Marchand M, Heynert KV, van der Most H, Penning WE (eds) Dealing with flood risk. Delft Hydraulics Select Series 1/2003. Delft University Press, Delft, pp 42-51

Sanchez-Arcilla A, Jimenez J, Valdemoro HI (1998) The Ebro delta: Morphodynamics and vulnerability, J Coast Res 14: 754-772

Smit B, Pilifosova O, Burton I, Challenger B, Huq S, Klein RJT, Yohe G (2001) Adaptation to climate change in the context of sustainable development and equity. In: McCarthy JJ, Canziani OF, Leary NA, Dokken DJ, White KS (eds) Climate change 2001: impacts, adaptation and vulnerability, Cambridge University Press, Cambridge, pp 877-912

Sterr H (2004) Assessment of vulnerability and adaptation to sea-level rise for the coastal zone of Germany. J Coast Res, in press.

Stive MJF, Roelvink JA, DeVriend HJ (1990) Large-scale coastal evolution concept, International Conference on Coastal Engineering, ASCE, New York, pp 1962-1974

Suanez S, Provansal M (1996) Morphosedimentary behaviour of the deltaic fringe in comparison to the relative sea-level rise on the Rhone delta. Quat Sci Rev 15: 811-818

Tol RSJ (2002a) Estimates of the damage costs of climate change. Part I: benchmark estimates. Environ Resour Econ 21: 47-73

Tol RSJ (2002b) Estimates of the damage costs of climate change. Part II: dynamic estimates. Env Resour Econ 21: 135-160

Tol RSJ, Klein RJT, Nicholls RJ (2004) Towards successful adaptation to sea level rise along Europe's coasts. J Coast Res, accepted.

Tooley, M, Jelgersma, S (Eds.) (1992) Impacts of sea level rise on European coastal lowlands. Blackwell, Oxford

Turner, RK (2005) Integrated environmental assessment and coastal futures. In: Vermaat JE, Bouwer LM, Salomons W, Turner RK (eds) Managing European coasts: past, present and future. Springer, Berlin, pp 255-270

Turner RK, Doktor P, Adger WN (1995) Assessing the costs of sea-level rise. Environ Plann A 27: 1777-1796

Turner RK, Lorenzoni I, Beaumont N, Bateman IJ, Langford IH, McDonald AL (1998a) Coastal management for sustainable development: analysing environmental and socio-economic change on the UK coast, Geogr J 164: 269-281

Turner RK, Adger WN, Lorenzoni I (1998b) Towards integrated modelling and analysis in coastal zones: principles and practises. LOICZ Reports & Studies No. 11. LOICZ-IPO, Texel

Vaughan DG, Spouge J 2002: Risk estimation of the collapse the West Antarctic ice sheet, Clim Change, 52: 65-91

Viles H, Spencer T, 1995. Coastal problems, Edward Arnold, London

WASA Group (1998) Changing waves and storms in the Northeast Atlantic. Bull Am Meteorol Soc 79: 741-760

Waterman RE, Misdorp R, Mol A (1998) Interactions between water and land in The Netherlands. J Coast Conserv 4: 115-126

Watkinson A, Arnell N, Conlan K, Coker A, Gill J, Ledoux L, Nicholls R, Sear D, Tinch R (2003) Assessment of environmental impacts of future flood risk. Foresight Flood and Coastal Defence Project preliminary paper, Office of Science and Technology, Department of Trade and Industry, London (downloadable at http://www.foresight.gov.uk/)

WCC'93 (1994) Preparing to meet the coastal challenges of the 21st century. Report of the World Coast Conference, Ministry of Transport, Public Works and Water Management, The Hague

Wigley TML, Raper SCB (1993) Future changes in global mean temperature and sea level. In: Warrick RA, Barrow EM, Wigley TML (eds) Climate and sea level change: observation, projections and implications. Cambridge University Press, Cambridge, pp 111-133

Willows RI, Connell RK (eds) (2003) Climate adaptation: risk uncertainty and decision-making UKCIP Technical Report. UKCIP. Oxford (downloadable at http://www.ukcip.org.uk)

Woodworth PL, Gregory JM, Nicholls RJ (2004) Long term sea level changes and their impacts. The Sea Vol 11, Wiley, New York, accepted.

12. Assessment and monitoring requirements for the adaptive management of Europe's regional seas

Laurence D. Mee[1]

Abstract

The continued decline in Europe's coastal and marine ecosystems poses a major challenge to policymakers and enforcers. Adoption of the 'ecosystem approach' offers an opportunity to develop more integrated policies within rational system boundaries. However, policymakers often express concern regarding how this approach should be delivered. An adaptive management scheme is offered that works towards publicly understandable ecological quality objectives in steps that are within pragmatic political timeframes. The scheme must be accompanied by monitoring of relevant system indicators if it is to be effective. It is currently being tested in several projects sponsored by the Global Environment Facility, including the Black Sea Environmental Programme.

Introduction

There is growing evidence that Europe's marine and coastal systems are suffering widespread and significant degradation. Marine and coastal systems are inherently complex and often display non-linearity between socio-economic pressures and resulting state changes. Past management has been (and in many places continues to be) predominantly reactive in nature. Furthermore, the governance of marine areas remains very fragmented, both between countries and sectors, nationally and internationally. The environment of European Seas and their fisheries are often managed independently of one another and, until recently, little regard has been given to the state of the marine environment in the process of economic development of terrestrial catchment areas. Though evidence-based management is likely

---
[1] Correspondence to Laurence Mee: l.mee@plymouth.ac.uk

J.E. Vermaat et al. (Eds.): Managing European Coasts: Past, Present, and Future, pp. 227–237, 2005.
© Springer-Verlag Berlin Heidelberg 2005.

to remain the predominant paradigm for marine and coastal areas, there is growing acceptance of the ecosystem approach and a more precautionary approach to decision-making. The current paper examines adaptive management as a means to deliver the ecosystem approach and focuses on the assessment and monitoring requirements of an effective adaptive management strategy.

Emerging concepts and policy drivers

The ecosystem approach

A useful working definition of the ecosystem approach has been developed by the Convention on Biological Diversity (CBD 1998): "The ecosystem approach is based on the application of appropriate scientific methodologies focused on levels of biological organization which encompass the essential processes and interactions amongst organisms and their environment. The ecosystem approach recognizes that humans are an integral component of ecosystems."

The ecosystem approach has the following key features:

- Management objectives as societal choice;
- Management decentralised and multi-sectoral;
- Appropriate temporal and spatial scale;
- Conservation of ecosystem function and resilience;
- Appropriate balance between conservation and use;
- Management within system limits;
- The outward vision (respect interconnectedness) and long-term vision (change is inevitable);
- Broad use of knowledge, scientific and traditional;
- Incorporation of economic considerations (costs and benefits, removal of externalities, etc.);

The European Water Framework Directive (EC 2000) has the potential of delivering the ecosystem approach as it defines systems within natural boundaries (catchments or 'river basin districts') rather than political ones. From the perspective of Europe's seas however, its value is limited to the coastal zone since it sets marine boundaries at one nautical mile from the coast. For this reason, the European Commission has recently announced (EC 2002) the development of a 'Strategy to Protect and Conserve the Marine Environment'. This embraces the ecosystem approach and sets boundaries that correspond with the current European seas Conventions (Baltic, NE Atlantic, Mediterranean, Black Sea).

Adaptive management

Adaptive management, originally termed "adaptive environmental assessment and management", (Holling 1978) recognizes the need to experiment with complex systems in order to learn from them. Within this paradigm, it is recognised that the level of scientific uncertainties in natural systems are often too great to permit long-term management decisions based upon conceptual modelling or knowledge of a limited part of the system. Sometimes described as 'learning by doing' (Walters 1997), adaptive management employs the best available multi-disciplinary knowledge to construct a dynamic conceptual model to examine scenarios for how the system might behave under different management regimes. It then encourages managers to adopt the most favourable pathway for a limited period; closely observing the outcome through carefully focused monitoring. At the end of this initial learning period, the model can be further refined and new management objectives set.

Adaptive management offers a practical means of integrating knowledge over social and economic as well as ecological scales (Walker et al. 2002). It can accommodate unexpected events by encouraging approaches that build system resilience. Despite these advantages however, its explicit application has been rather limited and many attempts at application resulted in disappointment. McLain and Lee (1996) have attributed this to excessive reliance on linear systems models and the lack of attention to non-scientific knowledge and knowledge-sharing policy processes (insufficient stakeholder consultation and buy-in). Managers that are presently working in a command-control situation may also have difficulties coming to terms with the more flexible adaptive approach with its relatively large monitoring requirement. Despite these initial difficulties, adaptive management is gradually becoming accepted as a valuable tool for delivering the ecosystem approach.

An adaptive management strategy for European seas

Conceptual model of the strategy

A strategy for applying adaptive management to marine systems is illustrated in Figure 1. This strategy is currently being tested in a number of GEF transboundary waters projects (including the Black Sea) and its application in the Baltic is also being studied. Its components will be briefly described below:

Component 1: Initial assessment

The purpose of the initial assessment is to:

- Gather relevant information;
- Convert available data into objective information in order to define the nature and impact of the environmental problem and to establish priorities for further action;
- Add value to historical data through new interpretations;
- Make complex information available to a wide range of stakeholders in an understandable form.

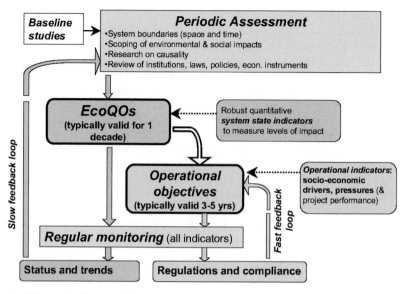

Fig. 1. Schematic diagram of a strategy for adaptive management in marine systems (see text for details)

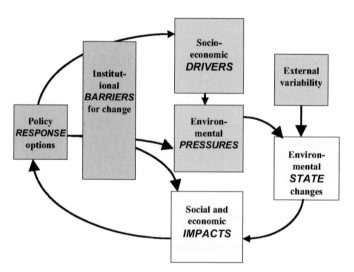

Fig. 2. Modified D-P-S-I-R model used for causality studies and scenario building

The assessment begins with a rather simple scoping study that can be conducted through a facilitated dialogue with stakeholders (including scientists) and a Delphi exercise to assess priorities based upon the severity of impacts. Scores for both ecosystem impacts and socio-economic impacts can be used applying agreed pre-defined criteria (Mee and Bloxham 2002). Available baseline data (see Hilborn and Walters 1981) helps to distinguish between natural and anthropogenic changes. A scaling exercise also takes place to define the problem-specific system boundaries (spatial and temporal).

Research is conducted on causality using the Drivers – Pressures – State changes - socio-economic Impacts – Responses model (Turner et al. 1998), modified to consider institutional barriers to change (see Figure 2). It is often convenient to conduct the analysis by building causal chains by successively answering the question "what is the cause?" starting from the state change itself and working backwards to the socio-economic drivers (or 'root causes'). An analysis of governance (including laws, institutions, finance, public participation) provides information on the barriers that must be overcome in order to develop effective policy responses. Application of the DPSIR model helps to define scenarios upon which to plan management decisions.

Component 2: The definition of ecological quality objectives and state change indicators

The EcoQOs can be regarded as a statement of 'vision' of how the stakeholders would like to see the state of the system in the future. They provide the long-term goal for adaptive management. The objectives however, are themselves based on human values and as functional participants in the ecosystem, we cannot be outside observers. As information, knowledge and wisdom grow, the EcoQOs themselves will tend to change and the adaptive management model has to be flexible enough to allow this to happen.

EcoQOs should be measurable environmental status goals that are clearly understandable by a wide range of stakeholders. They should be discussed with the stakeholders and where possible, developed with their full participation. They should also reflect key attributes of the system that can be quantified and they are often (but not always) set against a baseline clearly established in the initial assessment.

Examples of such EcoQOs, defined in the North Sea Ministerial Declaration process (NSMD 2002) are illustrated in Table 1. The important feature of the North Sea EcoQOs is that on one hand they address public concerns but on the other, lead to a cascade of technical requirements for policy actions, indicators and monitoring. This contrasts sharply with the 'blanket monitoring' approach of the 1970s and 1980s that tended to result in large databases that were difficult to use in a dynamic way.

Having developed EcoQOs, one of the major current challenges is to find robust system indicators, particularly those of emergent properties of the system (rather than measures of its consistent components). Such indicators have occasionally been described as measures of ecosystem 'health' (Schaeffer et al. 1988, an 'arcane' concept according to Ryder 1990).

The development of indicators for problems with multiple consequences is particularly difficult. This will be illustrated for the cases of eutrophication and chemical pollution.

Table 1. Examples of EcoQOs developed as part of the North Sea Ministerial Conference process

Quality element	EcoQO
Seal population trends on the North Sea	No decline in population size or pup production of $\geq 10\%$ over a period of up to 10 years
Proportion of oiled common Guillemots among those found dead or dying on beaches	The proportion of such birds should be 10% or less of the total found dead or dying, in all areas of the North Sea
Changes/kills in zoobenthos in relation to eutrophication	There should be no kills in benthic animal species as a result of oxygen deficiency and/or toxic phytoplankton species

An example of a suite of potential system indicators is illustrated for the case of eutrophication in Figure 3. The figure illustrates indicators based upon pressure (nutrient loadings), trophic effects and catastrophic state changes. The pressure indicators are poorly specific (and cannot be employed in isolation or without a system model) but are potentially able to provide an early warning of the phenomenon. Trophic effect indicators on the other hand may also provide an 'early warning' but are slightly more specific, especially when several tools are used together. Indicators such as diatoms/non-diatom ratio and fodder/non-fodder zooplankton ratios are based on emergent system properties and reflect the inability of the system to deal with excess primary production. Finally, indicators of hypoxia, decreased demersal fish catch and benthic mass mortality reflect catastrophic state change; a stepwise loss in system resilience (Scheffer et al. 2001). These are the most commonly used indicators of eutrophication but unfortunately provide information too late for response.

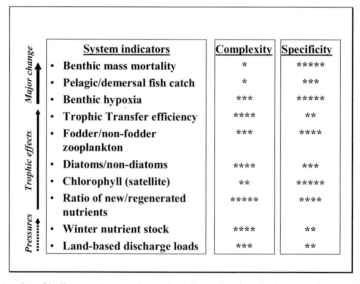

Fig. 3. A suite of indicators that can be applied for evaluating the impact of eutrophication

A similar situation arises with other indicators of state change. Pollution for example is often detected by mortality. Early warning indicators have been devised such as molecular biomarkers that reflect damage to individual cells by specific groups of pollutants but may be difficult to relate to community level changes. On the other hand, community health indicators are not very specific for individual pollutants. It is necessary for biologists to reconcile themselves to the unlikeliness of encountering 'magic bullet' indicators for most pollutants and a suite of measurements is normally needed. To date however, most chemical pollution studies still focus on the concentrations of substances known to pollute but our capacity to detect potential effects of thousands of unmeasured compounds and degradation products remains very limited (Mee and Bloxham 2002). This reactive approach introduces a potentially dangerous time lag between the introduction of a new pollutant and its inclusion in a statutory monitoring programme. Clearly more work is needed to find viable alternative indicators or to convince national agencies to employ those that already exist.

Component 3: Operational objectives and their indicators

Operational objectives are used to define the pragmatic steps towards achieving agreed EcoQOs. They are best devised through a stepwise process that starts with a facilitated 'brainstorming' activity between relevant stakeholders (McCreary et al. 2001). This activity seeks to examine all practical options for achieving the EcoQOs and the most promising options are submitted to feasibility studies. The options should not be limited to the 'technical quick-fix' but should include measures to address the problems as close as possible to their social and economic root causes. Actions nearer the root causes are more likely to be lead to sustainable change but are often less politically palatable. The improvement of intersectoral cooperation, capacity building, environmental education and more effective compliance with existing regulations, may prove to be more effective than end-of-pipe engineering in many circumstances.

Ultimately, the decision on which of the available options will be pursued is a political one. Sound advice will inevitably be tempered by political pragmatism. Whatever the outcome (assuming that it represents a step towards the EcoQO) it will require clear short-term objectives and appropriate indicators. The objective might for example, consist of an agreement to reduce an input to the system by a fixed percentage over a well defined timeframe (e.g. to decrease nitrate loads by 20%). This of course assumes a reliable baseline and a monitoring programme to test compliance. A decision to set such an objective is in itself an adaptive management experiment where the outcome must be carefully evaluated before moving forwards. A description of appropriate objectives and indicators is beyond the scope of the present article. However, the indicators would often refer to the reduction of pressures (or in some cases socio-economic drivers) in the DPSIR scheme. They may also include project performance indicators (management indicators demonstrating how efficiently a project is delivered).

Component 4: Monitoring schemes and feed-back mechanisms

In the present article, I have stressed the importance of monitoring within the adaptive management framework. The monitoring scheme should be clearly focussed on relevant system indicators. Those related to the EcoQOs constitute the basis of 'status and trends' reports that inform periodic assessments. These, in turn, lead to new or revised EcoQOs. The periodicity of the revision process would typically be about ten years; systems tend to recover quite slowly from a reduction in human pressure. This cycle has been termed the 'slow feedback loop' in Figure 1. Unfortunately, this frequency is well below that of most political processes (e.g. elections to legislatures) and this always presents a major challenge for the delivery of sustainable development.

The indicators for compliance with operational objectives are also monitored on a regular basis. However, the frequency of measurement should be much higher as rapid change would be anticipated. Indeed, part of the political pragmatism used when setting operational objectives is to ensure that they can be achieved within a project cycle of 3-5 years. Thus a rapid feedback loop emerges (see Figure 1) and new objectives can be set within the normal term of office of a public official (enabling greater 'ownership' of the process).

Practical application of the adaptive management scheme

At this point, the obvious question is whether or not the interpretation of adaptive management presented here has been applied in a practical situation in European seas. Certainly, the development of EcoQOs is a major ongoing activity within the North Sea Ministerial Conference process and in the context of the Helsinki Convention. A major test of the process underway in the Black Sea will be described in more detail.

The benthic ecosystem of the North-western shelf of the Black Sea has already suffered catastrophic change as a result of eutrophication (Mee 1992). It receives the combined discharges of Europe's second (Danube) and third (Dnipro) rivers, draining a huge agricultural area covering 13 countries. During the 'green revolution' that began in the 1960's, massive increases in fertilizer use and intensive animal husbandry led to the discharge of large loads of nutrients to the Black Sea. This situation resulted in eutrophication, eventually leading to the collapse of vast red algal (*Phyllophora*) beds on the shelf. The disappearance of this keystone species led to the demise of the benthic system and its replacement by a seasonally hypoxic pelagic system. Following the collapse of centrally planned economies in Eastern Europe at the beginning of the 1990s, farmers were unable to apply large quantities of fertilizers, intensive animal farms closed and nutrient loads to the Black Sea gradually subsided.

The collapse of the benthic system on the North-western shelf was precipitous, beginning in the early 1970s. By 1990, the entire *Phyllophora* bed had disappeared. At the same time, summer hypoxia resulted in the creation of a 'dead zone' that gradually increased in size reaching a maximum in the late 1980s. The onset of hypoxia in 1973 corresponded with a nitrate fertilizer loading to the drainage basin of about 1.5M tons/yr (Mee 2001). By 1996, nitrate loads had decreased to about the same level and there were no reports of a 'dead zone' for that year. In 2001, an exceptionally stable, warm summer and high river levels increased the seasonal nutrient runoff again and hypoxic conditions returned.

The situation described above is presented diagrammatically in Figure 4. This illustrates the catastrophic collapse of the benthic system. The recent decrease in hypoxia has not been mirrored by a return of *Phyllophora* however. The new predominantly pelagic system is also resilient to change and it is difficult to predict when and if the benthic system will be re-established.

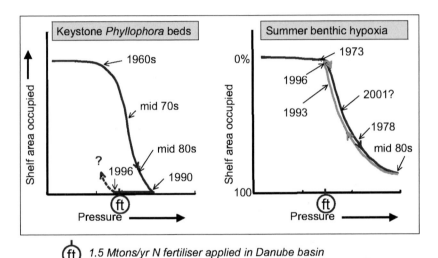

Fig. 4. The hysteresis effect on the Black Sea NW Shelf benthic system (conceptual model based on data in Mee 2001). Ft is the threshold point where collapse of the system began

In order to overcome the problem of eutrophication in the Black Sea, the coastal countries developed measures based upon the adaptive management paradigm. The 'vision statement' adopted was as follows: "The long-term goal in the wider Black Sea Basin is to take measures to reduce the loads of nutrients and hazardous substances discharged to such levels necessary to permit Black Sea ecosystems to recover to conditions similar to those observed in the 1960s."

Fortunately, data on the system in the 1960s is relatively abundant and it will be possible to develop a number of indicators to support this aspirational goal. The operational objective describes short-term targets: "As an intermediate goal, urgent measures should be taken in the wider Black Sea Basin in order to avoid that the loads of nutrients and hazardous substances discharged into the Seas exceed those that existed in the mid 1990s. (These discharges are only incompletely known.)"

Clearly from Figure 4, it is unsure whether or not this objective will be sufficient to return the system to its former state. However, the task of maintaining the relatively low nutrient loads of the mid-90s will not be an easy one, given the enormous pressure for renewed growth of the agricultural sector in Eastern Europe. The governments have set a six year target for revision of this objective and research monitoring has already begun to provide a better information base for setting new objectives in 2006.

Conclusion

This paper argues that adaptive management can provide a mechanism for delivering the ecosystem approach in Europe's regional seas. In order to do so however, an entirely new approach to monitoring will also be required. Emphasis must be on measuring the dynamic processes that characterise an ecosystem and on developing robust indicators that can be understood by a wide range of stakeholders. The approach described can be applied to a wide range of systems from river basins and enclosed seas to open sea areas. It is currently being tested in a number of projects supported by the Global Environment Facility, including its intervention in the Black Sea.

Acknowledgments

This work would not have been possible without a continuous and ongoing dialogue with my colleagues in the Marine and Coastal Policy Research Group and with practitioners and stakeholders. The stakeholder consultation to discuss the European Marine Strategy (Koge, Denmark, December 2002) provided the first opportunity to present this approach and the results of these discussions were further tested in the meeting on the sustainable development of Sweden's marine environment held in Waxholm in April 2003. My appreciation to Danish and Swedish Governments for inviting me to participate in these events.

References

EC (2000) Directive 2000/60/EC of the European Parliament and of the Council of 23 October 2000 establishing a framework for Community action in the field of water policy. Off J Eur Communities L 327:1-72

EC (2002) Towards a strategy to protect and conserve the marine environment. Communication from the Commission to the Council and the European Parliament. COM(2002) 539 final, Brussels

Hilborn R, Walters C (1981) Pitfalls of environmental baseline and process studies. Environ Impact Assess Rev 2: 265-278

Holling CS (1978) Adaptive environmental assessment and monitoring. Wiley, New York

McCreary S, Gamman J, Brooks B, Whitman L, Bryson R, Fuller B, McInerny A, Glazer R (2001) Applying a mediated negotiation framework to integrated coastal zone management. Coast Manage 29:183-216

McLain RJ Lee RG (1996) Adaptive management: promises and pitfalls. Environ Manage 20: 437-448

Mee LD (1992) The Black Sea in crisis: The need for concerted international action. Ambio 21:278-286

Mee LD (2000) Eutrophication in the Black Sea and a basin-wide approach to its control. In: Von Bodungen B, Turner K (eds) Science and Integrated Coast Manage. Dahlem University Press, Berlin, pp 71-94

Mee LD, Bloxham M (2002) Assessing the impact of pollution on aquatic systems at a global and regional level. Mar Environ Res 54:223-228

NSMD (2002) Bergen Declaration. Ministerial Declaration of the Fifth International Conference on the Protection of the North Sea. Bergen, 20-21 March, 2002. 35pp.

Ryder RA (1990) Ecosystem health, a human perception: Definition, detection, and the dichotomous key. J Great Lakes Res 16:619-624

Schaeffer DJ (1988) Ecosystem health: I. Measuring ecosystem health. Environ Manage 12:445-455

Scheffer M, Carpenter S, Foley JA, Folke C, Walker B (2001) Catastrophic shifts in ecosystems. Nature 413:591-596

Turner RK, Lorenzoni I, Beaumont N, Bateman IJ, Langford IH, McDonald AL (1998). Coastal management for sustainable development. Analysing environmental and socio-economic changes on the UK coast. Geogr J 164:269-281

Walker BS, Carpenter J, Anderies N, Abel N, Cumming GS, Janssen M, Lebel L, Norberg G, Peterson D, Richard R (2002) Resilience management in social-ecological systems: a working hypothesis for a participatory approach. Conserv Ecol [online] 6:14

Walters C (1997) Challenges in adaptive management of riparian and coastal ecosystems. Conserv Ecol [online] 1:1

13. Group report: Global change and the European coast – climate change and economic development

Emma Rochelle-Newall[1], Richard J.T. Klein, Robert J. Nicholls, Kevin Barrett,
Horst Behrendt, Ton H.M. Bresser, Andrzej Cieslak, Erwin F.L.M. de Bruin,
Tony Edwards, Peter M.J. Herman, Remi P.W.M. Laane, Laure Ledoux,
Han Lindeboom, Wietze Lise, Snejana Moncheva, Paula S. Moschella,
Marcel J.F. Stive, and Jan E. Vermaat

Abstract

The effects of climate change on economic development in the coastal zone cannot be ignored in future coastal zone management plans. This chapter reports the outcome of the group discussion centred round the three questions posed by Nicholls and Klein on how the coastal zone can be effectively managed in the future. The first question asked how we can marry together the human and natural values of a system in the upcoming decades. The results of the discussion highlighted the utility of using scenarios to obtain good management plans that take into account the three provisions of human safety, economic development and ecological integrity while still achieving a situation of sustainable development. The second question addressed the tools required to achieve these management goals and discusses the role of public participation and media communication. The third question asked what proactive strategies can be used to effectively manage the coastal zone in the 21st Century and an example of the Humber estuary management plan is given. It is proposed that a robust and flexible integrated coastal zone management plan is the only effective way to manage the coastal zone in a sustainable manner in the uncertain face of climate change.

---

[1] Correspondence to Emma Rochelle-Newall: emma.rochelle-newall@noumea.ird.nc

J.E. Vermaat et al. (Eds.): Managing European Coasts: Past, Present, and Future, pp. 239–254, 2005.

Introduction

Coastal zones represent the narrow transitional zone between the world's land and oceans, characterised by highly diverse ecosystems such as cliffs, beaches, dunes and wetlands. Many people have settled in coastal zones to take advantage of the range of opportunities for food production, transportation, recreation and other human activities provided here. A large part of the global human population now lives in coastal areas: estimates range from 20.6 per cent within 30 km of the sea to 37 per cent in the nearest 100 km to the coast (Cohen et al. 1997, see also Small and Nicholls 2003). In addition, a considerable portion of global economic wealth is generated in coastal zones (Turner et al. 1996). Many coastal locations exhibit a growth in population and income higher than their national averages (Carter 1988, WCC'93 1994), as well as substantial urbanisation (Nicholls 1995, Klein et al. 2002).

Natural climate variability is inherent in natural systems and it is upon this natural variability that human-induced climate change is overlain. Greenhouse gas concentrations have been growing since the Industrial Revolution and further-more, it is expected that these gases will continue to increase in concentration in the atmosphere through the 21st century, although this change will depend on a range of factors (e.g., Nakicenovic and Swart 2000). This is turn will lead to global climate change and sea-level rise with important impacts across Europe and the world (Houghton et al. 2001, McCarthy et al. 2001, Parry 2000, Nicholls and Klein, this volume). Therefore, adaptation to climate change, whatever its cause, is necessary for the future well-being of ecological and human systems. Matters are also complicated by other social and economic changes that also place stress on the coastal system and may interact with the effects of climate change (Holligan and De Boois 1993, Klein and Nicholls 1999, IGBP-LOICZ 2003).

Many of the effects of climate change, but by no means all, will present threats to the vulnerable coastal zone and also cause changes to both the inland catch-ments and open sea. Possible effects of climate change on the coast zone include:

- Changes in storm frequency and intensity, including surge tides;
- Increases in mean sea level;
- Changes in patterns of erosion and sedimentation;
- Increased flood risks;
- Increases in the temperature of land, air and water affecting the distribution of species, soil processes, etc;
- Changes in rainfall and run-off patterns and hence the flow regimes of rivers draining to the sea, which may in turn affect nutrient inputs, the salinity regime of estuaries, mixing patterns and flood risks;
- Changes in groundwater and saline intrusion;
- Changes in wind direction and velocity, which could affect mixing patterns and flooding;
- Changes in the distribution of human, animal and plant pathogens, and other health effects;
- Damage to archaeological sites and the historic environment;
- Changes in oceanic circulation patterns;

The effects of climate change and associated sea-level rise threaten economic sectors to a varying extent. The following socio-economic impacts were identified by McLean et al. (2001):

- Increased loss of property and coastal habitats;
- Increased flood risk and potential loss of life;
- Damage to coastal protection works and other infrastructure;
- Loss of renewable and subsistence resources;
- Loss of tourism, recreation, and transportation functions;
- Loss of non-monetary cultural resources and values;
- Impacts on agriculture and aquaculture through decline in soil and water quality.

It must also be recognised that not only is the coastal zone and the economical development therein impacted by climate change but that this economic development can also impact upon climate change and so care must be taken not to exacerbate the problems now beginning to manifest in our coastal zones. Moreover, it should not be forgotten that although humans are the main driver of ecosystem change, they are also are part of the ecosystem itself. Therefore, the response to climate change in coastal areas should be a twin-track process (Nicholls and Lowe 2004):

- Reduce greenhouse gas emissions;
- Adapt to the change that is inevitable.

The mechanisms causing climate variation and the effects that climate change then will have on the coastal environment are still not fully understood. Likewise socio-economic trends are dependent on many factors, including political decisions. The quantitative prediction of the future impacts on the coast in the longer-term (say next 50 to 100 years) is consequently fraught with uncertainty. Nicholls and Klein (this volume) identified that while the European Union has taken a leading role in reducing greenhouse gas emissions, preparation for adapting to sea-level rise and climate change in Europe's coastal areas is much more patchy. In terms of encouraging and developing adaptation to the effects of climate change on the coastal zone, Nicholls and Klein (this volume) posed the following three questions, which the group explored:

1. Protecting human use and sustaining the natural functioning of the coast: how can we most effectively marry these two often conflicting goals (especially given the long-term commitment to sea-level rise)?
2. Considering the available tools and methods for coastal management, what useful tools and methods exist for climate change issues and what new tools and methods are required to effectively manage the challenges of the 21st Century?
3. What is the appropriate role of proactive versus reactive (wait and see) adaptation policies in long-term management of the coastal zone, and how can proactive approaches be best facilitated?

The remainder of this paper summarises these discussions to provide suggestions on how to approach the management of the European coastal zone in the light of inevitable but uncertain change. Section 2 discusses how to balance human values with the natural worth of a system and gives some suggestions about how we can achieve a balance between these two often opposing elements. The use of scenario development is proposed and an example of such an exercise is presented. This section also introduces the concept of management of a dynamic coastal environment. Section 3 covers the tools and mechanisms now available to help us effectively manage the coastal zone. This section also presents the five priorities identified by the group as important requirements for the future management of the coastal zone. Section 4 addresses the role of proactive strategies and provides an example of a proactive management strategy now in place in the Humber basin, England. Finally, Section 5 provides some conclusions and proposes some future guidelines for more effectively managing the coastal zone in the future.

Human values versus natural systems

As discussed in Nicholls and Klein (this volume), there is a challenge to provide human safety and promote economic development without compromising ecological integrity. Although, ecosystems contribute to human safety provision and economic development, this contribution is often not quantified and therefore the benefits are not recognised in management. This is exacerbated by the lack of knowledge of the contribution of ecosystems to human welfare, such as the ability of salt marshes to reduce wave energy in coastal systems and their potential role as a natural buffer (Allen and Pye 1992, Allen 2000).

In the past under 'traditional development', economic growth development and human safety concerns have usually taken precedence over the preservation of ecological integrity, often resulting in a slow but steady decline in the environment and the services that it can provide (Figure 1). While the degree of loss of environmental integrity can be kept to a minimum in a future where there is a high degree of certainty of the effects of climate change on a system, this is not the case in a situation of low certainty. In such a scenario, the potential losses to the environment are large and may increase with the pace of economic development. In contrast, if a policy of sustainable development is followed, the degree of loss of ecological integrity is less and should be minimised, particularly in the situations where there is a high degree of certainty of the climate impacts. Under the situation of low certainty, the potential range of loss or augmentation of ecological integrity is wider, but it is still higher than that of the traditional development strategy, regardless of the level of economic development. This is also true for human safety, as ecological integrity maintains geomorphic features (beaches, salt marshes, etc.) that contribute to human safety, the third key point.

In order to move towards a more sustainable development of Europe's coasts, the group agreed that it is critical that (Figure 1):

1. Adaptive management approaches be adopted in which we maximise the possibility to 'learn by doing';
2. Sustaining and enhancing natural buffers become a priority, rather than depend solely on artificial defences;

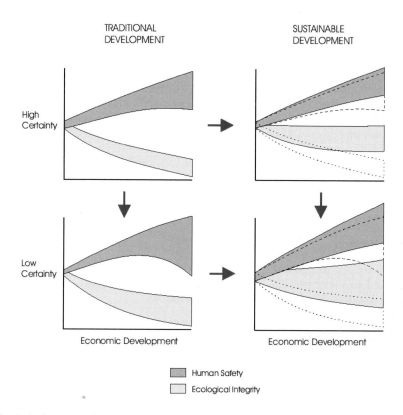

Fig. 1. A diagrammatic view of the evolution of human safety and ecological integrity with increasing levels of economic development under different development pathways and levels of certainty. The hatched lines on the right indicate the shifts that occur moving from a traditional development pathway to a sustainable development pathway

3. Strategic land use planning be implemented;
4. Disaster preparedness is considered in addition to traditional defences, so that we are prepared for all eventualities.

Scenario-based analysis

One way of addressing the three priorities of human safety, economic development and ecological integrity, given the uncertainty of the future, is to develop scenarios to define the range of possible outcomes of various contrasting projections of future situations (e.g. Parry 2000, UNEP, 2002). These should include various combinations of economic development and the degree of change expected. The example of scenario-based analysis used by the group is described below.

The context

Imagine yourself in the position of scientific adviser to the newly established coastal zone management agency of a European deltaic country. This country has a heavily developed coastal zone, where agriculture, tourism and industry are the most important sectors. In addition, it has one of Europe's largest ports, as well as valuable coastal habitats. More than half the country's population and capital assets can be found in the coastal zone. The newly established coastal management agency has the responsibility of developing and implementing a cost-effective integrated policy aimed at ensuring human safety, economic development and ecological integrity. The agency uses a 50-year planning horizon; it is aware that socio-economic change and climate change over this period will present a challenge to addressing these three priorities, although there is uncertainty as to the nature and magnitude of this challenge.

The scenarios

To allow for this uncertainty, the agency has developed three socio-economic scenarios and two climate scenarios, which provide six cases that you are asked to analyse. The scenarios can be summarised as follows:

Socio-economic scenario I: The coastal population continues to grow at its current rate for the first ten years and then gradually stabilises. Economic growth continues at its current rate until 2050. The main economic sectors remain the same.

Socio-economic scenario II: The coastal population and economy grow at significantly higher rates than is currently the case, which is associated with large-scale investment in infrastructure, intensified agriculture and a sharp increase in tourism.

Socio-economic scenario III: Economic growth in the coast stalls and people move away from the coast. Agriculture continues whilst industrial activity and tourism diminish. A marine national park is established.

Climate scenario A: Climate change will lead to a sea-level rise of 50 cm and an increase in summer precipitation of 20% by 2050. No other changes will occur.

Climate scenario B: Climate change will lead to a sea-level rise of 50 cm by 2050. In addition, storminess, summer and winter precipitation and river runoff will change significantly, but the magnitude and direction of these changes are uncertain.

The exercise

The group was divided up into three subgroups and each subgroup was asked to analyse two cases (one socio-economic scenario combined with the two climate scenarios) by answering the following questions:

1. What will be the combined impacts of the projected socio-economic change and climate change on human safety, economic development and ecological integrity?

Table 1. The main priorities identified in the group's scenario analysis

	Socio-economic scenario I	Socio-economic scenario II	Socio-economic scenario III
Human safety	• Maintain safety standards • Adaptive management constrained by need for emergency planning	• Hard engineering, high but affordable costs • Soft engineering in rural areas • Water buffer capacity in catchment • Good flood forecasting and emergency response	• Limited to urban centres • Reduced safety standards in rural areas • Policy of managed retreat, enhancing natural buffers
Economic development	• Uncertainty impedes development • Trade-off with ecological integrity	• Strong spatial planning	• Readjustment of agriculture to increased flood risk • Concentration in urban areas
Ecological integrity	• Threatened without management intervention • Could contribute to safety but uncertainty is large	• Supports recreation and amenities • Spatially reduced • Uncertain opportunities for wetland development	• Opportunities for improvement • Risk of exposure to historic pollution

2. Given the current state of scientific knowledge, what management strategies do you propose to address these impacts and ensure human safety, economic development and ecological integrity?
3. How can the current knowledge base for management be improved and what benefits will this bring in terms of ensuring human safety, economic development and ecological integrity?
4. What are the similarities and differences between these two cases and the other four?

The results of the exercise are presented in Table 1.

In the situation of high development (socio-economic scenario I), the proposed strategy is to locate the high value assets in concentrated regions and protect those regions with hard defences at the cost of sacrificing some of soft defences (e.g. polders). By encouraging marshes to grow in front of the sea walls, some ecological integrity can be maintained as well as providing a system of wave damping, but in this situation, it is clear that economic development and human safety are the first priorities at the cost of ecological integrity. Inland, upstream management practices such as the construction of above ground water storage and drainage systems that allow for controlled flow, as well as planning actions that are focussed towards tree planting and wetland development will increase the storage capacities in the watershed. These actions will augment the buffering capacity of the system and reduce the impacts of storms on the floodplains. One of the main differences between a fairly certain future climate and an uncertain future climate is that the costs will be higher in the uncertain situation as a much more interventionist design will be required. However, the high economic growth in this scenario means that society can pay for this highly interventionist style of coastal management.

In the second scenario, the less wealthy economy requires that there is a managed trade off between agriculture and ecological integrity. One strategy of dealing with this is to raise dunes and dikes to dissipate wind and wave energy, thereby using natural systems to create barriers against the sea. This requires, as in the first scenario, more knowledge in terms of future storm characteristics and then the marriage of this with coastal engineering to provide a safe solution.

In the third socio-economic scenario, that of lowest economic development but maintaining the highest ecological value, a widespread policy of managed retreat is proposed. This retreat would result in the reallocation of the populace into either urban areas or to regions above the flood line. The centralisation of population around urban areas of economic interest i.e. ports that are protected by hard defences, will allow for a relaxation of protection in other areas. This will result in shifts in agricultural practices in the coastal regions towards the use of islands, refuges, and towards a flood-adapted agriculture (as might have been found in Europe several centuries or more ago). Thus a high degree of ecological integrity will be conserved and this will enhance natural buffers. This third scenario, where funds are limited to deal with sea defences, means that there is most need to accept that there will be a risk of flooding and that the magnitude of risk will be dependant on the area characteristics.

The uncertainty in predictions of future situations means that any plan must be flexible. This means that the use of adaptive management strategies is imperative and therefore, more research on the behaviour of natural systems under climate change is essential. This can only be achieved by creating the knowledge base required to develop a more integrated understanding of functioning of the coastal zone. Furthermore, careful and controlled land use planning is critical to the success of any of these plans and it is evident that any planning action also needs to be flexible enough to adapt to the changing future. The development of a rapid feedback system that allows for adaptability within management plans will achieve this goal, a goal that is even more important in situations with a high degree of uncertainty where the degree of adaptability required is higher. Feedback systems need to be based on a rigorous monitoring, with an inbuilt feedback loop, which will allow for 'learning from mistakes'. Related to this is the requirement for the development of an efficient flood forecasting and warning system, including evacuation plans.

Exercises like this can result in the development of a series of priorities that can be tailored to a particular combination of requirements, including economic development, human safety and ecological integrity. The following types of priorities can be put forward as management options that are linked to the needs and resources of a particular situation (Table 1). By this means the scale of potential impacts can be evaluated and adaptation strategies identified that are flexible enough to deal with the outcomes of several scenarios. Thus, an insight is gained on a flexible way of accommodating uncertainty, as well as indicating future policy and research needs.

Managing the dynamic coast

The concept of managing a dynamic coast is inherent in each of the three scenarios outlined in Table 1. Traditionally, in coastal zone management terms, the coast has been viewed as a fixed feature and this has lead to "coastal squeeze" (French 1997, Parry 2000, Nicholls and Klein, this volume). As sea level rises, there is encroachment on to land, this causes change and leads to the development of new environments. However, if this encroachment is halted for one reason or another on the landward side, such as the construction of sea walls or buildings, the coastal zone becomes 'squeezed' between the rising sea and the human environment. This results in loss of habitat and a decline in ecological integrity. Sea level rise is not the only possible driver of coastal squeeze – sediment starvation and erosion area others, as all three induce a 'squeezing' when coastal migration landward is blocked. As previously noted, we need to prepare for a future in which climate changes at an increasingly rapid rate and so it is more appropriate that we move towards a framework that allows a dynamic definition of the coast. The more variability that is allowed within coastal management, the more we will be able to cope with climate-induced change in the coastal zone. Through the maintenance and enhancement of natural buffer zones, coastal squeeze can be reduced and even avoided (Hamm et al. 2002, Hansen et al. 2002). This can be achieved through policies of "set-back" where areas are created within which it is forbidden to do certain things, such as construct buildings. Secondly, just as we should not examine the coastal zone in isolation we also should not propose management actions that do not consider the whole ecosystem. Indeed, management practices in the coastal zone need to take into account the entire ecosystem, including the interests of those using, and influenced by, the river itself, the catchment (including the flood plain), as well as the marine part of the coastal zone. It is clear that a better understanding of the system will lead to a better coastal management which will in turn lead to a better understanding of the response of the system to stressors, thus creating a feed back loop that will enable managers to adapt their strategies to better manage the ecosystem.

Tools – analytical, management, and communication

To arrive at the solutions proposed in the previous sections, it is essential that we develop analytical and management tools that will enable us to attain a balance between the three priorities of human safety, sustainable development and ecological integrity. The problems of climate change are already manifest in the coastal zone of Europe and so it is necessary that we start now to deal with the problems. Clearly, we need to develop pro-active strategies to enable us to make the link between climate change and coastal zone management. This involves a fundamental change in ideology, leading to a translation of results from basic scientific research into results that can be applied in a management perspective.

The task of identifying which tools are required to accomplish this goal is complicated by the fact that different types of tools will be needed in accordance with the type of ecosystem to be managed. Furthermore, although, climate change is a global problem and requires a global solution, the impacts of climate change also occur at the regional, national and local scales. This creates a problem of scale that needs to be dealt with when proposing management options. It is clear that changes are required in international policy as well as at the national and local level and that there is a need for policy development. The Group identified four priorities:

1. Develop more integrative models that couple existing ecosystem models to hydrodynamic and geomorphological models. These coupled physical and biological ecosystem models should also include ecosystem resilience;
2. Integrate social science and natural science to address the long term economic impacts of climate change in the coastal zone. Development of predictive economics models are needed to provide robust methods of predicting the future in terms of economic and physical drivers;
3. Develop management approaches that are flexible enough to deal with uncertainty, whilst still maintaining a good degree of risk analysis resulting in the creation of more robust coastal systems;
4. Create a management system that involves all stakeholders in the decision making process.

There is also clear need to quantify risk and to communicate that risk to the public. In today's electronic, information oriented societies, the use of the media is essential to convey the message of researchers and managers to the general public. The general public are broadly aware of the threat of climate change and sea-level rise to the coastal environment. Using the media to aid in the visualisation of the problem will make it easier for the public to develop a deeper understanding and accept the potentially stringent measures that need to be taken. By doing this, information can be presented in a more accessible way that will allow the public to make more informed decisions. Indeed, one of the problems of accurate reporting lies in the uncertainty of the magnitude of the impacts of climate change on the coastal zone. However, it is possible to incorporate this uncertainty into reports of future expectations as has been very effectively demonstrated by meteorologists. The general public accepts that there is some degree of uncertainty in the prediction of the weather but nonetheless accept the general predictions and are willing to change their habits accordingly. However, to achieve this level of trust, we need to have strong predictability to get the public to believe in what we say. Alternatively we can assure public confidence and, thus, have their acceptance by finding a means to state the uncertainty more clearly. Effectively, this means that we as researchers need to find effective ways of presenting our results and projections in a "user friendly" manner. A suggestion could be found on the Tyndall site (e.g. http://www.tyndall.ac.uk/publications/fact_sheets/fact_sheets.shtml).

The role of the public in the decision making process is also important and the utilisation of the media is also an integral requirement for allowing decision-making processes to switch from a heavily top-down situation to a more integrated situation. Clearly, the media can be very important in determining the reaction of the public to management propositions and a well-informed local stakeholder will be more able to interact effectively with coastal management decision-making than a less informed stakeholder. This subject is discussed in more depth in Lise et al. (this volume). Briefly, a fuller participatory approach that switches from a wholly top-down system to a system balancing both inputs from stakeholders as well as central government is required. However, because climate change is a process occurring on long time scales (decades or longer), in addition to public participation, there is also a requirement for strong stewardship from government to balance short-term concerns.

Integrated assessment is another tool that is important for the development of a holistic strategy in the coastal zone. This is discussed in greater detail in both Nunneri et al. (this volume) and Turner (this volume).

Proactive strategies

From the previous sections it is clear that in order to effectively manage the coastal zone in the future we need to adopt a proactive approach that prepares for the significant but uncertain changes that are expected. This proactive policy needs to be based on feedback that allows for continued review and correction. It is also clear that because of the nature of the problems faced by coastal management in response to climate change that there is a requirement for a long term strategy incorporating strong planning, including land use.

The following idealised scheme is developed, with hindsight, from experiences of the Environment Agency's (UK) development of a long-term strategy for the investment in flood defences for the Humber estuary in North East England. In reality progress was less ordered and more complex than presented here, with a number of iterations. Some of the feedbacks are omitted for simplicity. There are no doubt other successful variations on this scheme and a number of the phases may be undertaken, at least partially, in parallel.

The estuary has the UK's largest port complex, an extensive floodplain, in which are located the homes of over 300,000 people and much industry, tourist and recreation facilities, and many archaeological and historic features. The estuary has developing fisheries and is of outstanding importance for wildlife (particularly waterfowl), which are protected under national and international legislation. The project on which this scheme is based has one principal organisation; it is led and funded by the Environment Agency (http://www.environment-agency.gov.uk), which has the responsibility for providing protection from tidal flooding. The stakeholders are involved in a "soft" partnership, which has a largely advisory role; they have no legal or financial commitment and are not responsible for the implementation. It is, however, wise to have agreed terms of reference and working procedures for such stakeholder steering group and similar fora.

The Humber is a large and complex estuary and its management is made more complicated by the fact that no single organisation is responsible for integrating all aspects of social, economic and environmental management. The Humber Strategies Manager represents the Environment Agency on the steering groups of projects led by other organisations (i.e. this person is a member of "soft" partnerships run by other "principals"). These partnerships are mostly to do with economic development and regeneration. An approach is, thus, developing whereby working together ensures that single-purpose plans fit together as far as possible into a clear, simple and integrated framework. An industry and nature conservation body (Humber INCA) has also been set up to encourage companies to help them to meet their environmental obligations and encourage environmental enhancements projects.

Other partnerships may be more difficult to develop. The UK Habitats Regulations 1994 require "relevant authorities" (organisations with statutory powers to manage activities within a European Marine Sites) to prepare a single management scheme for day-to-day activities (rather than plans or projects such as the shoreline management plan). These activities include flood defence maintenance works, recreation, fishing and dredging of navigation channels. Within the Humber basin there are 39 "relevant authorities". These include statutory agencies, local authorities, drainage boards, the Ministry of Defence, and water, harbour and navigation companies. This makes the "scoping" or "forming" phase more complicated as firm agreements need to be draw up for the partners, including on funding, project management, employment of project officers and/or consultants, and a disputes procedure. The responsibility for implementing the Humber Scheme is the task of individual authorities, although there will be collective monitoring of the results. In the case of the Humber, the relevant authorities have set up to the independent Humber Advisory Group to guide them and provide a consultative forum. The members of this "soft" partnership are representatives of estuary interests. It should be noted that there are examples of environmental enhancement projects based on hard "partnerships", which involve various combinations of public, private and voluntary sector organisations.

Partnership is a form of team working. Management theory identifies the following stages in team development:

- Forming: initial getting together;
- Storming: infighting for power, conflicting objectives, lack of clarity on roles and disagreement on how the partnership should work;
- Norming: recognition that progress will only be made by a more collaborative approach, constructive working out of shared objectives, roles and procedures; leading to:
- Performing: increasingly effective teaming working which delivers the outcomes.

An established partnership does, however, often require maintenance if it is not to get stale or complacent. It may slip back to the "storming" stage and then need to "norm" again. Risk management, quality assurance with an independent element external to the team, a good disputes procedure and judicious coaching by an experienced chair or other team member all help.

In several projects, "storming" or a dispute amongst stakeholders has led to the "forming" of a partnership, which has consequently passed smoothly to the "norming" and "performing" phases.

Improving the knowledge base

The previous sections present the results of the group discussions centred round the three questions posed by Nicholls and Klein (this volume). The focus of these questions was on how to reconcile the human and natural values of an ecosystem while considering coastal zone management under situation of climate change; on what tools and methods are available now and what is still required to effectively manage the coastal zone in the future, and on what proactive strategies can be implemented to achieve the management goals of providing human safety and promoting economic development without compromising ecological integrity.

It was clear from the discussion that:

- Europe's coasts are vulnerable habitats at risk from urban development, tourism and other economic activities. Climate change will exacerbate these threats;
- We must take climate change seriously and start to adapt now.

Climate change is inevitable and must be incorporated into any coastal zone management plan. From this standpoint the group then provided some suggestions on what strategies are needed to manage the coast in the future and what is needed to achieve the knowledge base necessary in order to attain those management goals.

- A long-term approach is necessary for managing the coastal zone although the prediction of climate and socio-economic change is a very uncertain process;
- Scenario analysis provides a powerful tool for examining the response to long-term but uncertain climate and other pressures (see Turner, this volume; Nunneri et al., this volume);
- Adaptive strategies should be based on sound science and a good knowledge of local circumstances with stakeholder and public involved incorporated into the process at the outset;
- Techniques such as cost-benefit analysis and multi-criteria analysis continue to be developed and provide valuable tools for strategy development and the assessment of options;
- There is a range of sustainable management measures that should be implemented to help adapt to the uncertainties of potential change in a robust and flexible manner so that human safety, economic well-being and environmental resources are all safeguarded;
- There is need for good environmental monitoring systems, targeted research and the willingness to adopt innovative solutions.

Empirical information on coastal adaptation to climate change is still scarce, although there is much more experience of adapting to climate variability (e.g. Klein et al. 2001; Tol et al. 2004). Continued impact and adaptation assessment, combined with fundamental research on coastal system response and economic, institutional, legal and socio-cultural aspects of adaptation, are required to understand which adaptation options might be most appropriate and most effectively implemented. In the past, many of Europe's coastal regions had extensive wetlands, dunes and related environments that acted to naturally regulate variations in water flow and nutrient loadings (e.g. Jickells et al. 2000; Boesch 2002). The natural sink and buffer capacities of these environments should be exploited within coastal management to ensure the ecological integrity, economic development and human safety concerns in the coastal zone (e.g. Nicholls and Branson 1998; Visser and Misdorp 1998). A holistic attitude, involving integrated modelling that incorporates both natural and social sciences, is also necessary to deal with the uncertainties of the future. Spatially explicit models of river basins that incorporate economic drivers and costs are already available (e.g., Costanza et al. 2002) and the development of coastal models of this type that are specific to the European context would be invaluable. Thus, the maintenance and enhancement of natural sinks and buffers in coastal zones combined with greater research into their effectiveness and limitations in ensuring human safety, economic development and ecological integrity, is clearly a necessary future goal.

References

Allen JRL, Pye K (1992) Saltmarshes: morphodynamics, conservation and engineering problems. Cambridge University Press, Cambridge

Allen JRL (2000) Morphodynamics of Holocene salt marshes: a review sketch from the Atlantic and Southern North Sea coasts of Europe. Quat Sci Rev 19:1155-1231

Boesch DF (2002) Challenges and opportunities for science in reducing nutrient over-enrichment of coastal ecosystems. Estuaries 25:886-900

Carter RWG (1988) Coastal environments—an introduction to the physical, ecological and cultural systems of coastlines. Academic Press, London

Cohen JE, Small C, Mellinger A, Gallup J, Sachs J (1997) Estimates of coastal populations. Science 278:1211–1212

Costanza R, Voinov A, Boumans R, Maxwell T, Villa F, Waigner L, Voinov H. (2002) Integrated ecological economic modelling of the Patuxent River watershed. Maryland. Ecol Monogr 72:203-231

French PW (1997) Coastal and estuarine management. Routledge, London

Hamm L, Capobianco M, Dette HH, Lechuga A, Spanhoff R, Stive MJF (2002) A summary of European experience with shore nourishment. Coast Eng 47:237–264

Hanson H, Brampton A, Capobianco M, Dette HH, Hamm L, Laustrup C, Lechuga A, Spanhoff R (2002) Beach nourishment projects, practices, and objectives—a European overview. Coast Eng 47:81-111

Holligan P, deBoois H (1993) Land-Ocean Interactions in the Coastal Zone (LOICZ). Science Plan. International Geosphere Biosphere Programme, International Council of Scientific Unions, Stockholm

Houghton JT, Ding Y, Griggs DJ, Noguer M, Van der Linden PJ, Xiaosu D (2001) Climate Change 2001 – The Scientific Basis. Cambridge University Press, Cambridge

IGBP-LOICZ (2003) LOICZ Future – Beyond 2002. Working document – draft version #10. International Geosphere Biosphere Programme, Land-Ocean Interaction in the Coastal Zone Project. (http://www.nioz.nl/loicz/)

Jickells T, Andrews J, Samways G, Sanders R, Malcolm S, Sivyer D, Parker R, Nedwell D, Trimmer M, Ridgway J (2000) Nutrient fluxes through the Humber estuary - past, present and future. Ambio 29:130-135

Klein RJT, Nicholls RJ (1999) Assessment of coastal vulnerability to climate change. Ambio 28:182-187

Klein RJT, Nicholls RJ, Ragoonaden S, Capobianco M, Aston J, Buckley EN (2001) Technological options for adaptation to climate change in coastal zones. J Coast Res 17:: 531–543

Klein RJT, Nicholls RJ, Thomalla F (2003) The Resilience Of Coastal Megacities To Weather Related Hazards. In: Kreimer A, Arnold M, Carlin A (eds) Building safer cities: the future of disaster risk. Disaster Risk Management Series No. 3. The World Bank, Washington, pp 101-120

Lise W, Timmerman J, Vermaat JE, O'Riordan T, Edwards T, De Bruin EFLM, Kontogianni AD, Barrett K, Bresser THM, Rochelle-Newall E (2004) Group report: Institutional and capacity requirements for implementation of the Water Framework Directory. In: Vermaat JE, Bouwer LM, Salomons W, Turner RK (eds) Managing European coasts: past, present and future. Springer, Berlin

McCarthy JJ, Canziani OF, Leary NA, Dokken DJ, White KJ (2001) Climate change 2001 – impacts, adaptation and vulnerability. Cambridge University Press, Cambridge

McLean R, Tsyban A, Burkett V, Codignotto JO, Forbes DL, Mimura N, Beamish RJ, Ittekkot V (2001) Coastal zone and marine ecosystems. In McCarthy, JJ, Canziani, OF, Leary, NA, Dokken DJ and White, KS (eds) Climate change 2001: impacts, adaptation and vulnerability, Cambridge University Press, Cambridge, pp 343-380

Nakicenovic N, Swart S (2000) Emissions scenarios – a special report of working group III of the intergovernmental panel on Climate Change. Cambridge University Press, Cambridge

Nicholls RJ (1995) Coastal megacities and climate change. GeoJournal 37:369–379

Nicholls RJ, Branson J (1998) Enhancing coastal resilience: planning for an uncertain future. Geogr J 164:255-343

Nicholls RJ, Lowe JA (2004) Benefits of climate mitigation for coastal areas. Global Environ Change, under review

Nicholls RJ, Klein RJT (2004) Climate change and coastal management on Europe's coast. In: Vermaat JE, Bouwer LM, Salomons W, Turner RK (eds) Managing European coasts: past, present and future. Springer, Berlin

Nunneri C, Turner RK, Cieslak A, Kannen A, Klein RJT, Ledoux L, Marquenie JM, Mee LD, Moncheva S, Nicholls RJ, Salomons W, Sardá R, Stive MJF, Vellinga T (2004) Group report: integrated assessment and future scenarios for the coast. In: Vermaat JE, Bouwer LM, Salomons W, Turner RK (eds) Managing European coasts: past, present and future. Springer, Berlin

Parry M (2000) Assessment of potential effects and adaptations for climate change in Europe. Jackson Environment Institute, University of East Anglia, Norwich

Small C, Nicholls RJ (2003) A global analysis of human settlement in coastal zones. J Coast Res 19:584-599

Tol RSJ, Klein RJT, Nicholls RJ (2004) Towards successful adaptation to sea level rise along Europe's coasts. J Coast Res, accepted

Turner RK, Subak S, Adger WN (1996) Pressures, trends, and impacts in coastal zones: interactions between socioeconomic and natural systems. Environ Manage 20:159-173

Turner, RK (2004) Integrated environmental assessment and coastal futures. In: Vermaat JE, Bouwer LM, Salomons W, Turner RK (eds) Managing European coasts: past, present and future. Springer, Berlin

UNEP (2002) Global Environment Outlook 3. Earthscan, London

Visser, J. Misdorp R (1998) Coastal dynamic lowlands – the role of water in the development of The Netherlands: past present and future. J Coast Conserv 4:105-168

WCC'93 (1994) Preparing to meet the coastal challenges of the 21st century. Report of the World Coast Conference, Noordwijk, 1-5 November 1993. Ministry of Transport, Public Works and Water Management, The Hague

14. Integrated environmental assessment and coastal futures

Abstract

This forward look analysis aims to shed light on a number of locally specific and more generic problems, which are likely to arise in European coastal and regional sea areas. The methodology deployed utilizes the DP-S-I-R scoping framework to identify significant environmental change driving pressures and their ecological economic and socio-political consequences. Future scenario analysis is then used to indicate the likely outcomes and policy options available, given the high degree of uncertainty present.

Introduction

In global terms, 60% of the population lives within a zone 100km wide along the seashore and many more within the drainage basins of the coastal seas. It appears that coastal-catchment economic development together with ongoing protection measures have grown out of control, and the consequent degradation or destruction of the coastal environment continues to increase. European countries are not immune from the consequences of this global change process. Interrelated global driving pressures such as, for example, urbanization, industrial development and mass tourism together with anthropogenically induced climate change, impact on regional and local resource systems with consequent local (yet generalised) management problems. However, while the problems have wider applicability they will often require co-ordinated responses by policy makers at the national or supernational scale (i.e. EU and beyond). The problem faced by policymakers is how to regain control and mitigate resource degradation to better conserve environmental systems and the socioeconomic activity that depends upon them. The

[1] Correspondence to Kerry Turner: r.k.turner@uea.ac.uk

J.E. Vermaat et al. (Eds.): Managing European Coasts: Past, Present, and Future, pp. 255–270, 2005.
© Springer-Verlag Berlin Heidelberg 2005.

future sustainable management of coastal resources is therefore an important policy goal for all governments of European countries with coastlines.

In any forward look it will be necessary to anticipate generic problems likely to arise or intensify within the established member countries of the EU, but also to recognize those relevant to the new accession states (NASs). Both the underpinning 'science' of these problems as well as policy and management options need joint multidisciplinary investigation. A number of general problem policy contexts can be highlighted:

- The future impacts of trade and economic development including, for example, increased dredging activities (both geomorphological and biogeochemical impacts) and the welfare consequences for natural and social systems; and the spread of invasive exotic species into local ecosystems.
- The environmental change impacts on fisheries and the implications of more extensive and intensive aquaculture developments.
- The need for and consequences of future coastal protection and sea defence systems and strategies (both immediately at the coast and along estuaries and into freshwater fluvial catchments).
- Water quality deterioration, future monitoring, evaluation and policy responses.
- Degradation and/or destruction of a range of natural habitats and ecosystems and policy responses.

Three particular characteristics of coastal zones complicate the management task: the extreme variability present in coastal systems, the highly diverse nature of such systems, and their multi-functionality and consequent high economic value. Coastal zones are important economic zones supporting livelihoods through flows of income derived from the utilization of the in situ natural capital stock and through global trading networks. However, simultaneously coastal areas are socio-cultural entities, with specific historical conditions and symbolic significances (Turner et al, 2001). This adds to their 'value' but is also problematic because their institutional domains can have transnational boundaries, which cross national jurisdictions and require international agreements and legislation. Institutions and regimes struggle because they are often not coincident with the spatial and temporal scales and susceptibility of biogeochemical and physical processes (the so-called scaling mismatch problem, Von Bodungen and Turner 2001). The resource management task is further compounded by the existence of multiple stakeholder interests and competing resource uses and values typically found in coastal zones.

Understanding the interactions between the coastal-catchment zones and environmental change cannot be achieved solely by natural science modelling and observational studies. Modelling and analysis of socio-economic, socio-cultural and political process is also a vital component of any decision-support system aiming to buttress coastal management institutions and practice. Integrated natural and social science must be the ultimate goal as advocated at the Earth summit in 1992 and reiterated in South Africa in the 2002 sustainability conference.

The principal objective of sustainable and integrated coastal management can be portrayed as the sustainable utilisation of the multiple goods and services (consumptive and non-consumptive uses and values) generated by coastal resources, together with the 'socially equitable' distribution of welfare gains and losses inherent in such usages. This social welfare accounting process is essential for designing sustainable development policies, practices and institutions. It presents a number of challenges: scientific understanding; evaluation of the environmental change process; and political acceptability in terms of greater inclusion/participation of stakeholders in the change management process.

In order to help resolve some of the methodological issues involved in coastal management, the so-called Driver Pressure State Impact Response (DP-S-I-R) framework can initially be deployed. Its usefulness as a scoping device, when combined with scenarios, has already been proved on the basis of work undertaken in different European coastal-catchments. The aim should be to set out a more complete methodology for integrated coastal management, together with an appreciation of the practical problems and constraints faced as such methods are actually deployed.

The requirements of a practical coastal management support system will include: databases, indicators and monitoring measures, scenarios, models and evaluation methods and techniques and will all pose their own problems. There is an 'interface' problem, when science and scientific understanding merges (or not) into decision making in the political arena. Questions posed by the policy process often need to be redefined before science can provide useful answers and some existing scientific knowledge is too specific to be of relevance to policymakers. Finally, it remains the case that there are some questions that science cannot yet answer. Policy targets and indicators then become very important but far from straightforward issues to address given the sustainability principles of:

- Economic and ecological efficiency (including the cost-benefit and polluter pays principles);
- Equity and fairness principle (including more inclusionary decision making and the subsidiarity principle); and
- Precaution, given the inevitable uncertainties that persist in coastal science and policy domains.

Embedded within the management process is the resource valuation problem per se. A range of valuation methods and techniques has been examined and their place and contribution to the overall policy analysis procedure has been assessed (Turner et al. 2001). To be judged an ultimate success, integrated coastal management as a process must unite government, civil communities, science and management and overcome or mitigate competing sectoral and public interests. It should interalia improve the quality of human communities who depend on coastal resources while maintaining the biological diversity and productivity of coastal-catchment ecosystems, and therefore the functioning of nature. Such a task can only be achieved incrementally over time and will be constantly challenged by complexity and uncertainty constraints. The future will always be shrouded by uncertainty and therefore accurate forecasting of coastal futures is not a feasible goal. However, it is possible to formulate scenarios, which can shed light on and offer insights into possible future environmental and socio-economic developments. The information generated by such 'futures thinking' can assist the policy process in a more efficient and effective search for appropriate projects, programmes and policies.

This paper first provides an outline of the main stages of an integrated environmental assessment process for coastal-catchment areas and in particular focuses on the formulation and deployment of 'futures' scenarios as a vital component of 21st century coastal management. Secondly, it uses the DP-S-I-R scoping device to highlight environmental change pressures, impacts and consequences for European coasts. Finally, a synthetic approach to scenario formulation and deployment is demonstrated in the European futures context.

Integrated environmental assessment

Managing resources across the catchment-coastal spatial scale and with long run sustainability policy objectives in mind, requires an appreciation of the full functioning of hydrological, ecological and other systems, together with the total range of valuable functions and functional outputs of goods and services provided. The management strategy and process must be underpinned by a 'valid' and 'reliable' decision support system i.e. a toolbox of methods and techniques encompassed by an analytical framework. A number of basic stages can be defined (Figure 1):

- Scoping and auditing stage – to set the management issues and problem in a proper context; the DP-S-I-R framework fulfils this role (Von Bodungen and Turner 2001);
- Identification and selection of complementary analytical methods and techniques – given the spatial and temporal scale issues involved and the mismatches between scales when science and policy issues are combined, the analysis is a non-trivial task requiring GIS, coupled natural science models, economic analysis, institutional and stakeholder analysis and scenario formulation and utilization;
- Data collecting, analysis and monitoring, supported by environmental change indicators and future environmental change scenarios; and
- Evaluation of project, policy or programme options – using methods such as cost-effectiveness, cost-benefit analysis and/or multi-criteria assessment.

The following sections set out a DP-S-I-R scoping exercise for Europe's coasts and relevant catchments, as well as a futures scenario exercise in order to highlight potential future problems for the original EU and its extended version and possible policy response options.

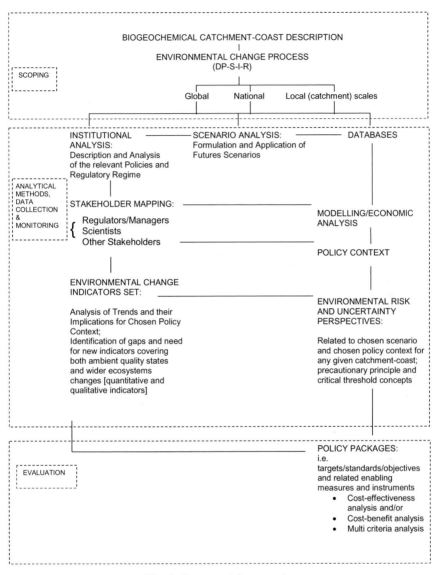

Fig. 1. Conceptual framework

DP-S-I-R analysis

The most important developments in Europe since the 1970s are the socio-economic and political institutional processes resulting from the expansion of the EU and the transition from centrally planned economics and societies in Central and Eastern Europe to more open market economics and societies already dominant in Western Europe. Thus although the three sub-regions of Europe – Western, Central & Eastern (out to the Caucasus and Central Asia) - have similarities there are also distinct and significant differences.

All the countries in the European region (i.e. EU countries and the 12 countries of Eastern Europe, the Caucasus & Central Asia) are under environmental pressure from generic driving forces within the globalisation process. But the impacts of these environmental changes are not uniformly intense and policy responses are also non-uniform across Europe. While all countries face the 21st century challenge of sustainable economic development only limited progress has so far been made in terms of more efficient resource utilisation (EEA 2003). While gross domestic products (GDPs) are no longer tied inexorably to increasing resource use (at least in Western Europe) the absolute level of material use is still probably unsustainable given foreseeable technological advances. The countries of Central and Eastern Europe face the additional dilemma of achieving Western European levels of socio-economic welfare via a growth process that needs to be constrained by sustainability principles and objectives. While per capita GDP grew on average from $9000 in 1972 to $13,500 in 1999, wide variations exist across Europe. In Western Europe the average was $25,441 in 1999 but only $3,139 in Central Europe and $771 in Eastern Europe.

Europe's population has grown by 100 million since 1972 and totalled some 818 million in 2000 (13.5% of the global total). The most significant demographic trends in Europe are the overall ageing of the population, the increasing number of households and the increasing mobility of people throughout Europe.

At the generic level, all European countries are engaged in growing and extensive trading activities. This is putting severe pressure on 'local' coastal areas, as well as possibly shifting other environmental cost burdens to other regions around the globe. Imports currently constitute around 40% of the total material requirement of the EU and they grew rapidly in the 1990s. Maritime transport increased by 35% in the EU between 1975-85 but has since levelled off. Nevertheless, it accounts for 10-15% of total SO_2 emissions and in the Mediterranean oil spill and related risks are high because of the volume of traffic. Some 30% of all merchant shipping and 20% of global oil shipping crosses that sea every year. Ports and associated industrial development are responsible for land conversion/reclamation, loss of intertidal and other habitats, dredging and contaminated sediment disposal, increased flood protection measures and also play an enabling role in the spread of invasive exotic species and 'local' biodiversity loss. The latter problem is a complicated one because of the causal mechanism comprising the introduction of double hull vessels, climate change and the flouting of legislation designed to protect local environments from illegal 'ballast' water dumping.

The increasing physical growth of the European economies manifests itself, among other ways, in massive new construction of buildings and infrastructure.

The spread of the built environment is having profound effects on catchment-wide processes, leading to increased flood risk, changes in sediment fluxes (and contamination risks) as well as habitat and biodiversity loss in the catchment-coast continuum. Direct physical alteration and destruction of habitats because of 'development pressure' is probably the most important single threat to the coastal environment (GESAMP 2001). But human-induced changes in the natural flow of sediments are also a contributory threat to coastal habitats. Deltaic areas such as the Po, Rhone and Ebro have suffered from sediment starvation as hydrological changes in catchments (dams etc) have cut sediment supplies. The built environment expansion has multiple causes, but two factors are especially relevant for coastal areas, transport and tourism.

Road transport is the dominant mode in Western Europe, not least because, despite tax escalators, fuel for road transport remains perennially cheap and fuel–efficiency provisions for vehicles remain voluntary. Rail networks are somewhat better developed in Central and Eastern Europe, but the 'lure' of the car/lorry poses a strong threat to any future transport strategy. Transport infrastructure and trade are strongly linked and so coastal habitats and ecosystems face fragmentation along with other areas close to main arterial transport routes. The official European investment agencies have biased their lending heavily in favour of favour of road projects e.g. between 1992-2000 the European Investment Bank gave 50% of its loans to road projects and only 14% to rail (EEA 2003).

Europe's coasts host around 66% of the total tourist trade, and in the Mediterranean, for example, arrivals are expected to grow from 135 million per annum in 1990 to as many as 353 million by 2025. Tourism's main environmental impacts are also generated via transport requirements, together with use of water and land, energy demands and waste generation. In some popular Southern European localities irreversible environmental degradation has probably already been inflicted. Tourism-related demand for passenger transport (pre 'nine eleven') has grown remorselessly, arrivals in South Western Europe grew by 91% 1985-2000. The environmental impacts are highly concentrated and seasonal within or close to resort areas. But lateral expansion along coastlines is also a common phenomenon and the construction of second homes (the Mediterranean and Baltic areas are prime examples) is a particular concern.

Europe's semi-enclosed and enclosed seas (with their limited water exchange) are particularly sensitive to pollution threats. Marine and coastal eutrophication from elevated nitrogen levels (riverine transport and atmospheric deposition) quickly emerged as a worrying trend, the impacts of which were exacerbated by the loss of natural interceptors such as coastal wetlands. Severe eutrophication has occurred in the Black Sea and in more limited areas in the Baltic and Mediterranean. A majority of European countries have made significant progress in combating point-source pollution of watercourses, estuaries and coasts (e.g. sewage treatment plants and industrial facilities). So discharges of heavy metal and organic substances into the North East Atlantic fell during the 1990s. Nitrate concentrations fell by 25% in the Baltic and North East Atlantic over the period 1985-98 and phosphate concentrations also fell in North Sea areas. Waste water treatment levels and discharges are still problematic in the Mediterranean, Adriatic and Black Sea areas. However, the issue of diffuse pollution at the catchment scale and beyond is a continuing threat. Agricultural activities and run-off are one of the contributors to this problem. Overall consumption of fertilizers in Europe has stabilised in recent years following a fall in the early 1990s in Central and Eastern Europe. Current levels of fertilizer and pesticide use are probably not environmentally sustainable and measures such as integrated crop management need to be adopted more widely (3% of the utilised agricultural area of Europe is under such integrated management). Irrigated cropland retains a significant share of the agricultural areas in Western, Central and Eastern Europe, ranging from 11% to 18% respectively. Irrigated land continues to expand in some Western European and Mediterranean areas. This type of production has serious water resource implications and also poses a major threat to wetlands. The Aral Sea catchment serves as an extreme warning as to what processes could unfold. Although the threats are varied some 31% of Europe's population now lives in countries that use more than 20% of their annual water resource (EEA 2003).

Most of the capture fisheries of Europe have now been over exploited and other substitute populations also denuded. But while fisheries production has declined, marine aquaculture in Western Europe has increased significantly. This development threatens the nutrient status of receiving waters and remaining wild populations if controls are not imposed.

Climate change is expected to cause significant impacts across Europe, but the south and the European Artic are possibly the most vulnerable areas (with the caveat that some 'local' areas are especially at risk for sea level rise e.g. The Netherlands and South East England) (Parry 2000). All areas face hydrological and water resource risk increases, which may then impact ecosystems and biodiversity, as well as human health. If the modelling predictions about changes in marine water circulation patterns turn out to be correct then there are significant negative implications for regional seas in terms of eutrophication and contamination risks.

The Caspian Sea provides an alarming example of the combined negative effect of temperature and precipitation change. It has recently risen by 2.5m causing severe flooding and damage costs. It is predicted that by 2020-40 an additional 1.2 - 1.5m increase is possible (EEA 2003). Europe's forests can play an important role in any climate change mitigation policy. The total area of forest in Europe is increasing and there is a future opportunity to diversify its service functions in order to provide watershed protection from soil erosion and floods and excess sedimentation, as well as carbon sequestration, recreation and nature conservation benefits. This also serves to reinforce the point that coastal zone management requires an appreciation of measures deployed within the relevant catchments if it is to be effective.

The concept of a more integrated coastal management approach has been advocated for more than a decade but so far full adoption has not been practised anywhere. In Western Europe, awareness has been raised but sectoral policies have not been radically modified, let alone integrated. More generally, only 15% of Western Europe is under national designation for nature conservation and 9% or less elsewhere in Europe. Table 1 presents an overview of a DP-S-I-R scoping exercise for European coastal areas. It remains the case that 85% of Europe's coast face moderate to high risk from economic development – related pressures and some 25% of the coastline is subject to erosion. Given these erosional trends and flood risks, hard engineering sea defences have been the traditional response. But these defences also serve to reduce sediment input to the coastal system, which intensifies erosion and the need for additional defences. Armouring the coastline in this way is essentially unsustainable on economic cost grounds alone.

Scenarios and coastal zone management

A scenario can be defined as a coherent, internally consistent and plausible description of a possible future state of the world (Parry 2003).

It needs to be emphasized that a scenario is not a forecast because it cannot assign probabilities to any particular outcome. Instead, scenarios portray images of how society and its supporting environment could look like given different sets of assumptions and consequent conditions. Early scenario planning was undertaken by various military agencies. This strategic approach was then adopted by multinational companies, such as oil companies in the 1970s, seeking to improve their decision making. The implicit rationale seemed to be to evolve better procedures for coping with future 'surprises', by forcing analysts to think laterally and radically (Hammond 1998). Scenarios typically contain qualitative storylines augmented by varying amounts of quantified data. They can be informed by relevant history but not conditioned by it, except in the case of so-called baseline or 'business as usual (BAU)' scenarios. The latter can be utilized as benchmarks against which to portray other possible states of the world and are completed with the aid of trend data. Table 2 presents a simple typology of scenario characteristics in terms of basic principles. In practice, scenarios will combine a range of features depending on their real world application and the scale at which they are pitched.

Scenarios are not precise future predictions but methods to aid decision makers in their efforts to cope with inevitable uncertainty. Scenarios may possess a variety of characteristics and can be deployed at different spatial scales and across different temporal scales (typically from 10 years to 100 years). They can be used to facilitate consensus, or negotiation, in situations where multiple competing stakeholder interests are at issue; or at least provide part of a more inclusionary process for decision-making. They can be focused on particular policy objectives and/or instruments and provide sensitivity assessments. Finally, they can portray the consequences of policy and strategies that incorporate radically different worldviews in a more visionary way. In this context the 'alternative' visions are most often reflected against a baseline (BAU) trend scenario.

Table 1. DP-S-I-R Framework: European seas and regional catchments

	Arctic	Azov Sea	Baltic Sea	Black Sea	Caspian Sea	Mediter-ranean	North Sea	Atlantic Coast
Main Driving Pressures:								
Climate Change (CC)	CC+	CC+	CC	CC+	CC+	CC (locally+)	CC (locally+)	CC
Built Env Expansion	-	-	BE+	BE	-	BE+	BE	BE
Trade, Ports & Related Industry (P)	-	-	P+	P+	P+	P+	P+	P
Tourism (T)	-	-	T+	T (locally+)	-	T+	T	T
Fishing & Aquaculture (F)	F+	F+	F+	F+	F+	F+	F+	F+
Agriculture (A)	A+	A+	A+	A+	A+	A	A	A
Environmental State Change /Impacts:								
Contamination (C)	C	C+	C (locally+)	C+	C+	C (locally+)	C+	C (locally+)
Eutrophication (E)		E+	E+	E+	E+	E (locally+)	E	E (locally+)
Biodiversity Loss (inc. invasive exotic impact) (B)	B	B+	B (locally+)	B+	B+	B (locally+)	B (locally+)	B
Sea level rise & coastal erosion (SE)	SE	SE	SE+	SE	SE+	SE+	SE (locally+)	SE
Policy Responses:								
International Agreements (IA)	IA	-	IA	IA	-	IA	IA	-
National Laws (NL)	-	-	Part. implemented	-	-	NL	NL	NL
Regional-local measure (RL)	-	-	RL	-	-	RL	RL	RL
ICZM	-	-	ICZM	-	-	ICZM	ICZM	ICZM

Notes: + = Very significant; - = minor to insignificant; Contamination = heavy metal accumulation, pesticide etc residue accumulation, oil and gas spills; ICZM = demonstration projects, training and evaluation; Source: Adapted from EEA (2003), Parry (2003)

Table 2. Scenario characteristics typology

Type 1	Forecasting scenarios: they attempt to encompass future alternative development paths from the standpoint of the current situation (time = t_o); they can also include expected of desired policy switches
Type 2	Backcasting scenarios: they take as their initial start pint some desired future (time = t_{1-n}) state of affairs or policy objective and then explore alternative strategies to maximize goal attainment.
Type 3	Descriptive scenarios: they set out a sequenced set of possible events in a neutral way.
Type 4	Normative scenarios: their sequences explicitly incorporate different interests, values and ethics.
Type 5	Quantitative scenarios: usually computable model-based exercises
Type 6	Qualitative scenarios: which rely solely on narratives
Type 7	Trend (BAU) scenarios: based on the extrapolation of current trends.
Type 8	Peripheral scenarios: which attempt to include surprises i.e. unlikely and/or extreme events and their consequences.

Source: Adopted from EEA (2000)

European coastal futures

The rapidly changing record of the last half century highlights the difficulties involved in forecasting the future decades ahead. But the underlying context for any 'futures' thinking for Europe must include an appreciation of the globalisation process and the implications of an expanding European Union and Single Market. Globalisation has brought a growing interdependence between financial markets and institutions, economies, culture, technology and politics and governance. So far, market forces and related social systems have become the increasingly dominant paradigm. For Europe, there is the additional dimension of change involving the potential inclusion of the countries of central and eastern Europe into the EU, together with their own internal transitions from centrally planned to market-based economies and societies.

If we follow the advice of scenario analysts, then we need to foster a process by which alternative worldviews (ecocentrism, technocentrism, weak sustainability or strong sustainability) and conventional wisdom are challenged and clarified, in order to focus on critical issues (Gallopin and Raskin 1998). We can envisage a future in which globalisation and liberal market conditions and values continue to evolve, or one in which coordinated and concerted collective actions are agreed upon by national governments and implemented by increasingly influential international agencies. Alternatively, self-interest and protectionism may become the dominant characteristics of a future society, with Europe gradually fragmenting rather than unifying. But even more radical change is a plausible future and the strong sustainability and 'deep green' visions may also become realities.

There is no shortage of candidate scenarios to choose from and the sub-section that follows outlines a hybrid approach which borrows from a set of scenarios previously formulated to investigate the impact of climate change technological advances and environmental consequences in a range of contexts (Parry 2000;

Lorenzoni et al 2000, OST 1999, EEA, 2000). The aim is to first provide a set of basic contextual narratives within which to set four somewhat more specific scenarios (UNEP 2002) with relevance for coastal-catchment areas across the European sub-regions (Western, Central and Eastern). The critical issues, which are, highlighted mirror the pressures, impacts and responses set out in Table 2 within the DP-S-I-R framework.

The narrative contexts and the scenarios are framed by two orthogonal axes, representing characteristics grouped around the concepts of societal values and forms of governance. The values axis provides a spectrum from individualistic, self-interested, consumerist, market-based preferences through to collectivist, citizen-based communitarian preferences, often with a conservationist bias. The vertical axis spans levels of effective governance from local to global. The four quadrants are not sharply differentiated but rather are bounded by overlapping transitional zones not distinct boundaries. Change occurs as certain trends and characteristics become more or less dominant across the different spheres of modern life-government, business, social, cultural and environmental – see Figure 2.

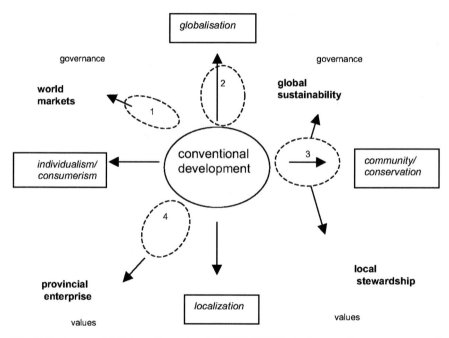

Fig. 2. Environmental Futures Scenarios (to 2080). NOTE: the four quadrants are separated by 'zones of transition' not distinct boundaries. Source: OST/DTI, Environment Futures, Foresight, OST London http://www.forsight.gov.uk. See also Lorenzoni et al (2000); and UNEP (2002)

Taking the four contextual background conditions first, *World Markets* is dominated by globalisation, which fosters technocentric and often short-term societal views. Expectations about an expanding EU and Single Market are born out and economic growth remains the prime policy objective. Environmental concerns are assumed to be tackled by a combination of market-incentive measures, voluntary agreements between business and government and technological innovation. Decoupling of the growth process from environmental degradation is assumed to be feasible, not least because ecosystems are often resilient. Weak sustainability thinking is favoured and 'no-regret' and 'win-win' options are the only ones pushed hard by regulators. Rapid technological change, sometimes unplanned, will be the norm, as will trade and population migration. Private healthcare, information technology, biotechnology and pharmaceutical sectors of the economy, for example, will thrive, while 'sunset' industries will rapidly disintegrate (e.g. heavy engineering, mining and some basic manufacturing). The internal and external boundaries of state will retreat producing a more hollowed out structure (Jordan et al 2000). National governments will struggle to impose macroeconomic controls as transnational corporate power and influence escalates. Multilateral environmental agreements will prove problematic and prone to enforcement failures.

Under *Global Sustainability*, there would be a strong emphasis on international/global agreements and solutions. The process would be by and large 'top down' governance. Trade and population migration would still increase but within limits often tempered by environmental considerations. EU expansion would be realised but social inequities would receive specific policy attention via technology transfer, financial compensation and debt for nature swaps/agri-environmental programmes.

Provisional Enterprise would be a much more heterogeneous world, EU expansion might stall and a slow process of fragmentation (economically and politically) might be fostered. A protectionist mentality would prove popular and economic growth, trade and international agreement making prospects would all suffer.

Local Stewardship would put environmental conservation (ecocentrism) as a high priority. A very strong sustainability strategy would be seen as the only long term option. This strategy would emphasise the need for a reorientation of society's values and forms of governance, down to the local community scale. Decentralisation of economic and social systems would be enforced, so that over time local needs and circumstances become the prime focus for policy. Economic growth, trade, tourism (international) and population migration trends would be slowed and in some cases reversed.

With this backdrop in mind, the four scenarios 1,2,3 and 4 can be roughly located in Figure 2, with the arrows indicating the general direction of change over decadal time. Scenario 1 is almost a trend/baseline scenario. The policy goal of maximise GDP growth is achieved via an extended single market system stretching into central Asia. New accession states are given transitional status to ease their progress into market-based systems. The relatively weak enforcement of environmental standards in these countries fosters short run profitability but may hinder long run resource use efficiencies. Rapidly growing volumes of trade and travel increase the level of economic interdependence in Europe, but social cohesion remains somewhat weak, as people strive to satisfy individual consumerist preferences. Scenario 2 imposes sustainability constraints via a 'top down' governance process but also encourages citizens to 'think global and act local'. Scenario 3 allows for a much more radical paradigm shift in societal values and organisations, environmental conservation and social equity rise up the political priority agenda. In Scenario 4, protectionism breeds growing disparities across the sub-regions of Europe. Inequality and possible conflicts spawn a relative isolationist response at the nation state level. So now we turn to the implications for the future coastal zones in Europe, given the different scenarios.

Under all scenario conditions pressures on coastal ecosystems are seen to increase, either through direct exploitation of coastal resources, including local use changes and an increase in the built environment at the coast; or through changes in related catchments associated with the spatial planning of development and transportation policies, changes in agriculture policy, especially trade regimes and reform of the Common Agricultural Policy (CAP). Another striking feature of the scenario analysis is that the impact of climate change does not vary significantly across scenarios until around 2030 – 2050 because of delays in the response of the climate system.

If *Scenario 1* conditions prevail then trade, economic development and the migration of people across Europe all increase significantly. This increase in general economic activity will outweigh improvements in energy efficiency stimulated by carbon taxes and technological change. Public transport will remain underdeveloped as car transport remains the dominant mode of transport. Air traffic also grows due to trade activities and international tourism. Emissions of air pollutants and greenhouse gases (GHGs) continue to rise because of the dominant effect of the energy and transport sectors of the European economies. It is also the case that effective policies to reduce CO_2 and other GHGs are slow to evolve or remain dormant. By the 2030s sea level rises of an average 10cm will have occurred with increased flooding risks and defence costs. The extent of the built environment at the coast and across adjacent catchments expands rapidly in Western Europe but is more stable in Central and Eastern Europe if population declines continue. The number of people living in areas of increased freshwater stress also rises as already vulnerable areas in the south of Europe face a deteriorating situation and other areas become newly stressed.

The Mediterranean coast is a particular pressure and stress problem through a combination of urban growth with inadequate waste water treatment facilities, tourism growth and increases in intensively farmed croplands close to estuaries. While the CAP is reformed this is achieved via a stronger imposition of international market pricing regimes. Some funds are made available to provide short run support for farmers, particularly in the new entrant countries of central and eastern Europe but land is still abandoned in these regions. Elsewhere land abandonment is restricted to small areas, for example, in upland areas. Overall, only the most efficient farmers survive by intensifying production and embracing GM technology, with consequent diffuse pollution and other environmental risk increases.

To sum up the coastal zone consequences under scenario 1, tourism impacts escalate leading to local environmental problems such as salinization and eutrophication of coastal waters. Second homes expand in almost all areas but are particularly problematic in the Baltic and Mediterranean areas. Sea level rise exacerbated by climate change begins to pose major difficulties and other catchment flooding events are exacerbated by the expansion of the built environment, from the 2030s onwards. Policy responses are somewhat restricted. At the international level environmental agreements prove to be difficult to negotiate and only partially effective if and when implemented. Coastal areas are therefore by and large left in the hands of local authorities and local regulators. This forum of governance offers unpredictable results and falls well short of ICM principles and practice.

Under *Scenario 2* conditions, Europe begins to resemble more of a federal state and is characterised by strong environmental and other regulatory agencies (e.g. European Environment Agency), which promote sustainability. Transport and energy sector growth is constrained by a range of intervention policies. Internationally the World Trade Organisation adopts an environmental mandate to complement its existing remit. Nationally, green belt policies and designations like Natura 2000 are strengthened and ecotourism principles are supported. CAP reform is dominated by switches to funding for a range of agri-environment schemes. The more proactive 'environmental' strategy succeeds in stabilising air pollution and GHGs emissions, largely because of energy efficiency gains and extensive switching to non-fossil fuels, and declines are possible from 2030 onwards. Sea level rise difficulties remain to be solved but policy responses are more flexible e.g. managed realignment schemes.

Areas under severe water stress remain more or less constant or fall slightly in some regions as irrigated agriculture is abandoned. Overall, Western Europe coasts are moved closer to ICM regimes, elsewhere basic coastal management measures are put in place and historical zoning plans revitalised. The Water Framework Directive is fully implemented as are Regional Seas Agreements.

Under *Scenario 3*, a combination of strong framework policies designed to ensure sustainability principles can be put into practice and most significantly attitudinal and lifestyle changes in society in general generate significant falls in air pollution and GHGs emissions, beginning in the 2020s climate induced sea level rise is still a problem but is tackled almost exclusively via 'soft engineering 'measures and relocation schemes with compensation for sufferers. Public transport networks are encouraged and succeed in reducing the dominance of the motor car (which itself become significantly 'cleaner' and 'more efficient') Local tourism activities flourish at the expense of international tourism and 'package' holidays. The built environment expansion is halted, except for some areas in Western Europe where development pressures remain particularly strong and a major reconversion of lost habitats is stimulated (more designated sites, agri-environmental schemes, managed realignment etc) to ensure increases in biodiversity. The total area under severe water stress is constant or declining as demands (especially from agriculture and mass tourism) are blunted by pricing policies and changing consumption patterns i.e. decline in meat eating. Policy responses at the coast embrace ICM principles but are more effectively enabled because of the existence of voluntary partnerships across stakeholders and other participatory arrangements at the local level. This 'bottom up' activity serves to complement the full imposition of the EU Water Framework Directive and Regional Seas Agreements.

Under *Scenario 4*, the expanded EU itself may fail to materialise. A fall in overall economic activity, trading activity and tourism is likely. Because of this relative economic stagnation at the micro level (most severe in Central and Eastern Europe) overall air pollution and GHGs emissions remain stable. Sea level rise remains a problem and there is a lack of resources to invest in both mitigation and adaptation measures. The numbers of people living in water stress areas increases as new areas join the vulnerable category. Less than full implementation of legislation like the Water Framework Directive bring forth the risk of water resource conflicts and more extensive water contamination problems. Coastal zones in Western Europe remain under built environment/economic development, tourism, port expansion and other infrastructure growth pressures. In central eastern Europe, conditions are more stable but do not improve, while in eastern Europe coastal zones could become militarised zones with restricted areas, except for port facilities.

References

Von Bodungen B, Turner RK (2001) Science and integrated coastal management. Dahlem University Press, Berlin

European Environment Agency (2000) Cloudy crystal balls: 'an assessment of recent European and global scenario studies and models'. Environ Issues Series No 17, EEA, Copenhagen

European Environment Agency (2003) Europe's environment: the third assessment. Environ Assessment Report No 10, EEA, Copenhagen

Gallopin GC, Raskin P (1998) Windows on the future: global scenarios and sustainability. Environment 40:7-11; 26-31

GESAMP (2001) Protecting the oceans from land-based activities. GESAMP Reports & Studies No 71, UNEP, Nairobi

Hammond A (1998) Which world? Scenarios for the 21st century. Earthscan Publications Ltd, London

Jordan AJ, Lorenzoni I, O'Riordan T, Turner, RK (2000) Europe in the new millennium. In Parry, ML (ed.) Assessment of the potential effects of climate change in Europe: The Europe ACACIA Project. University of East Anglia, Norwich, pp 35-45

Lorenzoni I, Jordan A, Hulme M, Turner RK, O'Riordan T (2000) A co-evolutionary approach to climate change impact assessment: Part I. Integrating socio-economic and climate change scenarios. Global Environ Change 10:57-68

Office of Science & Technology (1999) Environmental futures. Report for the UK's National Technology Foresight Programme. DTI/Pub 4015/IK 399 NP, VRN 99647, London

Parry M (2000) Assessment of potential effects and adaptations for climate change in Europe: The Europe Acacia Project, Jackson Institute, University of East Anglia, Norwich

Turner RK, Bateman IJ Adger WN (2001) Economics of coastal and water resources: valuing environmental functions. Kluwer, Dordrecht

UNEP (2002) Integrating environment and development: 1992-2002, UNEP, Nairobi

15. Group report: Integrated assessment and future scenarios for the coast

Corinna Nunneri[1], R. Kerry Turner, Andrzej Cieslak, Andreas Kannen,
Richard J. T. Klein, Laure Ledoux, Joop M. Marquenie, Laurence D. Mee,
Snejana Moncheva, Robert J. Nicholls, Wim Salomons, Rafael Sardá,
Marcel J.F. Stive, and Tiedo Vellinga

Abstract

A prototype scenario assessment was carried out with help of a DP-S-I-R framework to provide an outline forward look at the European coastal areas. Impacts of
change were assessed for the following major sectoral or cross-sectoral drivers:
climate change, agriculture/forestry, urbanisation, tourism, industry and trade,
fishery and shellfish fishery, and energy. The present situation was tabulated prior
to an outline of the impacts of three scenarios, i.e. (1) a world market perspective,
(2) global sustainability and (3) environmental stewardship. From twelve identified impact categories, three were judged to be of particular significance in the
present situation: habitat loss (including coastal squeeze); changes in biodiversity;
and the loss of fisheries productivity. A group of three impacts – eutrophication-
related effects, contamination-related effects and coastal erosion – were all judged
to be of moderate importance in most areas. All others were allocated only local
importance. The analysis suggests that the major current drivers will still play the
dominant role, augmented by climate change. Drivers and impacts intensities usually increase under the perspective of a more globalised world (scenario 1) and
usually decrease through better management, mitigation and adaptation measures
of scenarios 2 and 3. Under scenario 1, the eastern countries (Black Sea and Baltic
Sea areas) are particularly prone to eutrophication and contamination impacts, as
well as habitat and biodiversity loss, due to expansion of mass tourism eastwards,
together with intensification of agriculture and aquaculture. Under scenario 2
more stringent regulations and management reduce environmental impacts. Under
scenario 3 impacts are reduced through decentralisation, although this may also
result in sub-optimal management (local fragmentation).

---

[1] Correspondence to Corinna Nunneri: corinna.nunneri@gkss.de

J.E. Vermaat et al. (Eds.): Managing European Coasts: Past, Present, and Future, pp. 271–290, 2005.

Introduction

The use of scenarios for scoping relevant emerging issues is one way of mitigating the problems of the time lag lying between the initiation of scientific analysis and the use of scientific knowledge (findings) in decision-making and management. A research project analysing Drivers and Pressures at the coastal zone can take 3-4 years to properly identify and highlight the main issues. In order to be usable in Decision Support Systems (DSS), scientific findings need an additional 3-4 years, once the project is finished, for capturing the attention of the policy-makers and incorporation in the decision making procedure. This transfer of 'usable knowledge' to 'used knowledge' (from research to application) will then take altogether an average ten-year time span. In other words we have to start now the relevant research for 2010 and beyond now. The use of scenarios can shed light on which fields are likely to generate future problems and needs for management (and management studies). Moreover scenario assessment is an appropriate tool for integrating social and natural sciences (as well as engineering sciences) because it facilitates analysis of interwoven processes typical of intensively populated regions subject to multiple uses such as coastal areas. In this context scenarios are an essential component of integrated assessment studies. The primary purpose of the presented prototype assessment was to provide an outline forward look at the European coastal areas as the EU extends its membership and adjusts its socio-economic and political structures and strategies under the on-going globalisation driving pressures process. The global environment has come under increasing stress over the last 40 years and the European environment (from the Atlantic coast to the shores of the Caspian Sea) has been heavily impacted throughout this period of environmental change. A key question for preparing a future management strategy and vision is therefore: what are the current Drivers and Pressures in European coastal areas and what are the long term consequences from a foresight perspective? A further step in the integrated assessment focuses on possible future changes with respect to the projection of the current situation (baseline scenario) other feasible futures and the impacts that those "alternative" futures are likely to bring about.

Methodology

Based on the scenarios assessed by Turner (this volume), the aim of this session was that of identifying present and future driving forces for development of the coastal zones in different sea basins and their consequent impacts on the coastal zone functions and ecosystems. An integrated perspective has been chosen in order to utilise both natural science and social science findings. The methodology adopted for this foresight exercise is comprised of a scoping framework, the Driver-Pressure-State-Impact-Response (DP-S-I-R) approach (Turner et al. 1998, EEA 2003) and the use of qualitative future scenarios. Following the DPSIR scheme, a methodological approach for integrated assessment (IA) starts from the analysis of the current state of the system. Then successive steps look for causality chains, identifying drivers and pressures, which may change the present state (the current pressures and drivers might be different from the future ones), the political alternatives, as well as institutions and mechanisms for policy and management responses. The coastal system is characterised by dynamic and discontinuous (both natural and anthropogenic) processes (e.g. change of water inflow can both be due to natural causes, e.g. changing rainfall, or to anthopogenic activity, e.g. upstream river damming), some of which may have their origin in the riparian countries lying far away from the coastal zone (e.g. eutrophication related to nutrient input through agricultural land use in the catchment). Furthermore a very strong social component, characterised by multiple stakeholders, sometimes with competing interests and resource uses, superimposes upon and interacts with the natural processes taking place in the coastal zone. Such a context requires an integrated assessment methodology, coupling social and natural sciences, in which an essential role is played by stakeholder participation. The stakeholder mapping itself (who the stakeholders are and how they interact with each other) is not a trivial exercise. The capacity for mapping stakeholders, as well as initiating and developing participatory approaches varies across Europe, due to cultural, traditional, economic and social factors (e.g. people willingness to participate). The consolidated experience in the Northern countries contrasts with the Southern and Eastern countries, in which integrated assessment research is still in its fledgling stages. Every country has institutions (either Universities and research centres, or governmental agencies) that deal with participatory processes. Since the results are pre-formed and shaped depending on the kind (scientific or administrative) of structure initiating and leading the participatory process, this might call, on one hand, for assigning official responsibilities regarding integrated assessment studies to specific institutes (this would imply a leading role of some central extranational authority). Or, on the other hand, the need might be fore more harmonised approaches and procedures, in other words, a more watertight methodology and set of criteria. This though would imply at least at a certain level, a 'one size fits all' vision of participatory processes in a very heterogeneous Europe.

The management process engaging the stakeholders can, as a result of both socio-cultural and economic differences, result in different forms of dialogue, varying from a minimum, i.e. consultation via information provision (stakeholder involvement), to a maximum, i.e. full participation in decision making. The capacity to undertake participatory approaches varies considerably throughout Europe; the tendency is at most towards consultation rather than full participation in the Southern and Eastern European countries. This feature stresses the short-term need for a 'reduced form' assessment that can be made more sophisticated over time as finances and human resources allow (Timmerman and Langaas 2003). Moreover, in the context of integrated assessment, the responsibility of riparian countries for the coastal environment still has to be enabled through effective co-operation of institutions lying in the catchments and in the coastal areas.

The stakeholder dialogue is meant to mitigate stakeholder conflicts, either in a top-down or in a bottom-up approach, depending on the kind of society, needs and awareness of the general public. In other words a fully bottom-up participatory approach will be more likely to happen in some scenario worlds than in others.

The advantages of active participation of stakeholders is briefly summarised as follows:

- It enriches assessment by offering insights from different points of view;
- It offers a basis for confronting all conflicting issues and needs that are the fundamental cause of dissent and lobbying, thus helping to look for win-win solutions;
- Arguing parties may agree on the choice of a third party as an 'objective evaluator' of different management strategies;
- People actively involved in taking a decision will be more likely to help in implementing it.

This participatory approach can be operationalised by some pragmatic steps (see also Mee this volume; and O'Riordan this volume):

- Involved people (stakeholders and the general public in a fully participatory approach) express their desire about how 'their' future coastal environment 'should' look (structure, composition of the ecosystem) and what the coastal zone 'should' offer to them, also from the economic point of view (good/service outputs): this desirable future environment is 'the vision';
- Once a future vision has been agreed, then a set of indicators and methodologies for describing the current state of the system should be defined and agreed;
- Based on the current state of the system, what steps are necessary towards the desirable system? What structure (governance, governmental agencies) are available for directing systems towards this vision? A key point is a step-by-step path for reaching the vision. Even though the target may be long term (e.g. ecosystem integrity), intermediate, attainable shorter-term steps should be taken in order to achieve at least a majority acceptance by stakeholders.

A respectable assessment procedure should be adopted to confirm that the adopted policies are really leading in the chosen direction and in order to allow a timely policy change if otherwise. It also enables a periodic re-evaluation of whether or not the vision was an appropriate one. One of the advantages of this approach is that it adapts to changing perceptions, to newly available information and to the lessons learned from implementing the previous step.

A limitation of participatory approaches can be highlighted – they assume in principle the achievement of a consensus among the parties. This might not be the case in reality and time constraints may require top-down intervention deliberating management solutions within the operational time frame. Given this caveat, depending on societal (present and future) values and perceptions, a full-participatory, bottom-up approach might be seen as a too "idealistic" and ineffective instrument for supporting decision-making. Therefore, due to socio-economic, time or cultural constraints it might not be acceptable in every (present or future) society.

The scenario analysis is just one component of an overall integrated environmental assessment, but it can play an important role. In the first instance, scenario analysis can help inform the baseline studies, which initiate the IA system. In this sense scenarios are part of the methodological techniques deployed to scope the main coastal change pressures, drivers and impacts, as well as policy issues and conflicts that are generated. But scenarios can also be used later in the assessment process, to test whether a coastal management approach leads towards a set of agreed common objectives embodying the vision. The scenario can now play its second role as a sort of 'success prerequisites' checking device. Different policy packages and approaches (i.e. top-down, bottom up, mixed approaches), work better in one socio-economic, political, cultural and environmental context than in another. Scenarios can then illuminate the best matches between different policy approaches and the general status of the world evolving into future, thus allowing robust management solutions to be identified.

In the foresight exercise, the baseline trends and three further scenarios for the possible future development of the European coastal zones have been assessed in a round table discussion by a heterogeneous group of professionals with extensive and varied experience of coastal systems and their management problems.

The foresight exercise: DP-S-I-R across Europe

The group of experts that undertook this prototype exercise was biased in the sense that the mix of expertise available spanned a number of relevant disciplines, but was restricted in its geographical coverage (i.e. it became weaker the further East the appraisal areas extended), for this reason the Caspian Sea, which was initially included in the number of considered basins, has been subsequently removed. The group also only had a limited amount of time available for its deliberation and therefore choose to simplify and aggregate when complexities were severe constraints. In this context the regional differences present in all basins and particularly noteworthy in spatially extensive coastal areas (e.g. the Mediterranean, the Atlantic coasts), have been aggregated into an average assessment for the whole basin that might be very different from local or even national realities for some specific coastal areas. Therefore, this analysis has most value at the basin or larger scale. With these caveats in mind the group first set out to identify the main drivers of change in the contemporary (or past 30 years) systems in Europe's coastal areas. The driving forces were then translated into environmental pressures and consequent impacts with human welfare consequences (the DPSIR approach). Table 1 summarises the initial brainstorming exercise for connecting drivers and pressures to a set of the most likely significant impacts.

A key modelling and analysis characteristic of coastal zone change is the multiple links to the relevant catchment and catchment pressures such as, for example, nutrient, sediment and water flux changes triggered by land-use changes and other pressures. The catchment –coast continuum reality is another strong reason for the adoption of an integrated assessment approach, aiming to investigate, collect and *ideally* harmonise the conflicting interests not only in the coastal zone, but also between catchment and coast. The most relevant drivers determined in Table 1 are:

Table 1. Major Impacts related to the main Drivers and Pressures. This table is a result of a brainstorming exercise which aimed at a selective determination of major drivers and impacts. There is no pretension of completeness

Pressures/Drivers	Impacts
Climate change	Erosion, biodiversity loss, increased/changed floodrisk, altered species composition
Agriculture and forestry change	Eutrophication, contamination, biodiversity/habitat loss, subsidence, salinisation, altered sediment/water supply
Urbanisation and infrastructure change	Coastal squeeze, eutrophication, contamination, habitat loss/fragmentation/human disturbance, subsidence, altered sedimentation, increased flood-risk, salinisation, altered hydrology
Tourism development	Seasonal/local impacts, beach "management" (e.g renourishment), habitat disruption, loss of species, increased water demand, altered longshore sediment transport, loss of local cultural values
Industry and trade expansion	Contamination, exotic species invasion, dredging, sediment supply/erosion
Fisheries/aquaculture expansion	Species loss/overexploitation of stocks, impact on migratory species, habitat loss, species introduction/genetic pollution, contamination, eutrophication
Energy exploitation and distribution	Habitat alteration, altered water temperature, changed landscape/amenity, subsidence, contamination, accident risk, noise/light disturbance

1. *Climate change*: rising air and sea temperature and sea level, and changing rainfall and weather conditions (associated with flooding) represent a threat to which the coastal areas are particularly subject (Parry 2000, see also the group report by Rochelle-Newall et al. this volume);
2. *Agriculture and forestry*: changing farming practises and cropping regimes directly affect agricultural landscapes and simultaneously alter catchment inputs of run off water, sediment, nutrients and contaminants into soils and riparian waters, all of which eventually finds its way into estuaries and coastal waters;
3. *Urbanisation and infrastructure*: the expansion of the built environment can be at the expense of habitat and biodiversity and cause disturbance of ecosystem processes, but it can also alter the hydrological properties of catchments and interrupt the natural dynamics of coastal processes, including armouring of the coast; Moreover, it is a major source of contaminant through direct effluent discharge, storm waters and diffuse sources;
4. *Tourism*: "mass tourism" in particular is closely connected with further urbanisation, with local and seasonal maximum inputs of contaminants and nutrients but also with disturbance of sensitive species and progressive change in local cultures;
5. *Industry and trade*: industrial discharges in the catchment reach the coastal areas through the river systems, while ports and harbours contaminate coastal wa-

ters and increase the risk of accidents (oil-spills, illegal ballast water discharge) and the "import" of exotic species;

6. *Fisheries and aquaculture*: Over fishing and the unrestricted use of particular fishing techniques can generate very disruptive impacts, particularly destruction of the bottom habitat and can endanger target and substitute species; aquaculture is strongly connected with water contamination (e.g. antibiotics) and possible genetic pollution through animal escape;

7. *Energy*: demand for and generation of energy can affect the coastal zone in different ways, of which the most evident are the alteration of riverine hydrography through damming (and consequent alteration of water and sediment input to the coastal zone), disturbance of wildlife (e.g. cases of altered birds' flight by wind mills), altered water temperature, toxic spills including radioactive discharges, etc.

Table 1 illustrates the multi-causality of some impacts, such as habitat and biodiversity loss and highlights the implicit difficulties facing the management system designed to deal with these issues.

Although many impacts are interwoven in a cause-effect chain (e.g. eutrophication and altered species composition), broadly speaking four groups of significant impacts can be categorised according to particular ambient environmental states:

1. Coastal morphology, hydrodynamics and sediment quality:
   - Altered sediment and water supply to the coastal area, connected erosional processes;
   - Subsidence of coastal regions due to groundwater extraction or other economic activities, which makes those areas more prone to flooding events, a situation which can be further exacerbated by climate change (e.g. the Po delta and the Ebro delta areas: Bondesan et al. 1995, Jimenez et al. 1997);
   - "Coastal mismanagement", i.e. construction of coastal defences and connected beach renourishment driven by local interests or carried out without in-depth studies of the interwoven coastal processes, altering (especially longshore) sediment transport patterns;
   - Contamination of sediments requiring disposal or treatment or representing a threat by resuspension processes.
2. Hydrology:
   - Salinisation of groundwater;
   - Altered water temperature;
   - water contamination through catchment-based or estuarine-based activities such as ports and harbours (including marinas);
   - increase in turbidity.
3. Ecosystems:
   - Changes in biogeography (including migration routes, spawning and wintering grounds) due to climate change; this is rapidly becoming a major concern in the North East Atlantic;
   - Eutrophication;
   - Contamination (toxicity/bioaccumulation);

- Altered species composition potentially resulting in loss of biodiversity (e.g. due to introduction of exotic species, introduction of new species and "genetic pollution" through escape from aquaculture establishments, over fishing); biodiversity loss is also intimately connected with eutrophication, contamination and habitat loss;
- Habitat loss, fragmentation or disruption due to human activities;
- Disturbance through light and/or noise (e.g. birds are attracted by light);
- Coastal squeeze between retreating coastlines and (hard) coastal defences (due to drivers such as urbanisation);

4. Human culture and values:

- Altered landscape and loss of amenities;
- Loss of historic cultural values or symbolic nature values;
- Increased accident risk, also including contamination events;
- Increased flooding risk and damage, potentially including fatalities.

After having determined what are in principle likely to be the most relevant Drivers, Pressures and Impacts, the current significant drivers and pressures operating across the different coastal regions of Europe were identified. In order to do this, Table 1 was re-shaped into a matrix (see Table 2), which gives more emphasis to the impacts, and multi-causality effects. The impacts themselves are also further categorised in terms of geographical location and very roughly scaled in terms of intensity. Finally, the contemporary trends are re-examined in terms of different future states of the world, based on three scenarios assessed by Turner (this volume).

The question then becomes what scientific challenges do these coastal futures pose? What can we learn from ELOISE and other existing science and what do we still need to know? Equally, what social science, policy and management challenges do they pose? What social science findings can we already use and what more do we need to investigate through research (or through a 'learning by doing' approach)? By trying to answer these questions potentially successful policy options can be identified together with the barriers to implementation that remain to be tackled.

Results: Identifying significant drivers and pressures

The outcomes of the group discussions are summarised using a matrix approach (Tables 2 and 3). Given the very high uncertainty characterising the assessment of future scenarios, a simple three-value-scale (insignificant, significant, very significant) seemed to be appropriate as a first approach.

Coastal DP-S-I-R matrix: Contemporary trends and recent past experiences

A number of primary drivers of environmental change in Europe were initially identified, including climate change (Table 1). So far the consensus position was that this driving force had imposed limited effects across all sub-regions chosen for the study. Changes in the agricultural and industrial sectors of the European countries have generated significant pressures and a range of negative impacts in the Black Sea (Mee 1992, Zaitsev and Aleksandrov 1997) and Mediterranean regions, but these effects, while not negligible, were less significant for the other regions. One exception to this generalisation might be the role that agricultural change played in the eutrophication of the Baltic, although in this case a significant proportion of the nitrogen introduced into the system was via atmospheric deposition with sources located in industry and transport movements outside the Baltic drainage basin (Gren et al. 2000). The huge and very destructive hypoxic event that caused severe damage around Denmark in 2002 was a direct consequence of agricultural nutrients transported in the unusual summer flooding that swept through Central Europe. This has resulted in a new perspective on the issue in the Baltic – a combination of increasing extreme events (not yet clearly verified) and large residual agricultural loads. There is certainly great concern within the Baltic community regarding agricultural loads. The expansion of the energy sector in all countries (exploitation, processing and supply, distribution) has set up a series of impacts spread extensively across the entire region but the coastal effects were judged to be particularly significant in the Baltic Sea with its oil and gas exploitation and growing intensity of shipping, and in the Mediterranean with its high density of shipping routes and movements.

In Table 2, a trial run to relate drivers and impacts in an interdisciplinary way is shown, although the tendency might be that of falling into a "sectoral trap". An example of a such shortcoming may be that agriculture might also bring about fisheries loss (via eutrophication). This impact could be mentioned separately in the table but it is not, and this might cause confusion. Only after 'knowing' that eutrophication often (but not always) leads to fisheries loss or alteration, would it be possible to work out this impact. Similarly for fisheries as a driver, eutrophication is not only caused by discharge of nutrients from aquaculture but by the excessive removal of predators.

Four drivers (not internally ranked) were judged to be of most importance in the characterisation of the baseline conditions in Europe, namely:

1. *Urbanisation and transport* (i.e. expansion of the built environment). This driver was a ubiquitous one across all regions and furthermore stimulated changes and impacts in both the catchments and related coastal areas (and is linked to the second and third members of this set of primary drivers). It especially affected habitat and amenity loss across all the considered basins (e.g. EEA 1998);
2. *Tourism* was the second member of this set of primary drivers. The impacts related to this driver were of major significance for the Mediterranean, the Baltic and the Black Sea. Although the visits in the Mediterranean probably are twenty fold those to the Black Sea, the latter often result in a higher unit impact (Laurence Mee, pers. comm: approximately 150 million verus 5-8 million annually). The impacts

of huge increases in the volume of tourism in Europe and the consequent expansion of the built environment in the resort localities and elsewhere (e.g. transportation facilities), have been exacerbated by the lack of adequate planning controls or the lack of proper enforcement of controls (Simeoni and Bondesan 1997). The Mediterranean region is a particular example of this phenomenon where the situation is further complicated by a growing trend of tourist arrivals from the Eastern European countries. The problem of second homes is a major issue in the Mediterranean and a possible emerging problem in those parts of the southern Baltic and Black Sea close to the accession countries (Sardà & Fluvià, 1999). Land price escalation in coastal areas and problems of solid waste disposal are also noteworthy problems in this context. Policy responses have also been mixed with the possible abandonment of the tourist ecotax measure in the Balearics and the fostering of ecotourism facilities and practices in some of the less intensively developed Greek islands. Along the Baltic coastlines HELCOM plans to install a special land-use planning zone of 3 km, and Poland has expanded the coastal area subject to restricted use to an up-to-500 m "technical belt" (previously 100 m) and a 2 km "protective belt". (Act on Marine Areas of the Republic of Poland and Maritime Administration of 21 March 1991 with later amendments, Ordnance of the Board of Ministers of 29 April 2003 on the minimum and maximum width of the technical and protective belt and the method of determining their boundaries);

3. *Expansion of industrial activities and trade* (plus related infrastructure) was the third major driver, which was a ubiquitous factor across all regions. A diverse range of impacts from habitat loss and degradation to changes in species composition and the introduction of invasive exotic species and contamination-related impacts via dredged sediments and the transportation of oil and gas highlight the variety present in this set. The number of species, including exotic phyto- and zooplankton blooms recorded in Black Sea increased dramatically during the late 1980s and 1990s (Moncheva et al. 2001; Zaitsev and Osturk 2001). The alterations of the Black Sea phytoplankton structure, mainly related to eutrophication and the superimposed alien invasions, exert modifications in the historical predator-prey interactions of the food-web and fuel the disruption of the marine environment and economy of the affected areas (Moncheva and Kamburska 2002);

4. Finally, the *expansion of fisheries and aquaculture* represent the fourth main driving force and again is omnipresent throughout the pan-European area, with fisheries at a near collapse in a number of seas. The causal mechanism lying behind this productivity loss impact is a complex mixture of ecology, politics, human psychology and financial livelihoods (Farrugio et al. 1993). As an example, in 1988-1989 an estimated six-fold reduction of anchovy stocks with respect to the early 1980s was assessed as the result of both overfishing and the introduction of the ctenophore *Mnemiopsis leidy* in the Black and Azov seas (Zaitsev and Mamaev 1997, Gucu 2002). A further trend has emerged with control of fisheries and aquaculture passing from local interests into national/international control. Within these monopolistic market conditions, short termism is an ever present danger and the prospects for stock conservation are less then optimistic. The drive for maximum output is fuelled by the internationally competitive market forces.

Table 2. Drivers, pressures and Impacts: Present trends assessed for the Baltic Sea (B, b), Black Sea (BS, bs), Mediterranean (M, m), North Sea (N, n) and Atlantic coast (A, a). Capital letters symbolize a very significant impact, small letters a significant impact, no letter means no significant impact

Drivers	Impacts											
	Eutrophication related impacts	Contamination related impacts	Coastal squeeze and habitat loss	Species composition	Salinisation	Land subsidence	Flood risk change	Decreased sedimentation/coastal erosion	Increased sedimentation	Landscape/amenities	Fisheries loss	Loss of water resources
Climate change	B,BS, m, n		b, n	m, N, A			b, bs, m, n, a	b, n, m				
Agriculture/forestry	B,BS, m, n	bs,n			M,bs	m, n	M					bs, M
Urbanisation	b, BS, m, n	b, BS, m, n	B, BS,M,N,A	b	B, bs, M	m,n	m					
Tourism	b, bs, m	m, bs	b, BS, M, n, a	m	B, m			m		b, BS, M, M, n, a		
Industry and trade	b, bs	b, bs, m, n, a	b, bs, m, n	b, BS, m, n, a				b, m, n		b, BS, n, a		
Fishery and shell fish fishery	b, m, a	a	B, BS, M, N, A	B, BS, M, N, A							B, BS, M, N, A	
Energy	b, bs, m, n, A					n		BS				bs

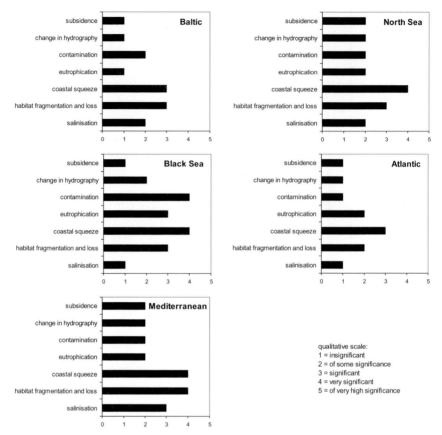

Fig 1. A more "visible" comparison of the impacts caused by tourism throughout the considered basins under present conditions. The scale of the impacts goes from 1, insignificant, to 5, very significant, 3 represents a significant impact

The selected drivers represented in Table 2, will result in a series of often interrelated impacts. The driving forces are not independent, but are all part of a joint process of globalisation and change. Some twelve impact categories were identified, each often encompassing a group of similar or related effects. An examination of the impact grouping revealed that three categories were judged to be of particular significance:

- Habitat loss, including coastal squeeze;
- Changes in biodiversity in terms of (impoverished) species composition;
- Loss of fisheries productivity (in some cases to the point of extinction at least at the economic level).

Other impacts were judged significant, but were not as extensively distributed as the three primary impact categories. A group of three impacts – eutrophication-related effects, contamination-related effects and coastal erosion – were all judged to be of moderate importance in most areas. Salinisation effects, land subsidence (due to water, oil or gas extraction), increasing flood risk, reduced sediment supplies, landscape, amenity and cultural losses together with incremented stress on water resources, were sometimes significant at a local level (location specific), or were judged minor problems.

As tourism was recognised as being one of the major drivers (see Sarda et al. this volume), it was chosen for a more in-depth assessment of some current tourism specific impacts on a five-value-scale. The resulting assessment is shown in Figure 1, which provides a more immediate comparison for the different sea basins.

At present, the Black Sea seems to be the most severely affected basin, through coastal squeeze, contamination, habitat loss and eutrophication (Mee and Topping 1999, Lancelot et al. 2002, Kideys 2002). In this regard, the comparison needs some caution, as it would require some estimates of the geographic extent of the human footprint, while it is always difficult to get a balanced perspective because the places people (even experts) are familiar with are often the most frequented ones. For example, there is very little tourism along the Black Sea coast of Turkey which is nearly 1000 km long. Certainly the Turkish coast suffers from many other problems such as urbanisation (direct discharge of solid and liquid effluent), heavy fishing pressure and problems of transport infrastructure (hundreds of km of road built on top of the beach), although this is not as much driven by tourism as by development and globalisation. Thus the coast of Bulgaria and half of Romania are impacted by tourism but this represents only ca. 350 km of coastline. Similarly part of Russia and Ukraine are impacted but it is a combination of impacts that pushes the system beyond thresholds. In this context the legitimate objection may arise, that a 'five' for habitat loss in the Mediterranean can be hardly comparable with a four or five from eutrophication from tourism (a relatively minor source of nutrients) in the Black Sea or elsewhere. This Delphi exercise therefore does not claim completeness, but aims to be a first exploration of some holistic way of representing and comparing present and future issues.

The Mediterranean follows, with problems associated with coastal squeeze, fragmentation and habitat loss as well as altered hydrological conditions causing salinisation. The Baltic Sea faces similar issues with the addition of contamination. The North Sea and the Atlantic coast appear to be less severely impacted by tourism than the other selected regions.

Scenario analysis

Once the baseline trend condition overview was completed, the group moved on to a consideration of possible, plausible coastal futures. The considered time-frame was the time-span until 2050. Three scenarios were selected relating to a (1) world market perspective, (2) global sustainability and (3) environmental stewardship. In scenario 1, it is assumed that the globalisation process (and the connected short-termism) continues and perhaps intensifies. In scenario 2, steps are agreed at governmental levels (national or international) to address environmental concerns and introduce sustainable development strategies in a top-down process. In scenario 3 local groups and interests are given a much greater role to play in management through full participation bottom-up approach. A more detailed description and elaboration of the scenarios can be found in Turner (this volume).

The group was then split into smaller teams each addressing a separate regional sea/catchment area - Baltic, Black Sea, Mediterranean, North Sea/Atlantic coast – and exposed the baseline data to the three scenarios in turn. Table 3 shows the results of this assessment.

Some of the more interesting insights that emerged from this scenario process are as follows:

Under *scenario 1*, tourism retains its primary driving pressure role, as mass tourism continues to expand eastwards in Europe and reverse flows also continue to increase. The second home problem becomes an even more extensive problem with threats not only to the environment, but local community, culture, amenities and lifestyles. Habitat losses connected with consequent urbanisation and coastal squeeze also escalate.

Other characteristics of scenario 1 are the possible flight of some industries currently located in Eastern Europe to other parts of the globe, while simultaneously these eastern areas experience an intensification of their agricultural and aquacultural sectors. The consequences are increased eutrophication and contamination-related impacts (GM crops and increased herbicide applications playing an important role). Water demand for agriculture also increases and catchments become even more affected by damming. Port complexes increase in size and number in many areas with consequent habitat loss and biodiversity changes through more importation of invasive exotic species and the additional driver of climate change, which is expected to be more significant than during the 20th Century (De la Vega-Leinert et al. 2000).

The Mediterranean area sees eutrophication effects intensify as the result of the combined effect of climate change increases in the built environment and mass tourism. It also suffers increased water stress, erosion and flood risks. Eutrophication is also a major problem in the Baltic, North Sea and along the Atlantic coast, where fish farming also brings about increased contamination and biodiversity risks.

Under *scenario 2*, the effects of some of the primary pressures are reduced in their intensity as regional/international standards and agreements are rigorously enforced ("command and control") or even overcomplied with. Contamination and eutrophication-type impacts are reduced but not eliminated. Coastal squeeze pressures are also moderated by a management strategy which operates at an appropriate (larger) spatial scale (e.g. Leafe et al. 1998) and provides compensation for stakeholders who lose out (in, for example, management realignment schemes). But tourism patterns prove more resistant to radical change, although ecotourism principles are more widely adopted and new facilities are better designed with landscape amenity in mind. Other impacts such as sewage and solid waste disposal are subject to much more stringent regulation so eutrophication impacts are reduced. Ports and harbours expansion is not halted, especially in eastern Europe, although control of ballast water discharge and dredging activities are more strictly and effectively enforced. Better water resource planning and management helps to reduce the number of water stress regions and the levels of stress experienced. Species composition change continues as invasive species continue to spread aided by climatic change factors. In the energy sector, there is a rapid switch to renewables, with offshore wind farms being a particular feature of Baltic and North Sea areas. The expansion of the built environment is only marginally reduced and switches in modes of transport are also slow to take place, because of additional resistance to change among the public. Loss of habitat is stabilised, however, as nature conservation designation increases.

Under *scenario 3*, mass tourism on an international scale all but disappears, the second home problem stabilises and eventually disappears. Fisheries and water resources overexploitation is gradually reduced as local markets and lower overall production levels become the norm. The expansion of the built environment and habitat loss are halted and even reversed, as decentralisation is adopted as the main policy objective. Coastal squeeze is consequentially also reduced in principle, but localisation could potentially mean fragmentation as processes with large spatial scale dimensions might not be properly managed by local vested interests. In this context it is doubtful that small-scale communities would be able to afford the major engineering works (e.g. to protect cities such as Venice). Some of the historical habitat loss may prove irreversible. .

A specific development in the North Sea and Baltic Sea is determined by the potential use of renewable energy (e.g. for Germany: BMU 2001; Kannen et al. 2003), and especially highlighted by the current plans for large offshore wind farm construction (Kannen this volume). In Table 3 this development-path is shown by additional impacts on species composition and landscape/amenities in all three scenarios if compared to the energy baseline scenario.

Climate change impacts are not significantly changed under any of the scenarios until about 2050 when the more forcible mitigation strategies adopted in scenario 2 start to take effect on GHGs emissions and in scenario 3 when adaptive strategies are more widely accepted and adopted. It might be argued that there would be political feed-backs. For example, coastal squeeze could well undermine scenario 3 as communities become desperate to conserve property rights and heritage in the face of global change and they may move towards scenario 2 solutions.

Conclusions

The role and challenge of integrated assessment in managing coastal zone issues has been discussed, together with its limits and its differences throughout the European countries. In this context, given the different capacities and the manifold kinds of institutions carrying out this analysis, a need for homogeneous approaches and a common methodology, as well as essential guidelines for the "less expert" countries has been highlighted.

The double role of scenarios within the integrated assessment (IA) procedure is, on the one side, to initiate the IA by scoping the main drivers, pressures and impacts, as well as policy issues and conflicts that are generated (baseline studies) and, on the other side, they can be used for testing effectiveness of selected coastal management strategies for reaching a desired situation (the "vision").

Table 3. Scenario assessment of drivers and impacts compared with the baseline scenario (first row of each driver). As before: Baltic Sea (B, b), Black Sea (BS, bs), Mediterranean (M, m), North Sea (N, n) and Atlantic coast (A, a). Capital letters symbolise a very significant impact, small letters a significant impact, no letter means no significant impact

Drivers	Eutrophication related impacts	Contamination related impacts	Coastal squeeze	Habitat loss	Species composition	Salinisation	Land subsidence	Flood risk change	Decreased sedimentation/coastal erosion	Increased sedimentation	Landscape/amenities loss	Fisheries loss	Loss of water resources
Climate change	m		b, n	n	m, N, A			b, bs, m, n, a	b, n, m				
scenario 1	m		B, BS, n, a	BS, b, m	BS, B, N, A			b, BS, M	B, M, N, A	b			M
Scenario 2	m		B, BS, n, a	BS, m, n, a	b, BS, m, N, A			BS, m, N, A	B, M, N, A	b			M
Scenario 3	m		B, BS, n, a	BS, m, n, a	b, BS, m, N, A			BS, m, N, A	B, M, N, A	b			M
Agriculture /forestry	B,BS, m, n		n	bs, -n		m,bs			M				bs, M
Scenario 1	B, BS, m, n	bs	n	B, m, n	b, n	m	m, n						BS, M
Scenario 2	b, BS	bs		b			m, n						Bs
Scenario 3	b, bs, n	b, bs			b								
Urbanisation	b, BS, m, n	b, BS, m, n	B, BS, M,N,A	B, BS, M,N,A	b	b, bs, M	m,n	m					
Scenario 1	B, bs, M	B, bs, m, m	B, BS, N, A	B, BS, M, N, A	b	b, bs, M	bs, m, n	M	BS, M	bs	BS, m		m
Scenario 2	bs, m		bs, M	b, BS, M		bs, m	bs, m	m	Bs, m		bs		
Scenario 3	bs, m		b, BS, M, a	b, bs, M		bs, m	m	m	Bs, m		bs		
Tourism	b, bs, m	m, bs	b, BS, M, a	b,BS,M, n, a	b, m	b, m			M		b, BS, M, n, a	M	
scenario 1	B, M	B, bs, m	B, BS, M, N, A	B, BS, M, N, A	b, bs, m	B, m	m		M		B, BS, M, N, A	M	BS, m
Scenario 2	m	bs	BS, m, n, a	b, bs, m, n, a		m			m		b, BS, m, n, a	m	bs
Scenario 3	b, m	b, bs	b, BS, m	b, bs, m	m	b, m			m		b, bs, m	m	

Table 3. Continued

Drivers	Impacts					
industry and trade	b, bs	b, bs, m, n, a	b, bs, m, n	b, BS, m, n, a	B, m, n	b, BS, n, a
Scenario 1	bs, m	B, BS, M, n, a bs		b, BS, m, n, a	b, m, N BS	b, BS, n, a BS
Scenario 2		BS, m bs	BS	BS, n, a bs	b, n	b, BS
Scenario 3		b, bs, m bs	Bs		b	b, BS
Fishery and aquaculture	b, m	a	B, BS, M, N, A			B, BS, M, N, A
Scenario 1	bs, m, A	A	b, BS, M, N, A		a	B, M, N, A
Scenario 2	a	a	b, bs, M, n, a			m, n, a
Scenario 3	a	a	b, bs, M, n, a			b, m, n, a
Energy				n	BS	bs
Scenario 1	bs	b, bs, m, n, A	b, bs, n	n		bs, N
Scenario 2	bs	B, bs, m, n, A	b, n			n
Scenario 3		b, bs	b, n			b

In a round table discussion the current Drivers and Pressures in European coastal areas and their long-term consequences from a foresight perspective have been assessed based on expert knowledge. A baseline scenario was compared with three different scenarios (intensification of the globalisation process, government driven sustainable development and decentralised management). The analysis was conducted at the scale of five European seas: the Baltic Sea, the Black Sea, the Mediterranean, the North Sea, and Atlantic Margin.

Four main current drivers were considered to play a major role in shaping the present issues in across all European coastal areas: urbanisation, tourism, industry and harbour expansion, fishery and aquaculture. Those ubiquitous drivers affect especially habitat and composition of species and are a source of contamination processes and loss of amenities. Particular emphasis was placed on mass tourism, which is currently highly impacting the Black Sea and the Mediterranean Sea, moderately impacting the Baltic and the North Sea, and locally impacting the Atlantic coast. From the economic point of view, fisheries loss due to over exploitation seems to be the trend throughout all basins. The current effects of other drivers (e.g energy and agricultural sector) may also be relevant, but only more locally (at the sub-basin scale). The effects of climate change are not estimated to have caused very significant impacts over the past 30 years.

Under the three future scenarios ("world market perspective with intensified globalisation", "top-down strategies for global sustainability" and "bottom-up environmental stewardship") the analysis suggests that the major current drivers will still play the dominant role, augmented by climate change. Drivers and impacts intensities usually increase under the perspective of a more globalised world (scenario 1) and usually decrease through better management, mitigation and adaptation measures under scenarios 2 and 3. Under scenario 1, the eastern countries (Black Sea and Baltic Sea areas) are particularly prone to eutrophication and contamination impacts, as well as habitat and biodiversity loss, due to expansion of mass tourism eastwards, together with intensification of agriculture and aquaculture. Under scenario 2 more stringent regulations and management reduce environmental impacts by designing environmental friendly tourist-infrastructures, controlling sewage and waste disposals and through a better environmental risk management. Under scenario 3 impacts are reduced through decentralisation, although this may also result in management fragmentation and locally-oriented solutions which are sub-optimal at scales above the regional level.

Any drive towards renewable energy, such as large scale offshore wind farms or use of tidal energy, might bring about impacts on the marine and coastal ecosystems as well as changes in the institutional and socio-economic system. However, the relevance of those impacts is difficult to forecast at the present early stages of their development.

Climate change impacts are uniform across the three future scenarios to 2050, after which mitigation or adaptation strategies start to take effect in scenarios 2 and 3.

In summary, this paper has shown how an approach making use of scenario development and analysis as used in the presented (tentative) exercise can provide insights about future problems, thus guiding and informing subsequent integrated assessment and contribute to improving coastal management over time.

References

BMU (2001) Windenergie auf See. Positionspapier Bundesministeriums für Umwelt, Naturschutz und Reaktorsicherheit der Bundesrepublik Deutschland zur Windenergienutzung im Offshore-Bereich. Bundesministeriums für Umwelt, Naturschutz und Reaktorsicherheit, Referat Öffentlichkeitsarbeit, Berlin

Bondesan M, Castiglioni GB, Elmi C, Gabbianelli G, Marocco R, Pirazzoli PA, Tomasin A (1995) Coastal areas at risk from storm surges and sea-level rise in Northeastern Italy. J Coast Res 11:1354-1379

European Environment Agency (2003) Europe's environment: the third assessment. Environment Assessment Report No. 10, EEA. Copenhagen

Farrugio H, Oliver P, Biagi F (1993) An overview of the history, knowledge, recent and future reserach trends in Mediterranean fisheries. Sci Mar 57:105-109

Gren IM, Turner RK, Wulff F (2000) Managing a sea: the ecological economics of the Baltic, Earthscan, London

Gucu AC (2002) Can overfishing be responsible for the successful establishment of *Mnemiopsis leidyi* in the Black Sea? Estuar Coast Shelf Sci 54:439-451

Jimenez JA, Sánchez-Arcilla A, Maldonado A (1997) Long to short term coastal changes and sediment transport in the Ebro delta; a multri-scale approach. Bull Inst Oceanogr Monaco 18:169-185

Kannen A (2004) The need for integrated assessment of large-scale offshore wind farm development. In: Vermaat JE, Bouwer LM, Salomons W, Turner RK (eds) Managing European coasts: past, present and future. Springer, Berlin, pp 365-378

Kannen, A, Windhorst, W, Glaeser, B, Ahrendt, K, Colijn F (2003) Zukunft Küste – Coastal Futures Project Proposal, Unpublished, FTZ, Büsum

Kideys EA (2002) Fall and rise of the Black Sea ecosystem. Science 297:1482-1483

Lancelot C, Martin JM, Panin N, Zaitsev Y (2002) The North-western Black Sea: a pilot sight to understand the complex interaction between human activities and the coastal environment. Estuar Coast Shelf Sci 54:279-284

Leafe R, Pethick J, Townend I (1998) Realising the benefits of shoreline management. Geogr J 164:282-290

Mee LD (1992) The Black Sea in crisis: A need for concerted international action. Ambio 21:278-286

Mee LD, Topping G (1999) Black Sea pollution assessment. Black Sea Environmental Series, Volume 10. UN Publications, New York

Mee LD (2004) Assessment and monitoring requirements for the adaptive management of Europe's regional seas. In: Vermaat JE, Bouwer LM, Salomons W, Turner RK (eds) Managing European coasts: past, present and future. Springer, Berlin, pp 227-237

Moncheva S, Kamburska L (2002) Black Sea plankton stowaways- why a crucial issue to safeguard Black Sea biodiversity, CIESM Workshop Monogr 20, Monaco

Moncheva S, Gotsis-Skretas O, Pagou K, Krastev A (2001) Phytoplankton blooms in Black Sea and Mediterranean coastal ecosystems subjected to anthropogenic eutrophication: similarities and differences. Estuar Coast Shelf Sci 53:281-295

O'Riordan T (2004) Inclusive and community participation in the coastal zone: opportunities and dangers. In: Vermaat JE, Bouwer LM, Salomons W, Turner RK (eds) Managing European coasts: past, present and future. Springer, Berlin, pp 173-184

EEA (1998) Europe's Environment: Second Assessment. European Environment Agency, Copenhagen. Office for Official Publications of the European Communities and Elsevier Science Ltd

Parry M (2000) Assessment of potential effects and adaptations for climate change in Europe, Jackson Environment Institute, University of East Anglia, Norwich

Rochelle-Newall E, Klein RJT, Nicholls RJ, Barrett K, Behrendt H, Bresser THM, Cieslak A, De Bruin EFLM, Edwards T, Herman PMJ, Laane RPWM, Ledoux L, Lindeboom H, Lise W, Moncheva S, Moschella P, Stive MJF, Vermaat JE (2005) Group report: global change and the European coast – climate change and economic development. In: Vermaat JE, Bouwer LM, Salomons W, Turner RK (eds) Managing European coasts: past, present and future. Springer, Berlin, pp 239-254

Sardá R, Fluvià M (1999) Tourist development in the Costa Brava (Girona, Spain): a quantification of pressures on the Coastal Environment. In: Salomons W, Turner TK, . Lacerda LD, Ramachandran S (eds) Perspectives on integrated coastal zone mangament. Springer, Berlin

Sardá R, Mora J, Avila C (2004) Tourism development in the Costa Brava (Girona, Spain) – how integrated coastal zone management may rejuvenate its lifecycle. In: Vermaat JE, Bouwer LM, Salomons W, Turner RK (eds) Managing European coasts: past, present and future. Springer, Berlin, pp 291-314

Simeoni U, Bondensan M (1997) The role and responsibility of man in the evolution of the Italian Adriatic coast. Bull Inst Oceanogr Monaco 18:111-132

Timmerman J, Langaas S (2004) The role of environmental information in European river basin management. RIZA, Lelystad, forthcoming

Turner RK, Lorenzoni I, Beaumont N, Bateman IJ, Langford IH, McDonald AL (1998) Coastal management for sustainable development: analysing environmnetal and socio-economic change on the UK coast. Geogr J 164: 269-281

Turner, RK (2004) Integrated environmental assessment and coastal futures. In: Vermaat JE, Bouwer LM, Salomons W, Turner RK (eds) Managing European coasts: past, present and future. Springer, Berlin, pp 255-270

De la Vega-Leinert AC, Nicholls RJ, Tol RSJ (2000). Proceedings of SURVAS expert workshop on: European vulnerability and adaptation to impacts of accelerated sea-level rise (ASLR) Flood Hazard Research Centre, Middlesex University, Middlesex

Zaitsev YP, Aleksandrov BG (1997) Recent man-made changes in the Black Sea ecosystem. In:Ozsoy E, Mikaelyan A (eds) Sensitivity to change: Black Sea, Baltic Sea and North Sea. Kluwer, Dordrecht

Zaitsev YP, Mamaev VP (1997) Biological diversity in the Black Sea-A study of change and decline. Black Sea Environmental Series, Volume 3. UN Publications, New York

Zaitsev YP, Osturk B (2001) Exotic species in the Aegean, Marmara, Black, Azov and Caspian Seas. Turkish Marine Research Foundation, Instanbul

16. Tourism development in the Costa Brava (Girona, Spain) – how integrated coastal zone management may rejuvenate its lifecycle

Rafael Sardá[1], Joan Mora, and Conxita Avila

Abstract

The Costa Brava is seen as an old mature tourist destination within Europe. Tourism has been the basic wheel of its modern development. However, as is the case with other mature destinations in the Mediterranean, and in conjunction with the development of tourism, the coast has been very heavily used and degraded. At the turn of the century, traditional tourism has stagnated, where not replaced by a strategy model based on secondary residences and with construction as the main economic activity, consuming large parts of the territory. In order to rejuvenate the industry, an agreement between all the stakeholders with interests in the coastal zone appears to be necessary since uncoordinated individual strategies in the past were only able to improve specific cases but did not solve the problem of overexploitation. In this context, ICZM provides a conceptual framework where individual strategies for resolving issues and promoting sustainable coastal development may be formulated. In this paper we show how the Costa Brava case may help to spread tools and practices for coastal managers in emerging tourism destinations, or in mature destinations with similar problems.

Introduction

Over the last decades, tourism has become an economic sector of great importance for many developed and developing countries. The Tourism Industry serves millions of people travelling internationally and, as travel and recreation are becoming more accessible to people, it is expected to be the world leading economic sec-

[1] Correspondence to Rafael Sardá: sarda@ceab.csic.es

J.E. Vermaat et al. (Eds.): Managing European Coasts: Past, Present, and Future, pp. 291–314, 2005.

tor during this decade. Tourism covers such diverse activities as transport, accommodation, entertainment and catering, and has indirect effects on many others. Although the World Tourism Industry was shaken by the terrorist attack of September 11th, the last western economic recession, and more recently, the SARS epidemic disease, which reduced its growth during 2002 and probably during 2003, the Industry has been averaging a 4% annual growth or more during the last 20 years is expected that this situation will continue to be the case in the near future. Because of these numbers, politicians and managers usually see tourism as an important source for revenue and employment. The "quick fix" attraction of foreign visitors and the increase in earnings provided by tourism obscured in many places medium- and long-term planning considerations and diminished the quality standards of products and services at final destinations. This well-known process raises, consequently, serious problems in the surrounding areas because of different types of impacts on the local cultural and natural environment. Solutions to these problems, when possible, come later in the process with efforts to reverse the degradation patterns promoted by an uncontrolled growth. On the other hand, sustainable development is today a well-established transnational concept, and is perceived as a compulsory prerequisite for any proposed activity, including tourism activities. Tourism, more than any other human activity, depends on the quality of the environment for its continued success. Therefore, sustainable development of tourism is today an essential prerequisite. At the beginning of this century, the managing tourism practices in a sustainable manner as an approach to maintaining and improving the natural environment while preserving local culture and heritage, is becoming fundamental both for tourism product quality and the economic development of the host territories.

According to the World Tourism Organization, "Sustainable Tourism should manage all resources in such a way that economic, social and aesthetic needs can be fulfilled while maintaining cultural integrity, essential ecological processes, biological diversity, and life support systems". Sustainable Tourism requires careful planning, and a move away from the patchy regulatory regimes of the past. It should emphasise the quality of the tourist product "the result of a process which implies the satisfaction of all the legitimate product and service needs, requirements and expectations of the consumer, at an acceptable price, in conformity with the underlying quality determinants such as safety and security, hygiene, accessibility, transparency, authenticity and harmony of the tourism activity concerned its human and natural environment (WTO 2002)", and ensure as far as practicable compatibility with the needs of the local systems in which it is based. To make progress towards the sustainable development of tourism, it is necessary to analyse the information and know-how related to previous good or bad experiences as well as to integrate new facilities and developments with the ambient environment (Bramwell et al. 1996).

Concerning coastal areas in Europe, tourism is, and has been, one of the main socio-economic drivers of environmental change. Europe is the destination for 58% of all international tourist arrivals (400 millions in 2001, WTO statistics 2002) with the coastal areas representing the most popular destinations. The Mediterranean Sea is still the world's most important tourist's resort; around 250 million travellers came to this region in 2001. These figures do not include millions of people travelling for tourism in their own countries, and which were expected to double by 2020 before the terrorist attack of September 11th (Mastny 2001). All these statistics lead to the conclusion that the economic importance of tourism in the European Union is large, representing 5% of total Gross Domestic Product (GDP). According to the Dobris Assessment (Stanners and Bordeau 1995), tourism is likely to become the largest single economic activity in the EU. Inside Europe, Spain is undeniably one of the world's leading tourist powers, ranking second in terms of international tourist arrivals in the entire world (49.5 millions in 2001) and having the 7,1 % of the world market share (WTO 2002). The resident population of Spain is around 40 million people. Comparing resident people with foreign visitor, we can notice the large importance of the tourism Industry and all their associated activities for the economy of the country. The Spanish Tourist Industry represents over 10% of the National GDP, accounts for annual revenues exceeding 30 billion euros, and generates more than one million jobs (MEH-MMA 1999). Although with some exceptions, most of these figures are still associated with the "mass tourism of sun and beach", and they are related to mature destinations in heavily pressured areas.

With more than 100 years of tourist experience, Spain is now making strenuous efforts to apply sustainability strategies in tourism. Spain has moved away from the days when anything seemed to be acceptable and tourist destinations were built without any regard to either their load capacity or impact on the local environment (MEH-MMA 1999). Today, we have official programmes to stimulate sustainable tourism initiatives (Excellence and Dynamic Development Plans). Also, there are important local initiatives at mature destinations (Patronato de Turismo Costa Brava-Girona 1997, Municipality of Calvià 1999, the latter being one of the most recognised in Europe). And, moreover, the tourism industry is gradually adopting sustainable policies and practices. Spain still has a way to go, since not everything is going in the 'correct' direction. But it is also true that there have been some very positive actions and the experiences gained should be transferred to other less mature tourism in other countries and emerging destinations to avoid old mistakes being repeated. Besides that, tourist activities have expanded to almost every location in the world, current trends indicate a dynamic growth of tourism in other emerging Mediterranean areas such as in Turkey, Croatia, Morocco, Tunisia, and Greece This paper aims to provide some recommendations to help in the promotion of sustainable tourism, using the example of the Costa Brava (Northwestern Mediterranean Catalan Coast) as a case of study.

The Costa Brava (see Sardá and Fluvià 1999 for a regional description) can be seen as an old and mature tourist destination. Although different types of tourist resorts can be found along this coast due to the diverse socio-economic history of its towns , most part of the 220 km stretch has been associated in the past with "mass tourism". The environmental implications of tourism, especially "mass tourism" in coastal areas of the Mediterranean Sea, are well known (Fraguell et al. 1998, Sardà and Fluvià 1999). The impacts generated by an uncontrolled rapid development of tourism followed a clear tourism cycle that can be seen everywhere (Butler 1980). This concept of the tourism as a life cycle has major implications for sustainability, because it is difficult to manage due to the absence of proper and clear policies, procedures and tools.

Tourism products, such as marketed and labelled destinations, tend to pass through a life cycle that goes from a period of euphoria with large territorial transformation and rapid environmental degradation, to a period when the product reaches its mature state and environmental awareness evolves. Then, in the interim, sometimes pollution problems become less important due to the investment in environmental equipment and infrastructures to maintain what is called "environmental quality standards". Finally, more subtle conservation problems arise including loss of symbolic and cultural landscapes and livelihoods. This tourism life cycle evolution is a repeated phenomenon and we should clearly analyse and communicate its consequences to avoid the repetition of similar mistakes elsewhere. Although for old mature destinations sustainable practices have been applied at the end of the cycle to rejuvenate its tourism industry, the introduction of the sustainable principles in the management of the territory and its tourism activities at the very beginning would be essential for those which are right now introducing these practices in their own territories. For both, mature and emerging destinations, Integrated Coastal management can facilitate the introduction of sustainable development of the regions.

Von Bodungen and Turner (2001) have argued that Integrated Coastal management should be something more than an attempt to reorganize coastal spaces and political systems for the purpose of facilitating investment penetration by governments and/or transnational corporations (see also Nichols 1999). Integrated Coastal Management should also facilitate Sustainable Development. These authors recommend that work should be done by natural and social scientists to help coastal managers and society by developing a strategy to identify the problems and their nature, to find solutions, and to formulate products that could be used as guidelines for valuation, assessment, and policy making. Since 1999, work has been carried out, in close co-operation with the Patronage of Tourism of the Costa Brava-Girona, and the Department of Environment of the Generalitat de Catalunya to develop principles and procedures (tools and methodologies) to foster future sustainable tourist development in the Costa Brava. Sustainable tourism means ensuring the future success of existing destinations and planning emerging and new destinations, with their long-term future development in mind (France 1997, Swarbrooke 1999, Fullana and Ayusao 2002). Present managing tendencies recognize the need to develop integrated management tools and self-regulation practices to support policy makers, local operators, and coastal managers in the task of integrating the economic development of tourism with environmental actions at appropriate planning levels. The main goal of this paper is to contribute to the dissemination of these tools and practices to those managers that are facing early developments in emerging destinations, or to those who are having enormous problems in mature ones.

Tourism in the Costa Brava

The Costa Brava (Girona, Catalonia-Spain) is one of the most visited Mediterranean destinations (Figure 1). Its coastal fringe is administratively divided in 22 towns included in 3 main administrative regional divisions or "comarques" (Alt Empordà, Baix Empordà and La Selva). These 22 towns cover 657,4 square kilometres and have a resident population in 1999 of 173,169 inhabitants. This population is enlarged every year by foreign tourists and day visitors as well as by other hispaniards that came from other Autonomous Communities of Spain. The base population of this region (the population formed by the resident population plus the annual average seasonal visiting population) was estimated in 446,337 inhabitants in 1999, almost 2.6 visitors by resident (Table 1).

Fig. 1. Map of the Costa Brava region

Table 1. Statistical data for the 22 municipalities present in the Costa Brava during 1999

Municipalities	code	Resident population	Base/ resident population	Total site accommodation	Shoreline artificialisation (%)	Protected over natural area (%)	Accommodation coefficient	Construction coefficient
Portbou	POB	1,617	1.24	185	20.8	65.00	11.44	0.10
Colera	COL	601	3.38	1,017	32.8	20.69	31.95	1.43
Llança	LLA	4,150	3.02	1,347	89.9	24.17	20.07	3.06
Port de la Selva	PSE	859	5.10	1,512	27.9	87.02	34.81	1.03
Cadaqués	CAD	2,000	3.12	2,070	26.6	60.27	77.25	1.15
Roses	ROS	12,991	3.34	8,507	48.0	52.95	44.51	2.31
Castelló d'Empúries	CEM	6,1087	6.94	10,446	30.2	65.05	16.86	2.20
Sant Pere Pescador	SPE	1,440	5.09	11,705	50.0	33.27	11.25	1.27
L'Escala	ESC	5,942	4.30	6,182	75.1	6.12	14.83	2.16
Torruella de Montgrí	TOR	8,236	2.20	15,034	28.5	2.68	30.43	2.89
Pals	PAL	2,005	4.82	7,925	76.2	2.72	27.08	4.11
Begur	BEG	3,500	2.90	1,528	61.3	21.60	21.29	2.15
Palafrugell	PAF	18,289	1.71	5,473	62.9	29.54	8.27	3.15
Mont-ràs	MON	1,596	1.41	1,304	0.0	24.28	1.07	2.44
Palamós	PAM	14,525	1.65	6,183	53.2	23.24	5.93	3.18
Calonge	CAL	7,035	3.81	8,477	100.0	42.76	19.26	3.67
Castell-Platja d'Aro	CAR	5,785	4.17	6,942	91.9	6.35	78.38	4.37
Sant Feliu de Guixols	SFE	18,420	1.60	3,637	65.0	36.16	14.06	3.40
Santa Cristina d'Aro	SAR	2,945	2.41	2,721	34.4	79.19	10.63	3.18
Tossa de Mar	TOS	4,407	4.11	13,989	39.3	81.71	168.73	1.46
Lloret de Mar	LLO	20,086	2.44	34,635	67.0	60.26	155.41	3.03
Blanes	BLA	30,653	1.65	13,837	84.5	34.10	14.88	4.23

A life cycle analysis of the Costa Brava Tourist product

The first indications of tourist activities in the Costa Brava dated from the middle of the XIX Century. They were associated with the demands for beach space for the high societal classes of Catalonia (Barbaza 1966, Goytia 1995). However, it was not until the late 1920's when initial residential areas were built in Lloret de Mar, Sant Feliu de Guixols, and S'Agaro. The "Tourism cycle" began with some immediate impacts as a consequence of the spread of tourism in this particular zone (Figure-2). During the 1930's tourism activities became evident as many people from the middle social class travelled around and foreign visitors, mainly Germans and British, started to be attracted by small tourism resorts such as Tossa de Mar or S'Agaro (Castell-Platja d'Aro). The new development of the tourism product enhanced a period of euphoria in which economic recovery and social modernisation obscured the future consequences. At the beginning of this cycle, tourism used to facilitate a change never dreamt of by the players involved in the process. The development of tourism in the area led to the positive restructuring and improvement of public municipal infrastructures, modernized the lifestyle of the local population, and increased their own assets. There was no time for worries, and a relevant group of Tourism managers founded the Tourism Patronage of the Costa Brava in 1935 (Cals 1982).

During the 1950's and 1960's, after the Spanish Civil War and the II World War, tourism activities were enthusiastically accepted and tourist resorts and activities increased drastically in the Costa Brava (Figure-2). It is at this time that the concept of "mass tourism" was born along with the Costa Brava destination and little attention was paid to the future implications of this tourist invasion (Fàgregas and Barri 1970, Goytia 1995). It was a time for development, the time in which this area became a very popular tourist destination in Europe. The extraordinary demand for tourism activities attracted many public and private investments, and in parallel, other economic industrial and agricultural activities were almost abandoned. Economic diversification was restricted and the Costa Brava, as well as a wider part of the Spanish Mediterranean area, became very dependent on the tourism industry and its associated components. However, tourists arriving "in mass" to a resort can destroy the very qualities that attracted them to the resort in the first place, and this well-known problem was evident during the 1970's in this region. Nevertheless, there was no natural rational mechanism to avoid pollution, frequent and excessive use of natural resources.

Although recognition of the tourism pressure on the environment began in the 1970's and several actions were observed ("Debate Costa Brava", Aragó 1996), the arrival of democracy to Spain and the need to improve the Spanish economic indexes lead to a period in which we followed the developmental patterns observed in previous decades. Based around a very old Coastal Law of 1969, unplanned growth and over-development, abusive building of second houses, often in sensitive natural areas, and unsustainable use of natural resources, were typical (Sardá and Fluvià 1999).

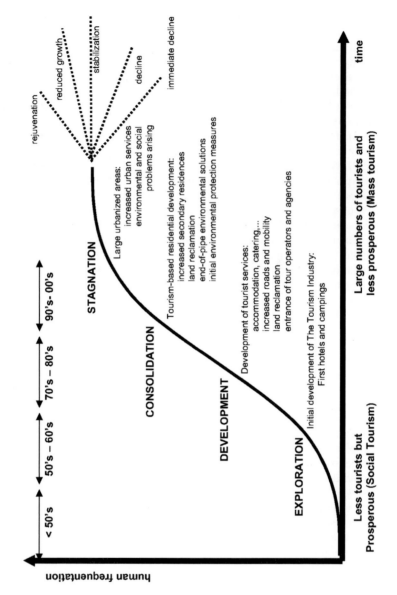

Fig. 2. Life cycle analysis of the Costa Brava traditional tourist product

Environmental awareness promoted by the globalisation of environmental probems brought the relationships between the economy and the environment up the priority agenda of politicians. Several important environmental policies were developed. In 1987, the Brundtland Report defined the concept of Sustainable Development giving much more relevance to other Mediterranean political regulatory measures such as the Mediterranean Action Plan (1975) or the Barcelona Convention for the Protection of the Mediterranean Sea (1976). In addition, a new Spanish Coastal Law was enacted in 1988 to protect the coast against unregulated development. The Rio Conference (1992) and the EC 5th Community Policy and Action programme in the field of Environment and Sustainable Development (1992) occurred more or less at the same time that the executive responsibilities for environmental policies in Catalonia were transferred from the Spanish Government in Madrid to the Generalitat of Catalonia, the Autonomous Government of Catalonia. The 5th Programme regarded tourism as an extremely important element in the economic and social life of the EU, a reflection of the legitimate aspirations of its citizens to enjoy other places, to get to know other cultures, and to engage in different activities or spheres of leisure outside their usual place of residence or work (MEH-MMA 1999). This new situation clearly started to change the way in which tourism and environmental issues were managed in the Costa Brava.

At the beginning of the 1990's, the Costa Brava appeared to be a very well consolidated destination and the of long-term impacts produced by past tourism activities were clearly recognized: a) environmental impacts: the excessive use of resources, and pollution, exacerbated by the concentration of visitors in time and space in destinations that were not geared to withstand such pressure; b) urban environmental impacts: abandonment or disappearance of old parts of towns, poor quality construction, and the transformation of old villages into tourist resorts; c) economic impacts: poor economic diversification, predominant a tertiary sector that must be adapted to a floating population with the necessity to be strong enough to support the seasonality of the demand; d) social impacts: with the development of conflict of interests and society disintegration by the division into the sum of individual interests; and, e) cultural impacts: with the loss of traditional activities and local customs. To reverse some of these problems, new instruments and tools were introduced: new end-of-pipe environmental solutions, conservation areas (natural parks), and environmental education. Finally, in 1997, the Tourism Patronage of the Costa Brava released its document, "The Action Plan of Sustainable Tourism in the Girona regions", a plan which emphasized tourism quality and the need for the introduction of sustainable practices inside the industry. Sixty years after its creation, the tourism industry realized that the future of the sector could only be assured through the sustainability of its own actions.

In a recent publication (Aragó 1996) the perceptions of 21 opinion leaders about problems arising from the past development of Tourism activities in the Costa Brava and possible solutions were discussed. Those leaders were asked to give their opinions on the most important improvement measures and potentials, as well as the most acute problems still to be solved (see Figure 3). The improvements in the quality of fresh- and sea-water as a consequence of end-of-pipe water treatment plants, and the protection of natural systems as a consequence of the in-

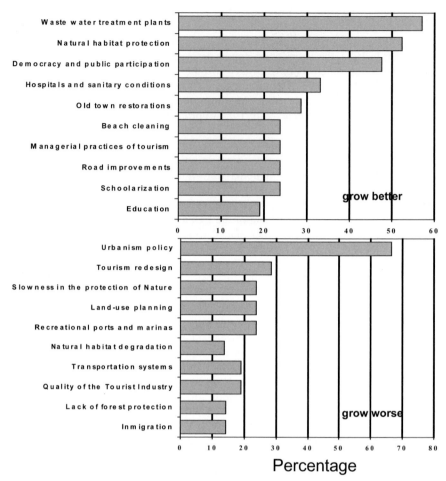

Fig. 3. Gains and losses connected with developmental pressures in the Costa Brava as a consequence of shoreline development through tourist activities during the period 1970's-1990's. The graph illustrates the percentage of responses of a group of interviewed opinion leaders

troduction of the "Pla d'Espais d'Interès Natural" in 1988, a Catalonian Act for the protection of Nature, were mentioned as clear improvements. The residential developments and the associated loss of natural habitats were seen as the most serious problems.

Present numbers for the tourism industry in the Costa Brava

The number of people visiting Spain is increasing every year (Figure 4). Foreign tourist arrivals have increased more than 100-fold since 1950. About 50% of for-

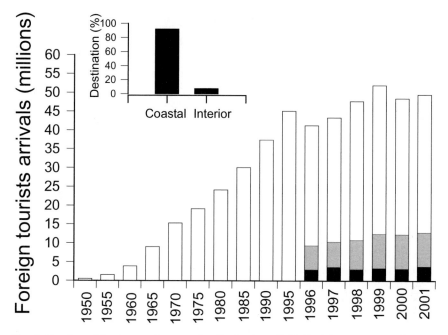

Fig. 4. Number of foreign tourists entering Spain since 1950. Each bar represents the mean annual number of foreign tourists that visited Spain (open bars), Catalonia (grey bars), and the Costa Brava (black bars) during the selected years. The percentage of foreign tourists who visited the coastal and interior zones during the last decade is plotted on the inset of the figure

eign tourists staying in hotel accommodation choose the Balearic Islands and Catalonia as their destinations, with the greatest concentration in Spanish Autonomous Communities situated on the Mediterranean Coast and the Islands, and the "mass tourism of sun and beach" as a the most representative tourist model (MEH-MMA 1999). Catalonia accounted in 2001 for the 27% of foreign tourist visitors to Spain (12.9 millions of foreign tourist excluding day-trippers, amounting to 7,6 millions). The Costa Brava represented roughly one third of these visits (31%). In addition, Catalonia received 4.7 millions of hispaniard travels coming from other Autonomous Communities of Spain (Catalunya Turisme 2002). Due to these figures, the Tourist economic sector contributed 9,9 % to the Gross Domestic Product of Catalonia in 2001 (IDESCAT 2002).

Between 1996 and 2001, the total number of foreign tourist arrivals to the Costa Brava increased by 24,5% (a 3,7% of annual increase, with high interannual variations). The total number of bed accommodation in the Costa Brava was reduced during the same period by 2,3 % for hotels and 1,6 % for campsites. The rotation of foreign tourists is higher now than in 1996, the time per visit has been reduced (8,9 days per visit for 2001), and the accommodation in secondary residences owned by foreign people, and apartments, is also high.

Table 2. Number of hotels and beds in the Costa Brava region

Location		1964	1986	2000
Alt Empordà	Establishments	130	133	149
	Beds	5,786	9,642	10,808
	Accommodation coeff.	34	32	26
Baix Empordà	Establishments	322	257	195
	Beds	14,261	14,591	16,208
	Accommodation coeff.	47	25	24
La Selva	Establishments	381	284	277
	Beds	20,849	38,746	42,211
	Accommodation coeff.	221	131	107

The profile of the tourist industry in the Costa Brava is summarised in (Figures 5a and b). The quality of hotel installations needs improvement. With an average of 2,3 star per bed (Figure 5a, top graph), the industry is trying slowly to improve the quality of these installations. The hotel quality variation has been positive during the period 1996-2000 (Figure 5b, top). This variation is associated with the disappearance of old installations and improved quality in the construction of the newer ones. However, for many small hotels and facilities, it is problematic to invest some of the earnings in modernization and upgrading of the quality of the services. An historical overview of the development of hotel accommodation capacity can be seen in Table 2. Although the number of beds has increased with time, the number of hotel facilities has been reduced since the 1960's. The reduction of hotel facilities goes in parallel with the reduction of the accommodation coefficient numbers (the number of tourist accommodation beds by each 100 resident people). Although for some locations such as Tossa de Mar and Lloret de Mar there are more than 1 tourist bed by resident bed (Figure 5a, middle), there is a clear tendency involving the reduction of the accommodation coefficient through time (Figure 5b, middle; Table 2). On the other hand, the construction coefficient (the number of primary and secondary residences built during the last 5 years by every 100 resident people) is increasing (Figure 5a, lower) and the tendencies are showing a maintained growth of around 8% during the period 1996-200 (Figure 5b, lower).

Seasonal fluctuation in tourist arrivals continues to be a characteristic feature of the tourism in this region, with almost 50 % of the tourist arrivals between June and September. This seasonal phenomena has enormous environmental implications as it obliges the administration to deal with urban infrastructure that is only used to full capacity for a few months during the year. This seasonal dynamics can be seen everywhere, in annual foreign tourist arrivals of Catalonia (Figure 6, bottom graph), in the structure of the seasonal population in the Costa Brava (Figure 6, middle), or even in the price of a bed in a tourist resort in La Selva Marítima (Figure 6, top).

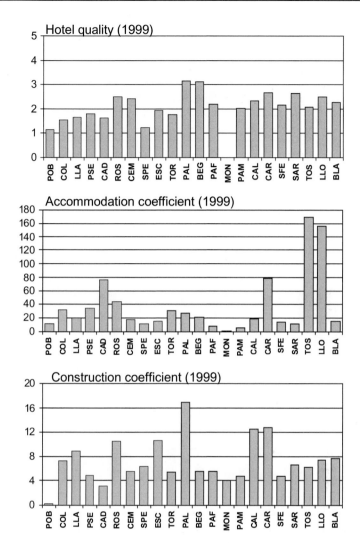

Fig. 5. Present situation of the Tourism Industry in the 22 towns of the Costa Brava expressed by some indicator numbers. Right graphs: Hotel quality (average stars by hotel bed). Accommodation coefficient (number of tourist accommodation beds by each 100 resident people). Construction coefficient (number of primary and secondary residences built during the last 5 years by each 100 resident people). Left graphs: Percentage of variation during the period 1996-2000 for the above indicators. Code for coastal towns follows Table 1

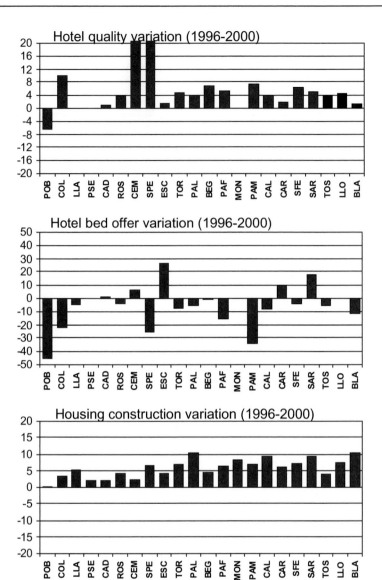

Fig. 5. Continued

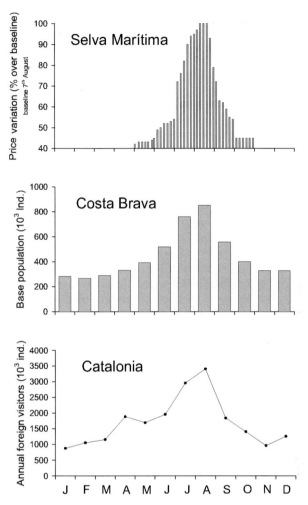

Fig. 6. Monthly seasonal variation for: Upper graph) the price of a bed in a tourist resort (percentage of variation since the maximum price, August 7th; adapted from Espinet and Fluvià 2001). Middle graph) the base population inhabiting the Costa Brava coastal towns. Bottom graph) annual foreign visitors in Catalonia (adapted from Catalunya Turisme 2002)

In conclusion, the growth of the traditional tourism industry in the Costa Brava, like in other locations of the Spanish Mediterranean Coast, has been reduced and it is believed that we have entered the beginning of an stagnation period. The traditional tourism industry is suffering today from the consequences of an historical uncontrolled and unplanned growth which has generated benefits in the short time, with absence of care for the natural and cultural heritage, and based on products of medium or low quality.

The need for development and ICZM process in the Costa Brava

Shoreline development in the Costa Brava has yielded a very complex situation in which two main circumstances can be seen together (Sardá and Fluvià 1999). There are very heavily pressured zones where shoreline development has been excessive and land reclamation, housing availability, road developments, environmental facilities, natural resource uses, beach and cove occupations, proliferation of recreational ports, marinas and other sea related recreational activities, energy and water consumption are still growing. However, on the other hand, there are other zones that were preserved in the past and which are in very good environmental condition, preserving the beauty of the old landscapes of this region. In this context, 48,4% of the entire shoreline of the Costa Brava has been artificialised (i.e. if human activities can be seen in its first 200 m of land) (Table 1). This evolution has been the consequence of decades of the application of market mechanisms with few interventions by regional and local politicians and coastal managers protecting the public interests. The communities changed economically, socially, an environmentally without a detailed analysis of future scenarios and consequences. At the change of the century there is a need for change, a need for a much more sustainable development of tourist activities, and we believe that an integrated coastal zone management process can help in this situation.

To tackle the issue of sustainable tourism development in mass tourism destinations as the Costa Brava, we first need to some: a) tourists are now much more concerned with the environmental quality of their holiday destinations and facilities, b) the strategy for sustainability in tourism fundamentally needs to be adapted to local destinations on which tourist base their choice of holiday, c) there is a higher demand of public participation in decision making, and a higher demand from people for protection measures for the territory, and d) there is a clear move in this zone away from a traditional holiday hotel-based tourism to a tourism based on secondary residences with higher demands on land.

With the general idea of the integration of tourism and environmental strategies into the sustainable development of the Costa Brava, an integrated coastal zone management process could be useful. This process could be structured in three particular actions (Sardá 2001):

- The development of sustainable plans for sectoral organizations. They are the main agents producing change, and for them these plans should be encouraged. It is necessary to work on the improvement of the environmental performance of private operators, and the awareness raising of tour operators and tourist themselves. The introduction of environmental policies, environmental management systems, and environmental practices in the companies should be recognized and rewarded;
- The need for the construction of a strategic regional coastal plan. This plan should follow an ICZM scheme to base daily decisions on the use of the territory. In this way information-based instruments must be developed and clear objectives and future scenarios should be formulated to underpin the managerial decisions of the Plan;

- The development of Local Agendas 21. Participation and cooperation between local authorities as well as discussion with their citizens, local organizations, and private companies are paramount to develop a local program according to the Rio Agenda 21.

The development of sustainable plans for sectoral organizations

As the major player in the tourism system, the private sector has a major influence in determining the extent to which impacts from tourism upon the environment will be either positive or negative. In 1997, the Tourism Patronage of the Costa Brava-Girona released its document, "The Action Plan of Sustainable Tourism in the Girona regions". One of the objectives was to promote environmental practices in the Industry. Although in the past private sector organisations such as tour operators and hotel and camping developers were solely interested in maximising their profits, they now started, slowly to move towards a much more proactive vision in the field of tourism, enhancing their environmental operations.

In the year 2001, we carried out a survey between the 338 hotels of the Costa Brava to evaluate the environmental perceptions and environmental policies of these private operators. A large questionnaire was distributed with the help of the Patronage of Tourism of the Costa Brava-Girona. We obtained a 20% response, a value that is similar to other questionnaires related to environmental issues (European Eurobarometer project, (Belz and Strannegard 1997), 17,6% of response; the 2001 Report of Environmental management on the Spanish Industry, Fundación Entorno, 2002, 22,3% of response). The average Costa Brava hotel is a small-medium enterprise, family-based hotel, 2,3 stars, open for 9 months of the year, and with more than 15 years of operations (from April to December). The main conclusions from the questionnaire survey were the following:

- The incorporation of environmental factors is low on the Costa Brava hotel sector's priority list. Activities questioned (conscientiousness, environmental evaluations, introduction of environmental policies, etc.) allow us to rate only 20-25% of the companies that replied to the questionnaire as having introduced some environmental strategies in their businesses; only 9% have an operative Environmental Management System (two of them having achieved an EMAS award, EC "Environmental Management and Audit Scheme"). This percentage ought to be balanced with the reply index obtained (around 19,23%), in the worse of cases meaning that only 5% of the hotels on the Costa Brava have introduced some environmental principals in their management;
- The sector seems to understand the need to improve both its environmental awareness and actions. But so far progress has been slow;
- There was some relevant and worrying lack of knowing among tourism professionals. While approximately 75% of the people surveyed claimed knowledge of environmental regulations (waste, water, noise, etc.), only 30% were aware of the Integral Intervention Catalan Community Act, that regulates the license of activities following the EC "Integrated Pollution Prevention Control Directive";

- In the cases where awareness is high, the implementation of Local Agendas 21 is well accepted among the hoteliers surveyed. Nevertheless there is a lack of knowledge of this basic tool in the context of Municipal Sustainable Development. Beach cleaning, tourist excellence programs, urban areas improvement, etc appear to be well valued. On the other hand, the urban sprawl of second homes and tourist eco-taxes are regarded suspiciously.

Although several environmental voluntary instruments and tools can be used by the private accommodation sector to advance towards a sustainable tourism, environmental auditing, environmental management systems, and eco-labelling, are for the most part restricted to large hotel chains. In the Costa Brava, were small business are more important, there is still a need to promote the use of these voluntary agreements. However, while only 2 hotels have been awarded the EMAS recognition, we have 5 campsites that had the award (there are 98 Campsites in the Costa Brava). Campsite accommodations are more proactive in environmental issues because they are much more nature related business and because campsites can demand higher prices if managed in a very sustainable way. Recently the European Community has set an EC "Eco-labelling Scheme" for Tourist accommodations, with the main criteria concentrating on limitation of energy consumption, of water consumption, the use of chemical substances, waste production; and the favouring of the use of renewable resources and of substances which are less hazardous to the environment. The European scheme follows a previous national eco-label scheme on Tourist accommodations done by the Generalitat of Catalunya, the "Distinctiu de Qualitat Ambiental". Today 2 hotels and 5 campsites have this ecolabel. Besides these good experiences, the accommodation tourist sector in the Costa Brava is still showing little interest in environmental concerns, and managers still have a lot of work to do in this particular issue.

Nevertheless, large tour operators such the two biggest European groups, the Touristik Union International (TUI) and the Thomson Travel Group, perform regular environmental audits in their operations. TUI, for example, consults with governments of host countries, international and national organisations with responsibilities in tourism and environment, regional and local authorities, business partners, and its own customers, to make them aware of good environmental practices. We expect to see more of these initiatives in the future.

The need for the development of an strategic regional coastal plan

The Economy of the Costa Brava region is highly dependent on the tourism industry. The need for a Strategic Regional Coastal Plan in the Costa Brava is becoming necessary as the incorporation of long-term considerations and the integration of the environmental factor, along with the participation of all stakeholders, would be required to reach sustainable tourism in the future.

We have been working with national and regional managers to develop some tools and methodologies for the evaluation and assessment of a future Plan for this region: a) a panel of indicators, b) the use of a Geographical Information System (GIS), and c) a set of different graphic package presentations (Sardá et al. 2003). In order to facilitate further management and planning, fourteen strategic indica-

Table 3. Strategic indicators used in the study by Sardá et al. (2003)

	Strategic indicator	Measure
1	Population density (Pr)	Inhabitants per km^2
2	Base population (Pb) over resident population (Pr)	Resident + average seasonal population over resident population
3	Impervious soil	Percentage over total municipal soil
4	Construction coefficient (Pb)	Built houses per 100 inhabitants in the last 5 years
5	Unemployement rate	Percentage over active population
6	Acommodation coefficient (Pr)	Hotel beds per 100 inhabitants
7	Hotel ratio quality over price	Average price by star at the peak season
8	Waste production intensity	Household and industrial wastes per unit GDP
9	Protected area	Percentage over total natural area
10	Coastal protection index*	Institutional index
11	Motorization coefficient (Pr)	Vehicles per 1000 inhabitants
12	Coastal fringe artificialisation	Percentage of coast artificialised over total
13	Beach quality	Institutional index based actually on sanitary conditions
14	Water depuration	m^3 per person per day

* This indicator was obtained from the "Pla de Ports Esportius" of the Generalitat of Catalonia (Recreational Ports Plan).

tors were selected (Table 3). These indicators incorporate a series of criteria; they are available for all the municipalities along the coast, they avoid duplication of information, they are easy to obtain, they compile information from the main economic sectors (tourism, construction and mobility) as well as from human population aspects, and environmental issues, and they are easily understandable.

Figure 7 illustrates the weight of four of these strategic indicators in a principal component analysis covering data on 22 Costa Brava towns. The location on the main axes follows a pattern of artificialisation as human density increases in these figures from right to left. Human density patterns (Figure 7a), and all their associated economic effects (construction, mobility, etc.), are inversely correlated with the index base population divided by resident population, a measure of seasonality of the population (Figure 7b), and with environmental protection (Figure 7d). See also Table 1 for town values. These two later indicators seem to be also in some way related with the average price by hotel star (Figure 7c). Hotel tourism investments in protected and less populated areas can be more profitable, giving higher revenues, than those made on less protected and highly populated areas. The price of hotel installations in less artificialised areas are higher for the same level of investment measured in terms of quality.

Therefore, using the developed environmental indicators, it is possible to classify the Costa Brava municipalities into three main groups. First, a group of towns that was created in the past around other commercial activities rather than tourism, and even if tourism has been important to develop these towns, there are other urban commercial, industrial and residential considerations that are important today. These towns are not classified as touristic towns by the Law 8/1987 of Catalonia because the annual seasonal population is not larger than the resident population and consequently, the index of seasonality of its population is lower than two (see

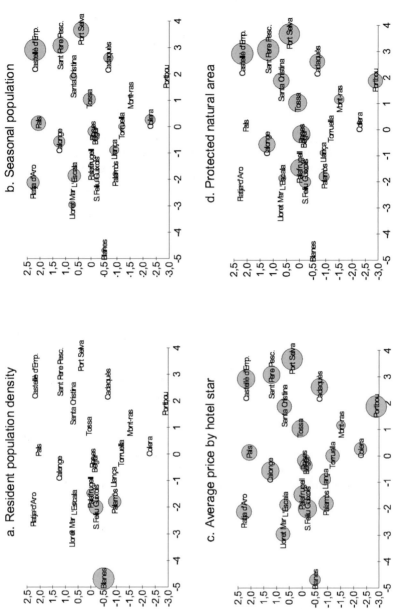

Fig. 7. Bubble representation of the weight of four of the twelve strategic indicators used in a PCA analysis to allocate differences betweens the Costa Brava coastal towns

Table 1). Second, a group of towns that was clearly associated with the tourism "boom" of the Costa Brava. Its growth was entirely dependent on this process and they did benefit in the 1960's and 1970's by the development of the industry in the region. And third, a group formed by the towns that were not initially considered in this development. These towns arrived to the 1990's with a larger portion of their territory not artificialised and, as a consequence, its natural territory had was potentially available for conservation. Most part of the towns belonging to this later group are included in the "comarca" of L'Alt Empordà. The further development of an integrated plan should emphasize these different realities, and it is obvious that the possibilities to rearrange the local activities for each town would depend on the group in which they are now included. However, although the actions and initiatives that could be derived from the development of such a Plan should be implemented at a local (municipal) level, the introduction of this local process in a structured regional plan should facilitate a harmonization of efforts and regional coordination.

The development of Local Agendas 21

Making tourism more sustainable requires also the involvement of all the stake-holders, including the local communities that will be directly affected by tourism's presence. As Agenda 21 has been accepted as the correct tool to allow social participation in the move towards sustainable development, its development through Local Agendas 21 is a high priority. Following these criteria, the World Tourism Organization, together with the World Travel and Tourism Council and the Earth Council produced their own Agenda 21 for the travel and tourism industry, governments, and others (WTTC et al. 1995).

In the Costa Brava, we are still in the process of developing the Local Agendas 21. Seventeen of the twenty-two towns of this region have signed the Aalborg letter, and nine of them are in the process of creating their own Agenda 21. However this tool has been applied in other well known cases such as the one in Calvià (Balearic Islands), one of the most outstanding cases of how to use this instruments to move towards the sustainability of mature tourist destinations (Ayto. Calvià 1999, MEH-MMA 1999).

Discussion

"Those of us who were not fortunate enough to spend our summers on the Costa Brava when we were children can only see these photographs as a dream that we almost experienced, something we barely touched with our fingers and which escaped us for ever". This phrase, extracted from a wonderful book about the Costa Brava "Rediscovering the Costa Brava", (Racionero et al. 1984) expresses clearly what shoreline development related to tourism issues meant for the natural and cultural values of this territory. If we are not able to manage tourism development in a region properly, irreversible processes related to land transformation and environmental degradation can quickly combine to the attractiveness of this region.

The evolution observed for the Costa Brava along its life cycle showed the movement from an initial tourism based on low volume high value principles (Social Tourism) to a one based on high volume low value principles (Mass Tourism; in parallel the development of tourism based on secondary residences which spread out everywhere. In many places on the Costa Brava, there is a clear perception of a loss of attractiveness, especially from repeat tourists and visitors.

Traditional tourism in the Costa Brava has reached a very consolidated, position. Several recommendations are given today to rejuvenate its position and to avoid further declining patterns: the move from "mass tourism" to other types of tourism "cultural-tourism, rural-tourism, eco-tourism, tourism for retired people ... to reduce also their seasonal effects, to implement a much more modern and conscientiousness management inside the tourist industry, and to protect the landscape, "mountain ranges run headlong into the sea, forming small, secluded pine coves and beaches that attracted foreign visitors at the beginning", and the cultural diversity of its towns and people. Recommended to tackle the problem.

In 1995, the Canary Islands developed its Canary Islands Tourism Regulation Act recognising the interdependence of economic and environmental systems, aimed to develop a future tourism of quality and at the same time affording a marked degree of protection of the island territory. In 1998, the Balearic Isles Regional Authority enacted two decrees regulating tourist establishments, and approved transitory measures limiting the opening of new tourism facilities and activities, including a tourist eco-tax. Nevertheless, all these strategies require an agreement between all the stakeholders intervening in the coastal zone, a longer-term perspective on policymaking, and a coordination of efforts between the general administration, the local coastal managers, the industry, and the participation of the society. Sustainable tourism is a goal, which can not be reached immediately, it is going to be difficult to readjust tourist activities given the history in this region, but the use of an ICZM process can really help. The development of an effective ICZM plan would be essential to achieve the goals for the future.

ICZM provides a conceptual framework within which individual strategies for resolving issues and promoting sustainable coastal development may be formulated (Richter et al. 2001). In the process of developing this effective management several steps needs to be addressed. The awareness of the problem that facilitates further dialogue and exchange of different views about the future, and then cooperation, coordination of efforts and actions, and the integration of all of them in a General Plan. We are still in the process of dialogue and cooperation in the Costa Brava. In order to avoid further degradation and to move towards a sustainable tourism, we will need to co-ordinate the needs of traditional tourism with the reality of the new tourism based on secondary residences, to find formulas in which both can coexist without diminishing the quality of the product. Maintaining the sustainability of tourism will require managing environmental and socio-economic impacts, establishing environmental indicators, and improving the quality of the tourism product, the Costa Brava product, either the well known tangible products such as installations and facilities, or the intangible ones, the natural and cultural heritage of the territory. ICZM can facilitate this move.

In a healthy western economic society such as the Costa Brava, it is still possible to promote the needed dialogue, to direct some income to the solution of environmental pollution problems or, even to protect new territory, or to restore degraded natural systems. However, when these processes are seen in developing regions of the world, damages are much more difficult to reverse (Mastny 2001). It becomes necessary to disseminate past experiences such as the ones observed in the Costa Brava to help local managers of emerging destinations.

The pace of tourism expansion is not expected to slow down. As developing countries grow richer more people can afford travel, tomorrow visitors are expected to come from Russia, China, and other eastern countries. The realisation that tourism may harm the natural, cultural and social resources, on which it is based, has generated increasing calls for its sustainable development. However, recognition of long-term environmental impacts generated by tourism, and appropriate regulatory and management actions are not easily realisable objectives. There are too many contrasting interests caused by the different sectors involved, and political circumstances make the tourism sustainable management complex. Besides, it is difficult to understand what the term sustainability means for a special territory. However, it is clear today that if tourism wants to contribute to the sustainable development, then it must be economically viable, ecologically sensitive and socially and culturally appropriate. Past experiences, such as the one observed in the Costa Brava, should cause us to think about the negative consequences derived from the fact that we were not able to incorporate long-term planning at the very beginning. The introduction of ICZM processes can be a very powerful tool to advance the cause of more sustainable tourism worldwide.

Acknowledgements

We are grateful to the organizers of the ELOISE Workshop to invite us to present a paper to the discussion on "developing coastal futures for Europe". Special thanks are due to Modest Fluvià, Anna Garriga (University of Girona), Josep Francesc Valls (ESADE Bussines School), Francesc Lopez, (Patronato de Turismo Costa Brava-Girona), and to Muntsa Solà, Sergi Taboada and Xènia Illas (CEAB-CSIC) for their help during the work that has been carried out. This paper is a contribution to the projects 2FD97-0489 and SEC2000-0836-C04.

References

Aragó N (1996) La Costa Brava, vint anys despres del Debat. Revista de Girona 179:60-93
Ayuntamiento de Calvià (1999) Calvià Agenda Local 21: la sostenibilidad en un municipio turístico. Ayuntamiento de Calvià, Mallorca
Barbaza Y (1988) El Paisatge humà de la Costa Brava. Edicions 62, Barcelona
Belz F, Strannegard L (1997) International Business Environmental Barometer. Cappelen Akademisk Forlag, Oslo
Bramwell B, Henry I, Jackson G, Prat AG, Richards G, van der Straaten J (1996) Sustainable Tourism Management: principals and practices. Tilburg University Press, Tilburg
Butler RW (1980) The concept of a tourist area cycle of evolution: implications for the management of resources. Can Geogr 24:5-12

Cals J (1982) La Costa Brava I el Turisme: estudis sobre la política turística, el territori i l'hotelería. Ed. Kapel, Barcelona

Catalunya Turisme (2002) Catalunya Turística en números 2001: dades bàsiques 2001. Catalunya Turisme, Barcelona

Fábregas i Barri E (1970) Vint anys de Turisme a la Costa Brava (1950-1970): cara i creu d'una època divertida. Ed. Selecta, Barcelona

Fraguell RM, Capellà J, Donaire JA, Muñoz JC, Ullastres H (1998) Turisme Sostenible a la Mediterrània. Ed. Brau, Girona

France L (1997) Sustainable Tourism. Earthscan Publications Ltd., London

Fundación ENTORNO (2001) Informe 2001 de la Gestión Medioambiental en la Empresa Española. Ministerio de Ciencia y tecnología. Madrid

Fullana P, Ayuso S (2002) Turismo Sostenible. Ed. Rubes, Barcelona

Goytia A (1996) Back to a sustainable future on the Costa Brava. In: Bramwell B, Henry I, Jackson G, Prat AG, Richards G, van der Straaten J (eds) Sustainable Tourism Management: principles and practice. Tilburg University Press, Tilburg, pp 121-145

IDESCAT (2002) Anuari Estadístic de Catalunya, 2001. Institut d'Estadística de Catalunya, Generalitat de Catalunya

Mastny L (2001) Traveling light: new paths for International Tourism. Worldwatch paper 159:1-88

Ministerio de Economía y Hacienda (MEH), Ministerio de Medio Ambiente (MMA) (1999) España: un turismo sostenible.Centro de Publicaciones de Ministerios, Madrid

Racionero L, Masagué A, Jordi S (1984) La Costa Brava recuperada. Ed. Lunwerg, Barcelona

Richter C, Burbridge PR, Bätje C, Knoppers BA, Martins O, Ngoile MAK, O'Toole MJ, Ramachandran S, Salomons W, Talaue-Mcmanus L (2001) Integrated Coastal management in Developing Countries. In: Von Bodungen B, Turner RK (eds) Science and Integrated Coastal Management. Dahlem University Press, Berlin, pp 253-274

Sardá R (2001) Shoreline development on the Spanish Coast: problem identification and solutions. In: Von Bodungen B, Turner RK (eds) Science and Integrated Coastal Management. Dahlem University Press, Berlin, pp 149-163

Sardá, R., M. Fluvià. 1999. Tourist development in the Costa Brava (Girona, Spain): a quantification of pressures on the Coastal Environment. In: Salomons W, Turner RK, Lacerda LD, Ramachandran S (eds) Perspectives on Integrated Coastal Zone Management. Springer, Berlin, pp 257-276

Sardá R, Mora J, Avila C (2003) Sostenibilidad y Gestión Integrada de Zonas Costeras. Documentos del VI Congeso Nacional de Medio Ambiente. Madrid

Sardá R, Mora J, Avila C (in revision) A methodological approach to be used in ICZM proceses: the case of the Catalan Coast. Catalonia

Stanners D, Bordeau P (1995) Europe's Environment: The Dobrís assessment. European Environmental Agency, EC-DG XI, Copenhagen

Swarbrooke J (1998) Sustainable Tourism Management. CABI Publishing, London

Von Bodungen B, Turner RK (2001) Science and Integrated Coastal Management. Dahlem University Press, Berlin

World Tourism Organization (2002) Tourism Highlights 2002

World Tourism and Travel Council, World Tourism Organization, Earth Council (1995) Agenda 21 for the Travel and Tourism Industry. London

17. Management of contaminated dredged material in the port of Rotterdam

Tiedo Vellinga[1] and Marc Eisma

Abstract

In the port of Rotterdam some 20 million m^3 of sediment is dredged each year. More than 90% of this material is not contaminated and can be disposed of at sea. The amount of contaminated dredged material has decreased due to, amongst others, the focus on the control of the initial sources of contamination by the Municipal Port Management. Further improvement of the quality of dredged material in the port is worked on in the Rhine Research Project II. Key concepts in this project are emission control, harmonisation of water and sediment policy and a shift towards integrated sediment management.

The part that is still contaminated is stored in a large disposal site, called the Slufter. If at all possible (given environmental, scale and economic considerations), the Rotterdam port authorities want to make beneficial use of the dredged material rather than store it in the Slufter. So, since 1992 sand is separated and recently some clay has been made from the dredged material. Since 2001 studies have been carried out into thermal immobilisation and into the possibilities for the actual use of the dredged material.

Introduction

The port of Rotterdam owes its leading position in Europe and the world to a major extent to its geographical location: on the North Sea and the Rhine and Meuse estuaries. This location implies large transports of sediments where sea and river meet, in and out of the port, both marine and fluvial. Marine sediments accumulate through tidal action mainly in the western port areas, whereas the eastern port areas are mainly influenced by fluvial sediments, transported by the Rhine (Figure 1).

---

[1] Correspondence to Tiedo Vellinga: t.vellinga@portofrotterdam.com

J.E. Vermaat et al. (Eds.): Managing European Coasts: Past, Present, and Future, pp. 315–322, 2005.

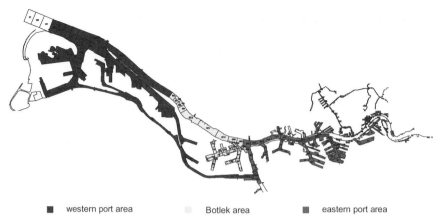

■ western port area   Botlek area   ■ eastern port area

Fig. 1. The port of Rotterdam

Most of the sediments to be dredged derive from the marine environment and only around half of the river sediment settles in the port. The other part of this fluvial sediment finds its way into the North Sea. Rotterdam Municipal Port Management (RMPM) is responsible for the nautical safety and shares responsibility for the quality of the environment in the port area. A sustainable clean and safe port is a basic condition for a healthy future for the port and surrounding region.

Commissioned by the managers of the port area and river (RMPM and the Ministry of Transport, Public Works and Water Management), to maintain adequate port facilities, 15 to 20 million m^3 of sediments are dredged per year.

The relocation of this dredged material to the North Sea, the preferred disposal option, is regulated by a set of chemical criteria, the so-called Sea/Slufter limits. Dredged material exceeding these limits, mainly sediments from the eastern port areas (and partly from the Botlek area), has to be disposed of in a confined site, the Slufter (Figure 2). However, nowadays most of the dredged material is returned to where it came from: the sea.

Fig. 2. The Slufter – a confined disposal site for contaminated dredged material

Sediment management in the port of Rotterdam

Rhine research project

Since in the seventies the contamination of the river sediment became apparent, the Rotterdam Dredged Material Management Program has been strongly focussed on the control of the initial sources of contamination, by way of the Rhine Research Project. Agreements were successfully reached with all major discharging companies concerning a radical reduction in their discharges. Over the past fifteen years, this approach has led to a significant reduction in point discharges and resulted in a significant improvement with regard to the quality of the Rhine water and consequently to the quality of the dredged material in the port of Rotterdam (Figure 3). Although for a number of contaminants the target values have not been met, the Rhine river management interests now tend to shift to other issues.

These developments made it necessary for the RMPM to start a follow-up: 'Rhine Research Project II'. The main objective of the 'Rhine Research Project II' is to ensure that all dredged material is sufficiently clean by 2015, in line with the concept of a sustainable port and region in which port activities take place. In other words, all dredged material should be clean enough either to be relocated in the North Sea or to be beneficially used.

The Port of Rotterdam is not in a position to solve the sediment and dredged material issues on its own. Collaboration with organisations that are involved in water and sediment management for the Rhine and the North Sea such as the International Commission for the Protection of the Rhine (ICPR) and OSPAR is an important implied objective.

Emission control, with a shift from point to diffuse sources

Several heavy metals, polychlorinated biphenyl's (PCBs) and polycyclic aromatic hydrocarbons (PAHs) are of concern with regard to the quality of sediments in the Rhine catchment area and are as well criteria for the relocation of dredged material to the North Sea. The question arises, how the contamination of dredged material will develop in future and whether it will reach levels that allow its relocation to the North Sea. With a model two types of scenarios were assessed for the time period until 2015. The 'business as usual' (BAU) scenarios take measures into account that are already agreed on or are 'in the pipeline', i.e. the implementation can most probably be expected. The 'Green' scenarios consider additional reduction measures that might be realised but largely depend on upcoming policies. Taking the present state as a starting point, the changes in modelled future inputs in the Rhine basin were extrapolated on the development of the quality of sediments in eastern parts of the Port of Rotterdam and compared to current Dutch quality criteria for relocation of dredged material to the North Sea (Figure 3).

Measures, accounted for in the *BAU* scenarios, are not expected to result in a substantial decrease in contamination of sediments/dredged material for most of the investigated substances. Additional measures (*Green* scenarios) could achieve more satisfying results towards reaching the Sea/Slufter limit values. However, even in the Green scenarios, defined target values will be still exceeded for the investigated compounds until 2015, with the exception of lead (not shown in Figure 3).

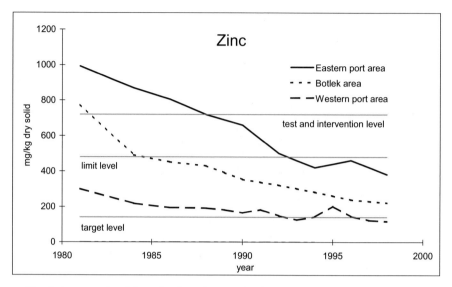

Fig. 3. An example of the reduction of contamination in the different areas of the port

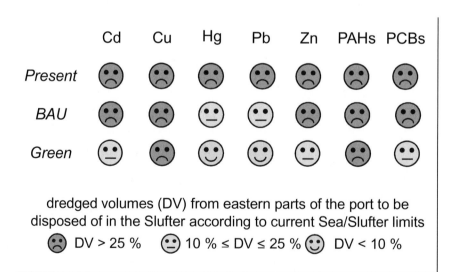

Fig. 4. Present and estimated future quality of dredged material in the eastern parts of the port of Rotterdam according to current Dutch criteria

Harmonisation of the Rhine – and North Sea policy

Also the related current and future policies and regulatory frameworks are investigated. For both the river Rhine and the North Sea and for the port of Rotterdam it is very important that policies are harmonised. Although it seems obvious that because of its nature water and sediment management should be integrated within the system of river catchment and coastal zone, in reality both policies do not seem to be linked. The lists of priority contaminants are very different. Upstream sediment can be returned to the system when it exceeds concentrations up to three times of what is actually present in the river. Downstream in the port it is not allowed to return the sediment with even the same quality. For the relocation of the dredged sediment in the marine environment permitting has been since 2003 also based on biological assessment. No such instrument is being developed for the Rhine sediment itself for the purpose of control of the initial sources of contamination. These sources include in this case also all the sleeping and seeping sources that are present in the Rhine catchment. Concerning the assessment of the quality of sediments and dredged material in the Rhine catchment area, 'new' substances can become of concern, which at present are not considered problematic.

The river catchment and the coastal zone should be treated as a continuum/one system.

Efforts to improve sediment/dredged material quality should encompass the assessment of biological, physical, chemical and economic factors, and the balancing of these gains and losses against political, economic and social welfare decision criteria. This implies the need for harmonisation of approaches and the integration of stakeholders in the decision-making process.

It is important that these basic principles be followed now when the focus of the EU water quality management is on the implementation of the European Water Framework Directive.

From dredged material to sediment management

At present, direct emissions from the river dominate the input for most contaminants into the North Sea. This underlines the need to view the Rhine catchment and the coastal zone as a continuum/one system with regard to achieving further reductions of contamination.

The decision on relocation or withdrawal (confined disposal) of sediments from the marine system should take two effects on the coastal ecosystem into account. On the one hand the withdrawal of slightly contaminated sediments will reduce inputs of contaminants (positive), on the other hand withdrawal of large amounts of sediments will upset the sediment balance in sedimentation areas (negative). In a modern impact analysis both aspects should be put in a context of questions about sustainable river management and sea level rise, with increasing coastal erosion and impacts on mud flats.

Following nature, the level of protection for the North Sea should be one of the leading principles for the river protection programs. The main objective of the management strategy should be to shift from dredged material management to sediment management.

Disposal and beneficial re-use of dredged material

Disposal of dredged material

Although there is a distinct reduction of contaminants in dredged material, disposal of contaminated dredged material is necessary. The Slufter, an enormous permanent disposal site for dredged material, came into use in 1987. RMPM and the Ministry of Transport, Public Works and Water Management own the Slufter. Due to the efforts put into source control, there is still room in the Slufter for contaminated dredged material (according to the present norms) up to 2015. As far as disposal is concerned, the policy of the RMPM has always been on the economical use of space in the Slufter. The main port function has to be assured, also in the future. That is safeguarding the continuing process of dredging and disposal of dredged material.

The Dutch policy

Contaminated dredged material is a national problem because of the spatial restrictions and environmental opposition policies and regulations are very critical towards the mere disposal of contaminated dredged material in a disposal site.

Fig. 5. Sandy dredged material

This is why increasingly measures (focussed on non-disposal) come into use. Important measures that have their influence on dealing with contaminated dredged material are:

- The levying of taxes on disposal of sandy dredged material (into use since 2002); if dredged material with more than 60% sand is put in a disposal site (without removing the sand), high taxes must be paid (€ 13/tonnes). In future maybe more dredged material may be subjected to the taxes.
- If treatment, instead of disposal, of the more contaminated material is carried out (with as a result a product that can be beneficially re-used), this is State-aided.
- Sand and clay can only be used in constructions if the contaminants do not exceed levels stated in the building Act. At this moment the building act is under revision because the act was not orientated on building materials coming out of dredged material, increasing the beneficial use potential.

The Rotterdam activities

The RMPM fully supports the Dutch policy in reducing the amount of contaminated dredged material that has to be stored. In this perspective the Rotterdam activities are mainly focussed on source control.

As far as the measures for disposal are concerned, from an economical point of view, the taxes for disposal can have large consequences for the Port of Rotterdam. Since the dredging and disposal is paid out of the harbour-dues, this can have consequences for the economical position of the port. Also most of the contaminated dredged material in the Rotterdam Port area is too clean to come into consideration for the state-aided treatment. In view of this it is important that, if beneficial use is concerned, a large-scale method with low costs is applied.

This is why, in 1992, the RMPM started with the extraction of the clean sand out of the material that was brought into the Slufter. Since 1993 sand separation has been common practice at the Slufter disposal site (Figure 4).

Also since the late nineties clay-fields were constructed at the Slufter site. At this moment a study is carried out into the use of alternating layers of sand and clay (out of dredged material) in a construction (road embankment or raise). The civil-engineering aspects are investigated. Maybe these so-called sandwich constructions can be used in the future extension of the port.

Another investigation carried out by the RMPM, some private companies and a company from the UK, is an innovative thermal immobilisation of dredged material. For the RMPM the focus is mainly on a recycling factory in the Port. Besides dredged material other additional substances are necessary for the process. Because of the treatment and disposal cost of these additional substances otherwise and because of the economical heat consumption, this method can be relatively cost-effective for dredged material and the additional substances. This method is still in the test stage.

Conclusions

The main objective of the RMPM is to ensure that all dredged material is sufficiently clean within the foreseeable future (2015), in line with the concept of a sustainable port and region in which port activities take place.

The basic principle of Rotterdam's policy is that it should be possible to relocate all dredged material from the port (back) into the water system, or alternatively, to use part of it on land without causing (hazardous) impacts.

The focus is mainly on source control but as long as there is contaminated dredged material economical use of the Slufter is eminent. For an economical sound Port, the focus should be on beneficial use from which the environment benefits the most. For the port of Rotterdam this means large-scale treatment methods for low costs (under the conditions are that there is a market for the products and that the rules and regulations do not prevent the use of the products).

References

Dredged Material in the Port of Rotterdam – Interface between Rhine Catchment Area and North Sea. GKSS research project site: http://coast.gkss.de/aos/dredged_material/

Rhine Research Project II (PORII). Havenbedrijf Rotterdam N.V. research project site: http://www.portmanagement.com/UK/Waterandweather/Waterquality/Index.asp

18. Integrated assessment for catchment and coastal zone management: The case of the Humber

Julian Andrews, Nicola Beaumont, Roy Brouwer, Rachel Cave, Tim Jickells, Laure Ledoux, and R. Kerry Turner[1]

Abstract

In the context of the Water Framework Directive (2000/60/EC), EU Member States are required to introduce water quality objectives for all water bodies, including coastal waters. Given the impact of catchment fluxes on coastal water quality, decision-making at the catchment scale is essential. This chapter investigates the use of integrated assessment as an overall decision-support process and toolbox in the Humber estuary.

Introduction

In the context of the Water Framework Directive (2000/60/EC), EU Member States are required to introduce water quality objectives for all water bodies, including coastal waters. Given the impact of catchment fluxes on coastal water quality, decision-making at the catchment scale is essential. This chapter investigates the use of integrated assessment as an overall decision-support process and toolbox in the Humber estuary. The context of this research is an ongoing European research project, EUROCAT, which is focused on the requirements of integrated catchment and coastal zone management and analyses the response of the coastal sea to changes in fluxes of nutrients and contaminants from the catchments. The first section of the chapter describes briefly a suitable decision-support framework and its main analytical steps. The chapter then describes the case study, the Humber catchment and estuary, exploring the management issues

---

[1] Correspondence to Kerry Turner: r.k.turner@uea.ac.uk

J.E. Vermaat et al. (Eds.): Managing European Coasts: Past, Present, and Future, pp. 323–353, 2005.

through the Driver-Pressure State Response (DP-S-I-R) scoping framework. The general methodology for scenarios is reviewed, before regional scenarios are derived, which describe three possible futures for the Humber. The chapter finally presents a policy analysis, including an abatement cost study, highlighting the issue of copper discharges to the estuary, and reports on how the scenarios can be used to investigate future fluxes and provide a consistent framework to evaluate potential policies to improve water quality in the estuary, in the context of the Water Framework Directive.

Integrated environmental assessment (IEA) for coastal zone management

When water resources, including coastal waters, are seen as components of a wider set of interrelated systems encompassed within catchment and watershed boundaries, more efficient management of water and related measures to protect the wider supporting ecosystems are all vital components of a sustainable development strategy. Managing water resources at the catchment scale requires an appreciation of the full functioning of hydrological, ecological and biogeochemical systems and the total range of valuable functions and functional outputs of goods and services that are provided. A sustainable approach to water management and pricing must therefore be based on a wide spatial and temporal appreciation of the landscape ecological processes present, together with the relevant environmental and socio-economic driving forces. Such a management strategy will need to be underpinned by a scientifically credible but also pragmatic environmental decision support system, i.e., a toolbox of evaluation methods and techniques, complemented by a set of environmental change indicators and an enabling analytical framework.

The support system should allow managers to identify a number of steps or 'decision rules' in order to use the framework in a given catchment. The main steps are listed below:

- Scoping and auditing stage - to scope the nature of the problem and the causes and consequences that are relevant. The so-called DP-S-I-R framework has proved useful at this stage (Turner et al. 2001);
- Identification and selection of complementary analytical methods and techniques - such as GIS, coupled natural science models, cost benefit analysis, etc;
- Data collection and monitoring via indicators of change and forecasting of future possibilities via environmental change scenarios;
- Evaluation of project, policy or programme options - using methods such as stakeholder analysis, cost effectiveness, cost benefit analysis and multi-criteria analysis.

The following steps and related information gathering and analysis procedures are recommended in the appraisal process:

Scoping and problem auditing

The DP-S-I-R (driving pressures-state-impacts-response) framework, originally developed by the OECD, is a useful device for the scoping of complicated management issues and problems. It can make tractable the complexity of causes of degradation or loss of water resources, habitat and species and the links to socio-economic activities, across the relevant spatial and temporal scales. It also provides the important conceptual connection between ecosystem change and the impacts of that change on people's economic and social well-being. Relevant indicators of environmental change can be derived (see below), and the loss of ecosystem function provision in terms of goods and services (direct and indirectly received) can be translated into human welfare loss and quantified in monetary and/or other more qualitative ways.

In this initial stage of the analysis an empirical description and explanation is also required covering the relevant policy context and regulatory regime that is to form the focal point of the research within any given catchment. The regulatory regime work where necessary, will need to encompass both regional/national and international regulations and designations and their implications for the catchment.

At the core of this analytical stage is the process of stakeholder mapping and the related identification (via the DP-S-I-R approach) of those impacts and consequences of environmental change which impinge on stakeholders.

Stakeholder mapping

For a given catchment it is necessary to identify the following:

- The different 'interest' groups within the catchment and outside (national and international) that are relevant to the policy issues and contexts being focused on;
- Existing stakeholder networks (or the lack of networks);
- Existing institutional arrangements and 'power' structures;
- The aggregate 'policy networks' (or the lack of networks) that serve to influence policy choice outcomes.

This stakeholder-related information should then be set against the relevant drivers and pressures of environmental change in the catchment (from the DP-S-I-R data) e.g. population growth and density changes, pollutants and contaminants trends and climate change. The findings should help, among other things, to highlight any distributional equity concerns (i.e. who gains who loses) and power relationships relevant to existing policies and future potential policy measures. The policy set should include any national, EC or other international regulations, designations and agreements. All this information will be relevant to the outputs from the futures scenarios and policy goals and measures research.

Scenario analysis

The future will always be shrouded by uncertainty and therefore accurate prediction is not a feasible goal. However, it is possible to formulate scenarios that can shed light on and offer insights about possible future developments. It is these scenarios, which can inform the policy targets, standards and measures packages relating to the policy issues chosen as the foci for the catchment-level research.

Indicators and critical thresholds

Given the degree of scientific and socio economic (economic, social, political and cultural factors) uncertainty that exists about current and future environmental change issues and consequences, the scenario-related perspectives are heavily influenced by the degree of risk aversion that may be adopted by different stakeholders and the government on behalf of the whole of society. The so-called precautionary principle is a reflection of this concern about uncertainty and the sort of decision-making approach that should be adopted. We turn to this concept next.

Different societal positions (perspectives) will encapsulate different approaches to the precautionary principle and the treatment of risk and uncertainty. The position taken over how to mitigate the uncertainty problem will also affect the indicators of change that are chosen and their interpretation. In many cases it will not be possible to determine single number outcomes for 'critical loads' of given substances, or 'critical thresholds'. Instead there will be standards/targets, which incorporate different interpretations of 'safety margins', necessary to avoid breaching uncertain thresholds. Relevant environmental change indicators may of necessity be fairly crude measures of trends (positive and negative) relating to substance fluxes, ambient quality states and wider ecosystem changes, augmented by a qualitative assessment.

Identification and selection of appropriate decision-making methods

Managed ecosystems will be in an almost constant state of flux as the natural processes and systems react to human management interventions, which in turn, subject to various lags, produce more policy responses i.e. a co-evolutionary process characterised by continuous feedback effects. It is therefore important to be able to assess the impact of alternative sets of management actions or strategies in order to judge their social acceptability against a range of criteria such as environmental effectiveness, economic efficiency and fairness across different stakeholder interests (including different generations). Evaluation methods and techniques have to be matched up to the chosen evaluation criteria.

Data collection, monitoring, and indicators

Official interest in quantifiable environmental indicators intensified during the 1990's as sustainability thinking came to prominence. They serve to reduce the complexity of environmental pressures, state changes and impacts and to increase the transparency of the possible trade-offs involved in policy options choice. Indicators do not, however, provide a panacea for scientific uncertainty and they also require suitable institutional structures to be in place to regularly collect and update the relevant background data.

Evaluation of project, policy, or programme options

A combination of quantitative and qualitative research methods is advocated in order to generate a blend of different types of policy relevant information. This applies to both the biophysical assessment of management options and the evaluation of the welfare gains and losses people perceive to be associated with the environmental changes and management responses. The main generic approaches that can form the methodological basis for strategic options appraisal are:

- Stakeholder analysis;
- Cost-effectiveness analysis;
- Extended cost-benefit analysis and risk-benefit analysis;
- Social discourse analysis;
- Multi-criteria analysis.

This chapter reviews how these different steps were applied to the Humber within the context of estuarine water quality management.

The Humber case study and the DP-S-I-R framework

The Humber catchment and the coastal zone

The Humber catchment covers an area of ca. 24,240km^2, more than 20% of the land area of England (Jarvie et al. 1997). It is home to 20% of the UK population, and a very significant proportion of the energy, industrial and agricultural production. Industrial, urban and agricultural development over the last few hundred years have adversely impacted the quality of the water entering the estuary from the rivers.

The macro-tidal Humber estuary is one of the largest in the UK. The area surrounding the Humber Estuary (generally referred to as Humberside) is mainly high quality agricultural land, with many thousands of hectares reclaimed from the estuary over the last few centuries (Murby 2001). As a result, it is estimated that over a third of a million people now live on areas of land below high spring tide level (EA 1999). Humberside ports handle 13% of the UK's sea-borne trade, with Grimsby-Immingham representing the largest port complex in the UK (DETR 2000a). In spite of extensive reclamation and coastal squeeze over several centuries, large areas of intertidal and coastal wetland habitat still exist in the estuary, supporting year-round bird populations, as well as species that use it during migratory passage and as a winter residence. The outer estuary is also of particular importance as a fish nursery area for North Sea plaice. Much of the Humber is designated under the Habitats Directive, and the entire Estuary has been proposed as a marine Special Area of Conservation, in recognition of its importance to nature conservation.

The Humber case study represents a good example of what can be called a "mature" environmental problem. Over the last few decades, as heavy industry in the catchment has declined, and regulations on emissions and inputs to controlled waters have become more stringent, the water quality of both catchment and estuary have improved. The "peak" loading impact to the Humber estuary in terms of nutrients and many metal inputs from the near-estuary and wider catchment zones was in the past. However, a legacy of contamination still exists in the sediments (Millward and Glegg 1997), and the loss of intertidal area has led to a severe reduction of the ability of the estuary to trap nutrients, which instead get exported direct to the North Sea (Jickells et al. 2000).

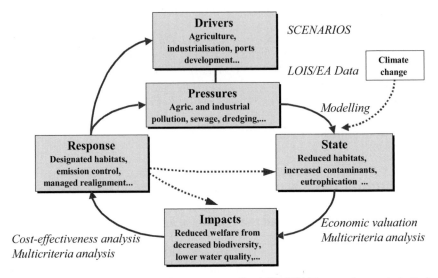

Fig. 1. The DP-S-I-R for the Humber (corresponding EUROCAT research steps in italics)

The DP-S-I-R framework for the Humber

The Driver Pressure State Impact Response (DP-S-I-R) framework has proved useful to scope sustainable development issues in coastal zone management. First developed for environmental reporting by the OECD (OECD 1993), it was further developed and adapted to the context of coastal zone management by Turner et al. (1998a). Figure 1 illustrates the DP-S-I-R framework applied to the Humber. At the root of environmental change are economic drivers, for example agricultural intensification, urbanisation, ports development, which in turn will create pressures: land conversion and reclamation, nutrient emissions, waste disposal in coastal waters, dredging. These pressures, along with physical factors such as climate change, will lead to changes in the state of the environment e.g. changes in nutrient concentration leading to increased risks of eutrophication, loss of habitat and species diversity. These physical changes will in turn have an impact on human welfare, for example through reduced fisheries productivity, health impacts, or reduced welfare from decreased biodiversity and poorer water quality. Environmental economic valuation measures these impacts on human welfare due to physical changes in terms of costs and benefits to society.

These welfare changes will provide the stimulus for management action, which will seek to control socio-economic drivers and consequent environmental pressures, thus creating a continuous and dynamic cycle with feedback loops.

Drivers in the Humber can be described as follows: (Cave et al. 2002):

Population growth and urbanisation

The total population of the Humber catchment numbers about 11.5 million (Office of National Statistics 2002). Of this, 6.1 million live in the Trent catchment, including 1 million in the city of Birmingham. The Ouse catchment has a population of 4.4 million, with the largest urban area being the Leeds-Bradford conurbation, totalling 1.2 million inhabitants. Nottingham, Leicester, and the West Midlands/Birmingham conurbation are drained by the Trent, the Leeds-Bradford area in West Yorkshire is drained by the Aire/Calder and the Sheffield/Doncaster area in South Yorkshire is drained by the Don (Figure 2). There is one major conurbation on the estuary itself, on the north bank at Kingston-upon-Hull, and several large industrial areas on the south bank, giving a total Humberside population of 0.9 million. There are also large rural regions, whose populations are currently experiencing high population growth, while the urban areas are showing a small decline.

Agriculture

The total agricultural area for the Humber catchment amounts to some 2.3 million hectares, from a total land area of just over 2.4 million hectares (source: DEFRA, UK Agricultural Census Statistics for 1999[2]). Of this, 1 million hectares is arable and horticultural land, growing mainly cereal crops, oilseed rape, and root crops, and the remainder is predominantly grazing land. Land use in the Ouse and Trent catchments is broadly similar, but with the Trent having almost double the urban area of the Ouse, while the Ouse has more extensive woodland, heath and bog.

Industry

In the catchment, traditional industries such as textiles and iron and steel have declined, while the chemical and petrochemical industries are thriving, as is the power industry. The estuary is now in a post-industrial phase, with some of the large polluting manufacturing plants on the estuary now closed down (e.g. tin smelter at North Ferriby), and others subject to more stringent effluent restrictions (e.g. titanium dioxide plants on the south bank).

---

[2] available from
http://www.defra.gov.uk/esg/work_htm/publications/cs/farmstats_web/default.htm

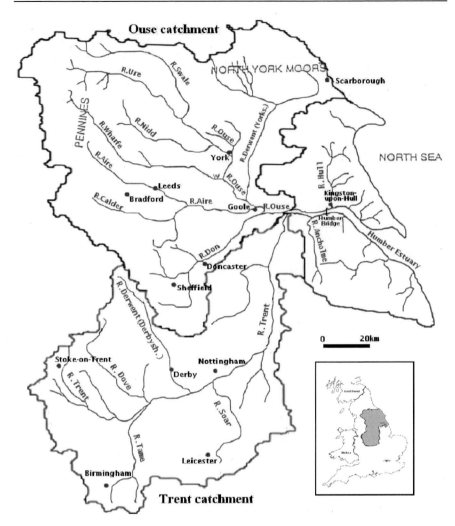

Fig. 2. Map of the Humber Catchment. (after Edwards et al. 1997). Inset map of part of the UK, showing the extent of the catchment in grey

Port development

The Humber ports (Goole, Grimsby, Hull and Immingham) handle 13% of the UK's seaborne trade (DETR 2000a). This has risen from 58 million tonnes in 1989 to 74 million tonnes in 1998. Up to 50,000 ship movements per year are handled by the Vessel Traffic Service Centre at Spurn Head, and the volume of ship traffic is expected to increase as the European Union expands over the coming years. Grimsby and Immingham overtook London to become the UK's leading port in 2000 with 52.2 Mt of freight traffic (DETR 2000b).

Climate change

In addition to the socio-economic drivers, an emerging additional driver in the Humber catchment for the foreseeable future is the necessity to cope with climate change. This includes more extreme weather events, which can lead to flooding in parts of the catchment (Longfield and Macklin 1999), and expected sea-level rise around the estuary. Major floods occurred in the catchment in 1986 1995 and 2000, with the Ouse catchment most affected. This type of event, both in the catchment and in the intertidal area could remobilise contaminated sediments.

An area of more than 800km^2 around the Humber estuary currently lies below the level of high spring tide, protected by extensive coastal defences. This includes urban/industrial areas, high-grade farmland, infrastructure such as roads and railways, and natural reserves e.g. wetlands. More than 280km of defences are currently maintained by the EA in the Humber area, and it is now realised that the long-term maintenance of all these defences is not desirable either on economic or environmental grounds.

Institutions and stakeholders in the Humber

International agreements

There are a number of international agreements and conventions that are relevant to the North Sea in general, and the Humber in particular. The most relevant aspects are highlighted here, and summarised in Figure 3. The Humber is one of the major contributors of fresh water from the UK to the North Sea, together with significant loads of nutrients and contaminants. As such, its outflow has been scrutinised under the terms of the Oslo and Paris Conventions and subsequent North Sea Conferences. The Convention for the Protection of the Marine Environment of the North-East Atlantic (OSPAR Convention) was opened for signature in 1992, and came into force in 1998 (see http://www.ospar.org/). The parts of the convention most relevant to the Humber are those that deal with the prevention and elimination of pollution from land-based sources to the North Sea, and with the prevention and elimination of pollution by dumping or incineration.

European legislation

A number of EU directives are also directly relevant to the HUMCAT study. Of particular importance is the Water Framework Directive (2000/60/EC), which is a major example of policy response addressing water quality issues at the catchment scale. Adopted in June 2000, it integrates previously existing water legislation, updates existing directives according to new scientific knowledge, and strengthens existing legal obligations to ensure better compliance (Kaika and Page 2002). Kallis and Butler (2001) point out that the directive introduces both new goals, and new means of achieving them (new organisational framework, and new measures). The overall goal is a "good" and non-deteriorating "status for all waters (surface, underground and coastal). Measures to achieve the new goals will be co-ordinated at the level of river basin districts, i.e. hydrological units and not political boundaries. Authorities should set up River Basin Management Plans, to be reviewed every 6 years, based on identifying river basin characteristics, assessing pressures and impacts on water bodies following future trend scenarios, and drawing on an economic analysis of water uses within the catchment (including a cost-effectiveness analysis of potential measures). Monitoring is also an essential component, determining the necessity for additional measures. Finally, an important innovation of the Directive is to widen participation in water policy-making: river basin management plans should involve extensive consultation and public access to information.

Although it does not target coastal zones specifically, the Directive does cover coastal water quality in its objective for good quality status, and provides a good example of integrated catchment management, addressing in particular the issue of diffuse pollution of coastal waters. A major part of the research within HUMCAT is relevant to the implementation of the directive: e.g. the scenarios analysing potential future fluxes of contaminants; the cost-effectiveness and multi-criteria analysis of potential policy measures to improve water quality, and the involvement of stakeholders throughout the project.

The Habitats Directive (92/43/EEC) also has major implications in the Humber estuary. The Habitats Directive is the main element of response of the EU to the Convention on Biodiversity (Ledoux et al. 2000). Together with the Birds Directive (79/409/EEC), it aims to create a network of designated areas (Natura 2000) to protect habitats and species of community-wide importance, on a biogeographical basis. It is, in effect, a "no-net-loss" policy, in so far as it requires all Natura 2000 areas to be protected from deterioration and damage. The Member States are required to take all appropriate steps to avoid the deterioration of those habitats and species for which protection is required. Under articles 6(3), a plan or project likely to have a significant effect on a Natura 2000 site must undergo assessment to determine whether it would damage the nature conservation interest of the site. If the plan or project is thought to impose a significant threat, it can only go ahead if (1) there is *no alternative solution;* (2) its implementation is of *overriding public interest*; (3) member states must provide compensatory measures which may include habitat restoration or recreation of the same type of habitat on the same site or elsewhere.

A significant number of habitat types listed in Annex II of the Directive are located in the coastal fringe (dunes, mud flats, coastal lagoons, coastal freshwater wetlands, etc.). In addition, the Habitats Directive specifically establishes Marine Special Areas of Conservation. The Habitats Directive can therefore be expected to have a major impact on the coast. In its strict interpretation, the compensation requirement for displaced habitats also applies to habitats lost through natural, or semi-natural causes, such as sea level rise and coastal erosion, which is likely to have far reaching consequences given the current climate change predictions. In the Humber, relevant authorities are anticipating this need for compensation and are planning ahead by recreating coastal habitats through managed realignment – realigning existing hard defences further inland thereby recreating intertidal habitats (Ledoux et al. 2003).

Finally, under the Urban Waste Water Directive (UWWD, 91/271/EEC), water companies will also have the responsibility of increasing waste water treatment to meet more stringent nutrient standards in receiving waters. The degree of treatment required is normally secondary treatment, although primary treatment might be allowed in "less sensitive areas". In this context, the EC has announced infraction proceedings against the UK government for failing to designate the Humber as a sensitive area. The degree of wastewater treatment required in the Humber will eventually depend on the result of this case.

Although it is not strictly speaking legislation, it is also worth mentioning that the European Union has recently developed a European strategy on coastal zones. The Strategy defines Integrated Coastal Zone Management (ICZM) as a "dynamic, continuous and iterative process designed to promote sustainable management of coastal zones (EC 1999 2000). It recommends: (i) promotion of ICZM within the member States and at the "Regional Seas" level; (ii) making EU policies compatible with ICZM; (iii) promoting dialogue between European Coastal Stakeholders; (iv) developing best ICZM practice; (v) generating information and knowledge about the coastal zone; (vi) disseminating information and raising public awareness. The philosophy underpinning the strategy is one of governance by partnership with civil society, with the EU providing leadership and guidance to support implementation at other levels.

A recommendation of the European Parliament and of the Council recommends that Member States should develop a national strategy, or where appropriate several strategies, following the principles of ICZM as described in the European Strategy. These strategies might be specific to the coastal zone, or be part of a geographically broader programme for promoting integrated management of a wider area.

National legislation

At the national level, the Humber Estuary Management Strategy (HEMS) identifies a long list of legislation affecting the HEMS area. The most recent and influential elements of this are (see also Figure 3):

- The Environmental Protection Act 1990 - established statutory provisions for environment protection purposes including integrated pollution control for dangerous processes;
- The Water Resources Act 1991 - consolidated previous water legislation in respect of both the quality and quantity of water resources;
- The Water Industry Act 1991 - consolidated legislation relating to the supply of water and the provision of sewerage services;
- The Environment Act 1995 - established the Environment Agency, and introduced measures to enhance protection of the environment, including further powers for the prevention and remediation of water pollution.

Stakeholder interests

The Environment Agency has the main responsibility for long-term water resources planning in England and Wales, and as such is responsible for water quality in the Humber. In 1992, the Department of Environment published guidance on

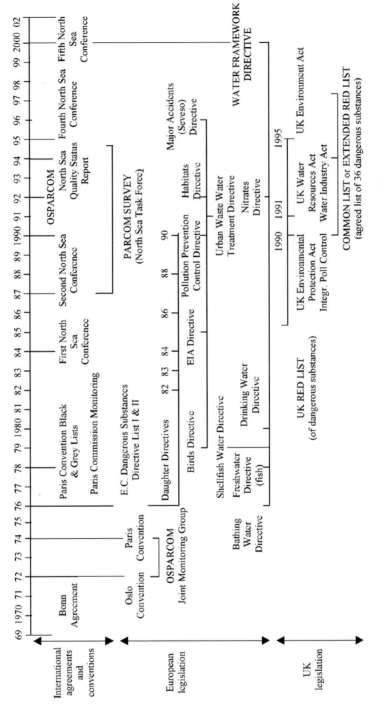

Fig. 3. Evolution of policy regimes relevant to the Humber and the North Sea

coastal planning (Planning Policy Guidance Note 20), advocating the creation of Estuary Management Plans, with the objective of bringing together decision-makers and stakeholders to adopt a strategic approach to estuary management. A management structure was set up during 1993 and 1994, comprising of a project officer, an executive steering group to guide the project, and a working group with which the Project Officer maintained regular contact, the latter two groups representing a range of interests. Extensive consultation and a close cooperation with the Environment Agency lead to the publication of a Humber Management Strategy, summarising the objectives for the long term management of the Humber, with the aim of guiding future activities through a voluntary framework of planning and management options, as a complement to statutory processes and legislative framework. Table 1 summarises the objectives from different stakeholders, as reported in the Humber Management Strategy (HEMS 1997).

Objectives listed in Table 1 suggest that there might be potential conflicts between types of interests. However, the way the objectives are formulated, it is also clear that some consensus has already been reached, and that there is common ground between the various interest groups.

Given the amount of consultation already done, the HUMCAT team chose to use this existing consultation exercise and adopt its findings, rather than start a whole new consultation process on general issues related to the Humber Estuary.

Stakeholder consultation was however carried out on a regular basis through setting up a Policy Advisory Board, comprising key stakeholders: the Environment Agency, British Associated British Ports, and the Royal Society for the Protection of Birds. Two years into the project, the group has now met six times, providing regular feedback on different stages of project development. Fuller consultation has also already taken place through the pre-existing Humber Shoreline Management Plan steering group, originally set up to maximise input from stakeholders into the Humber Strategy, and representing the main interests around the estuary. A workshop was organised to discuss potential policies likely to be adopted to improve water quality in the Humber estuary, and what main areas of interests should be taken into account for decision making.

Scenarios methodology

What are scenarios?

A scenario should provide a plausible context, but not a probabilistic forecast, for a possible future state of the world (Parry and Carter 1998). Scenario analysis has been under development since the 1960s across a number of different applications (Miles 1981; Kassler 1995; Hammond 1998). Scenario analysis usually proceeds through a series of spatial scales, involving a process of aggregation/ disaggregation until an appropriate policy/management option level is attained (Figure 4).

A growing body of research work has been exploring the application of scenario-based 'futures studies' as a means of obtaining a better understanding of the potential for societal adaptation to future environmental change (Gallopin and

Table 1. Stakeholder objectives in the Humber estuary (Source, HEMS 1997)

Type of interest	Objectives
Agriculture	• To foster viable and sustainable farm development to support rural communities;
Archeology and Cultural Resources	• To conserve and enhance the estuary's archaeological and cultural heritage, to ensure the maintenance of its special and diverse qualities and to secure its sensitive management and promotion;
Fisheries	• To support and promote sustainable exploitation of the fisheries of the Humber through appropriate regulation to protect estuarine habitats and a healthy food chain whilst recognising the value of the fisheries to local communities;
Flood defence and coastal processes	• To provide environmentally, technically and economically acceptable flood defences, developed through a strategic understanding of physical processes and interests on and adjacent to the Humber Estuary;
	• To produce a "state of the art" estuary Shoreline Management Plan" (ESMP) based on the current understanding of coastal processes, to a format compatible with the open coast Shoreline Management Plans (SMPs) to set a framework for the physical management of the estuary;
	• To develop an approach that would facilitate the construction of short term defence in a manner which meets the requirements of the Habitats Directive;
	• To ensure that the planning and implementation of flood defence strategies contribute to the sustainable development of the Humber Estuary and the delivery of biodiversity at a national level.
Industry and Commerce	• To create through a partnership a dynamic, diverse and environmentally sustainable economy that provides good quality employment opportunities for local people;
Integrated Pollution Control (IPC)	• To promote sustainable environmental management by working with developers, industrialists, farmers and the community in general, so as to ensure that natural resources are protected;
	• To improve land, water and air quality in the HEMS area without imposing disproportionate costs on industry or society as a whole.
Landscape	• To ensure that the special and distinctive qualities of the Humber landscape are protected and promoted, enhanced where appropriate and, where necessary, restored;
Nature Conservation	• To maintain and enhance the diversity and abundance of wildlife within the Estuary, especially the internationally important populations of birds;
Navigation and Port Development	• To ensure the continued growth and vitality of the Humber's Ports and Wharves and their related developments;
Sport, Recreation and Access	• To maintain, and improve the provision and availability of as wide a range of sport and recreational facilities as are compatible with the local environment of the Humber Estuary;
Tourism	• To maintain, develop and promote tourism on the Estuary in a way which ensures that all development achieves a suitable log term balance between needs of visitors, local communities, and the environment.

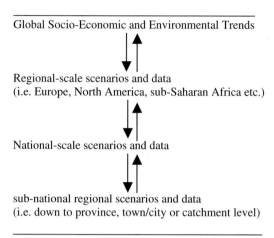

Global Socio-Economic and Environmental Trends

Regional-scale scenarios and data
(i.e. Europe, North America, sub-Saharan Africa etc.)

National-scale scenarios and data

sub-national regional scenarios and data
(i.e. down to province, town/city or catchment level)

Fig. 4. Scenario scaling

Raskin 1998). Scenarios can be used to inform present choices in the light of future alternatives, such as in the context of climate change mitigation and adaptation options (Lorenzoni et al. 2000a, 2000b, Parry et al. 2000). Thus a longer-term context can provide a framework within which to comprehensively evaluate short-term decisions, as well as providing the capacity to explore the consequences of surprise events ('side-swipes'). Plausible scenarios are potentially useful constructs for informed social learning among groups of stakeholders, as the different interests are exposed to different worldviews.

To summarise so far, scenarios are not precise future predictions but methods to aid decision-makers in their efforts to cope with inevitable uncertainty. They may possess a variety of characteristics and can be deployed at different spatial scales and across different temporal scales (typically from 10 years to 100 years) (Turner, this volume).

National scenario templates

The UK Foresight Programme sponsored a set of scenarios pitched at national level, to appeal to business and governmental audiences (OST 1999). The OST analysis is meant to cast light on future social, economic and environmental trends for the UK over the period 2000-2040.

The scenarios are framed by two orthogonal axes, representing societal values (ranging from consumerist, self-interested market-based preferences to collectivist and conservationist social preferences) and level of effective governance (from local to global) respectively (Figure 5). These axes determine four futures, which, for simplicity's sake, appear as independent possible states of world. In reality, the boundaries are fuzzy and the different states are differentiated because certain trends and characteristics become more or less dominant across government, business and public social contexts. World Markets, for example, is the equivalent of a baseline (almost trend) scenario. It is meant to portray conventional industrial/international capitalism continuing out into the foreseeable future. Sustainable development is interpreted in its 'weak sustainability' form (Turner et al. 1998b, Burbridge 2001). This scenario is characterised by a requirement to maximise total output (GDP), with widening income inequality. Environmental concerns are important but constrained to 'local' health and/or amenity concerns; international environmental agreements have relatively modest targets/standards.

Global Sustainability, on the other hand, is a scenario which contains the belief that environmental systems are often of infinite value and are the foundations of a sustainable economic and social system, not vice versa. Resource use efficiency/productivity can be radically improved via a mixture of regulation, economic incentive mechanisms and technological innovation. Sustainable development requires the redressing of global inequities of income and wealth, as well as efficiency gains. A move towards more globalisation of governance systems is supported.

The Provincial Enterprise scenario has a mixed technocentric and ecocentric makeup depending on the national resource base available and its configuration. Global and national governance are significantly downgraded in importance. Environmental concerns are local/regional but mitigation measures are constrained by budget limits.

Finally, in the Local Stewardship Scenario, social values coalesce around long-term requirements and the satisfaction of collective needs. Governance is concentrated on federal political systems, with the emphasis on deliberative and participatory/inclusive processes at the local scale. Key technologies become renewable energy and small-scale manufacturing processes. Environmental concerns are high but strong action is limited to local scales.

Regional scenarios

Three regional variants (Ledoux et al. 2002) were derived from the OST national scenarios and adapted to the Humber (Figure 5). The Business As Usual scenario (BAU) is the baseline scenario, corresponding roughly to the World Markets scenario at national level. It is a forward projection of the past 20 year trends in data, ignoring the recent sustainable development strictures. In this scenario, current legislation is only complied with in a formal way. For example, expected port expansion within the Humber estuary over the next 20-25 years would lead to a loss of 0.2% of the intertidal area in the Humber estuary, approx. 20 ha. In terms of flood defence around the estuary, hard defences would be maintained as far as possible, exacerbating the problem of coastal squeeze. Given the commitment of the UK to implement the Habitats Directive, compensation in the form of recreated habitats would have to be provided, but one could assume that this would be on the basis of minimum compliance. One would therefore expect a net loss of habitats in this scenario. Water quality objectives are likely to include exceptions for a variety of polluting industries. The standard of sewage treatment is likely to be relatively low (up to recently, there was no treatment for the large, direct sewage discharges in the tideway). At the catchment level, agriculture is likely to remain relatively intensive and based on technology (e.g. GM crops) to sustain high yields, leading to no net reduction in nitrate input to rivers. Contaminant concentrations would also remain at their current level.

In the Policy Target scenario, current and prospective legislative targets and objectives are all met on time, according to the EU schedule, with a genuine effort to comply and/or to overcomply with the objectives. The Habitats Directive is likely to be implemented in a genuine attempt to achieve zero habitat loss. This would involve compensation for loss of intertidal area by recreating equivalent or increased habitat in another area. For example, the environmental regulator is creating 80 ha of intertidal habitat at Thorngumbald, by moving back the flood defences. The scheme envisages providing compensation for flood defence works having an adverse impact on designated habitats and contributing some area towards alleviating coastal squeeze. This approach is close to the idea of "mitigation banks" or "land banks", whereby an extensive area of habitats is recreated ahead of development or natural loss, and which could facilitate the implementation of the Habitats Directive (Ledoux et al. 2000). In this scenario, there would be a net increase in intertidal area of 1200 ha by 2025, which would include compensation for losses through coastal squeeze. Sewage treatment around the estuary is likely to be of moderate standard, i.e. all sewage will receive secondary treatment prior to direct discharges, but there would be no tertiary treatment or phosphorus removal. In agriculture, application of fertiliser per unit area will be reduced through targeted policies, and will be timed to reduce the runoff to rivers. Overall reduction of the nutrient load from the catchment into the estuary would be approximately 50%, as foreseen by the OSPAR convention and the various international agreements on the North Sea. The current water quality standards would be met at all times for all contaminants.

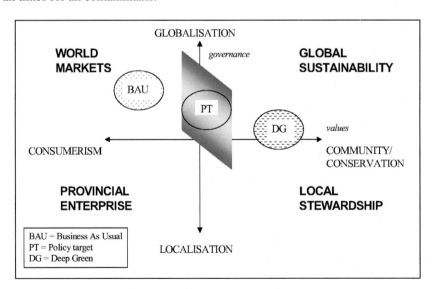

Fig. 5. From national contextual scenarios to regional scenarios for the Humber (modified from OST 1999)

In the Deep Green scenario, environmental protection is given maximum priority. It corresponds loosely to a state between the Global Sustainability and Local Stewardship national scenarios. This represents some environmental state beyond that which could be achieved if current policies were implemented. The economy is also likely to be more regionalised. A Deep Green scenario would involve substantial increases in intertidal areas, compensating for any new works or extension of existing installations, over and above coastal squeeze. The creation of mitigation banks in a formal and regulated setting might contribute to a strategic approach to an increase in biodiversity (Crooks and Ledoux 1999). A recent RSPB study (Pilcher et al. 2002) identified 2 858 ha with potential for intertidal habitat creation within the Humber Estuary. In this scenario, intertidal habitat would increase to 2500 ha, i.e. more than double the area in the Policy Target scenario. Agriculture is likely to become less intensive. Riparian zones will be created along most riverbanks bordering farmland to reduce inputs of nutrients to rivers. Environmental schemes such as reed bed treatment will be widely applied for secondary sewage treatment, rather than hard technology, and tertiary treatment will be widespread, removing nitrogen and phosphorus. In this scenario, the long run objective would be to approach "natural" background levels of nutrient and contaminant fluxes through the system, with due allowance for the historical contaminant legacy "locked" into sediments.

Data collection and sources of contaminants

The extensive research and data gathering (Table 2) that occurred in the past few years in the Humber enabled us to produce a good picture of what the main sources of pollution are.

Nitrate is the dominant source of N to the rivers, and the Trent and Ouse systems provide roughly equal shares of the nitrate load to the estuary. The ammoniacal N load from the Ouse was much higher than for the Trent in the 1980s, due to poorer sewage treatment in the Ouse catchment at that time. Virtually all catchment sewage now receives secondary treatment, e.g. the total population for reporting year 1999-2000 served by sewers for Yorkshire Water is 4.8 million (Ouse catchment, Yorkshire Water returns to OFWAT), of which 4.5 million are connected to plants with secondary sewage treatment. To date no specific efforts have been made to remove P in sewage effluent, although some reduction in load is assumed due to the decline in the use of P in detergents. Orthophosphate loads from the Trent are much higher than from the Ouse, reflecting the larger population in the Trent (Neal and Davies, in press). Most of the nitrogen input to the estuary (Table 3) is in dissolved form, and is exported to the North Sea as dissolved inorganic nitrogen. Most of the phosphorus input is also in dissolved form (Table 3), but is transformed by in-estuarine processes and is exported to the North Sea as particulate (Jickells et al. 2000). Only a fraction of the inputs are stored in sediments within the estuary.

Diffuse inputs to the rivers are the dominant source for arsenic, copper, lead and zinc in the Humber catchment (Table 4), although there is a very large source industrial of zinc direct to the estuary. Lead and zinc supply is dominated by the

Table 2. Research programmes on the Humber

Research Programme	Time Period	Institute/ Organisation	Project description / dataset
UK River Gauging Station Network	1960s-present	Environment Agency,	River Flow and catchment rainfall, National River Flow Archive
Land Ocean Interaction Study (LOIS) (UK contribution to LOICZ)	1992-1998	Natural Environment Research Council (NERC)	To quantify and simulate fluxes and transformations of sediments, nutrients and contaminants into and out of the coastal zone, extending from the catchment to the edge of the continental shelf
Humber Wetlands Project	1992-	English Heritage Univ's of Hull and Essex	Archaeology of Humber Wetlands
Lowland catchment Research (LOCAR)	1999-2006	NERC	Measurement and modelling of water and material fluxes through riparian and wetland habitats
Fate and impact of persistent contaminants in estuaries and coastal waters	1998-	EA, Univ. of Plymouth, Liverpool	Generic 3-D computer model, as a pollution information system
URGENT – Urban regeneration and the Environment	1999-2006	NERC Thematic Prog. in partnership with city authorities, industry and regulatory bodies	Integrating ecological, urban and environmental research across the geological, terrestrial, freshwater and atmospheric sciences
DEFRA/EA Estuaries Research Programme	1999-2000 2002-	EMPHASYS Consortium	Phase1 - Prediction of estuary morphology Phase 2

Ouse, and is largely in particulate form, a legacy of mining activity in the catchment. Arsenic and copper inputs are more equally divided between the Ouse and Trent systems. Copper supply is approximately 50% particulate, while 80% of the arsenic is dissolved in the river water (Neil and Davies 2003).

Sediment storage accounts for more than 90% of the present-day inputs of As, and 40% of the Pb. However, it appears that the bulk of the Cu and Zn entering the estuary are exported to the North Sea.

The current regulatory regime requires that measurements of the dissolved forms of the nutrients and contaminants above, as well as many other contaminants, are made 12 times per year (formerly 7 times) in the estuary and tidal rivers. The average of those measurements has to be below a set threshold (the Environmental Quality Standard, or EQS) in order for a 'pass' to be achieved. Sites in the inner estuary regularly fail to meet the EQS for copper, and occasionally fail to

Table 3. Estimates of nutrient loads to the Humber estuary, from rivers and direct industrial and sewage inputs. Data for 1980s are from NRA, (1993). River data from 1990-2000 are from Harmonised Monitoring Network Data (CEH, unpublished), direct inputs are from PARCOM data

(a) Nitrogen (total) kT yr^{-1}	1980s average	1990	1995	2000
Sewage (direct)	2.90	2.60	1.11	1.31
Industry (direct)	2.30	0.90	0.53	0.65
River *	54.00	31.34	34.87	39.10
Total	59.20	33.84	35.51	41.06
(b) Orthophos-P (total) kT yr^{-1}	1980s average	1990	1995	2000
Sewage (direct)	0.73	0.60	0.24	0.29
Industry (direct)	0.04	0.10	0.03	0.01
River *	7.45	3.85	3.65	3.98
Total	8.22	4.55	4.02	4.28

*River – these values include industrial and sewage inputs, as well as diffuse inputs.

Table 4. Metal contaminant loads to the estuary in 2000. Per cent data in brackets indicate the proportion of the total load that comes from point sources throughout the catchment. Load data are from Environment Agency monitoring data. Sediment burial data are from this work

Contaminant load	As T yr^{-1}	Cu T yr^{-1}	Pb T yr^{-1}	Zn T yr^{-1}
Total load to estuary	11.4	71.9	66.7	450.3
Total point source load	2.4	18.6	6.3	196
	(21%)	(26%)	(9%)	(43%)
Sediment sequestration	10.5	13.5	27.0	66.0

achieve the required EQS for dissolved oxygen. There are no EQS values set for contaminants in suspended particulate matter, nor in bed sediments within the estuary, despite the fact that the sediments are the habitat for benthic species which form part of the food chain and underpin much of the ecology of the estuary. Levels of contaminants in sediments and biota are monitored, however, and have shown significant reductions over the last decade.

Policy analysis

The general approach

In the context of EUROCAT, scenarios are useful to think about what type of management strategies would be best adapted to a variety of possible futures. In each scenario, there is a different objective for water quality improvement, and a different combination of policy options is likely to be implemented according to environmental objectives, but also depending on the general socio-economic and political context. Options might be related to: control measures on agriculture or industry, sewage treatment, cleaning up or removal of contaminated sediments, or managed realignment (the case of interest here involves realigning coastal defences further inland, thereby recreating intertidal habitats which can play a role in nutrient and contaminant removal). Practical measures are likely to be a combination of options. These options can be targeted at the level of the estuary or in the wider catchment.

Theoretically, all policy options would be analysed in each scenario, taking into account future conditions and preferences. In practice however, this is impossible to carry out, as it would imply too many assumptions about future price-elasticities, consumer preferences, general socio-economic conditions etc. The pragmatic approach taken here is to analyse in present time the policy options likely to be implemented in each scenario – i.e., what impact would these policy options have if they were implemented now. In a final step, the scenarios will provide a consistent framework to undertake a sensitivity analysis of how the outcome of the present time analysis would change under the three different possible future scenarios.

Policy analysis of water quality improvement options within EUROCAT focus on two complementary approaches: a "cost-effectiveness", and a "multi-criteria" analyses. The format of these approaches is described in detail in the following sections. Although the ultimate aim is to use the scenarios to explore impact of policies in different futures, as a first step, the scenarios are used to generate policy packages that would be likely to be used in each of the possible future worlds.

A cost-effectiveness analysis of reducing copper inputs to the Humber

One policy question is: how effective are available policy measures at reaching the environmental targets, and at what cost? This is a cost-effectiveness issue: what measures or combination of measures are able to reach a certain water quality target at least cost? In the Humber, this sort of analysis can help to answer the following questions: which economic activities generating pollution should be priority targets? Should policy measures target activities with direct input in the estuary or activities further up in the catchment? Are there any sub-catchments that are more obvious candidates for specific policies? This involves an economic analysis of abatement costs for different types of activities. This section first provides an example of cost-effectiveness analysis focusing on direct industrial inputs of copper in the Humber.

Copper pollution is one of the few current significant water quality issues in the Humber Estuary. Copper is a List 2 substance under the European Dangerous Substance Directive (76/464/EEC) and it is this designation that defines how copper is regulated in the Humber estuary. The Environmental Quality Standard for dissolved copper in estuarine waters is 5ug/l relating to soluble copper as an annual average concentration (Mance et al. 1984), and the Environment Agency's statutory duty is to ensure this standard is met. Copper concentrations in the Humber Estuary exceeded the Environmental Quality Standard (EQS) on several occasions in the period 1989 to 1995 (Beaumont and Tinch 2003), and as a result considerable investment has been made to reduce copper discharges to the estuary. Despite this reduction the dissolved copper concentrations in the estuary still exceed the 5 μg l^{-1} limit, albeit occasionally.

A cost-effectiveness analysis was undertaken to assess the potential methods of reducing copper inputs in the Humber. It is recognised that there are many sources of copper to the Humber Estuary (mainly contaminated sediments, sewage, and industry with point and diffuse inputs throughout the catchment, and direct inputs to the estuary). In the first instance, however, only the industrial point sources to the Humber Estuary were included in the analysis, with scope to include all the sources at a later date if the methodology proved successful.

Cost-effectiveness analysis in this context requires the derivation of abatement cost curves (Boer and Bosch 1995, Wickborn 1996, Maya and Fenhann 1994, Kram et al. 1995, Greer et al. 1997). Abatement cost curves provide an estimate of the cost to reach a required level of abatement, and also reveal the most efficient route to this discharge target. The description of costs in this format allows the ready evaluation of the impact upon the economy, and aids negotiations between the government, regulators and industries (Jung et al. 1996).

A drawback of the method applied is that it does not take into account the effects on the national economy, and the secondary impacts of the abatement technologies were also not included in the calculation of the abatement costs. Guidelines to standardise abatement cost studies have recently been produced (Wenborn et al. 1997, A.E.A. 1999) and these procedures were adopted in this study. There are 5 major steps in the derivation of the abatement cost curves:

1. Identification of all sources;
2. Collation of abatement techniques;
3. Total Costs and Abatement Potential;
4. Manipulation and standardisation of abatement cost data;
5. Production of the curves.

A database was developed which included the details of each abatement method, the current copper output of the industry (kg/year), the abatement level achieved (kg/year) following implementation, the investment cost, and operating costs. Investment costs are defined as the capital costs of the technology. The costs of implementing the technique over time are known here as operating costs.

There were a number of assumptions associated with the cost data. In many cases the investment has not been made specifically to reduce the copper concentration in the waste, however as the situation was rarely clear cut, to ensure consistency, the total cost of abatement was assumed to be targeted at copper. Another significant assumption was that secondary impacts of applying the abatement technologies were considered outside the scope of the primary analysis, but were taken into account when drawing conclusions. Abatement technologies can have beneficial side effects: for example, enhancing production, reducing running costs (including energy savings) and increased recovery of material. Health and safety benefits may also be enjoyed. There are also adverse effects related to the implementation of new processes, resulting in indirect costs. Older equipment is sometimes replaced prematurely, causing a financial loss.

All data was standardised to the baseline year of 1998. An annuity was calculated for each technology, permitting the comparison of measures on a cost/year basis. A variety of discount rates and time periods were used, as detailed in Beaumont (2000).

The abatement techniques in this case study were fully compatible and hence all combinations were used to produce the cost curve. Figure 6 depicts an aggregate cost curve for the most cost-effective application of available abatement measures, it is based upon a 4% discount rate and a 10 year time period. As expected the curve is stepped in nature and has a similar distribution to those documented in Riege-Wcislo and Heinze (1996).

The abatement cost curve can be used directly as a policy tool. Certain groups of abatement measures, or 'baskets', have drawbacks, and access to information about a similar abatement basket is useful. For example the most cost-effective method of reducing copper inputs by, say, 6000kg will be using a combination of measures that could be extrapolated directly from this curve. However, if one of the measures within this combination has a non-cost related adverse side effect, the curve can be utilised to provide information on cost-effective methods of attaining the same reduction (6000kg), whilst avoiding the application of the undesirable measure. Interestingly, the study also demonstrated that slightly less than 50% of the total abatement potential could be reached by the implementation of 9 techniques, costing less than 15% of the total costs.

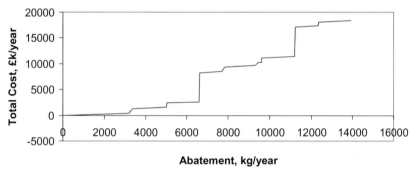

Fig. 6. Abatement Cost Curve for Copper Reduction in the Humber Estuary 1998 (Source: Beaumont 2000)

This analysis of the copper abatement techniques in the Humber region provided a valuable insight to the potential options for reducing copper discharges, and the barriers to this reduction. To undertake the study, several simplifying assumptions have been made, but the results still provide a generalised account of realistic pollution abatement options and their costs.

It is critical to note that the impact of waste discharges is not only dependent on how large they are, but where they are spatially located. Clearly, significant abatement at the mouth of the estuary will have less impact on the estuarine copper concentrations than similar abatement at the head of the estuary. Ideally the results from the abatement cost curve should be integrated with an understanding of the natural science of the estuary (Beaumont 2000). The discharge reductions in the estuary are only a small part of the cost-effectiveness analysis, as copper inputs also include catchment wide sources, historical sources (for example contaminated sediments) and sewage.

As a preliminary to a catchment-wide investigation, a broad-brush analysis can help identify the main sources of pollutants. Copper emissions in the estuary are used here again as an illustration. Figure 7 illustrates the total riverine inputs in the estuary (i.e. total copper inputs from the wider catchment), and compares them to direct inputs in the estuary, both from sewage treatment plants and industry.

Sewage and industrial inputs form only a small part of the riverine input to the estuary, and total riverine input of copper today is much greater than direct inputs to the estuary. A 'natural' background flux for copper of about 45 T yr^{-1} has been estimated using the fluxes for the Yorkshire Derwent, a non-industrialised, non-mining sub-catchment of the Yorkshire Ouse (Neal and Davies, in press). An estimated two thirds of the 'excess' of riverine input of Cu over industrial and sewage input can therefore be attributed to "natural" background flux, but given the historical mining and industrial legacy of the catchment, the remainder is likely to come from contaminated sediments in the catchment. Given the high costs of cleaning or removing these sediments, a mitigation option based on coastal managed realignment, for example, seems to offer some promise. The idea would be to create new intertidal habitat in the estuary to act as sinks for these sediments. An estimated 13.5 T yr^{-1} of the copper input to the estuary is currently buried by sedimentation on intertidal and sub-tidal mudflats.

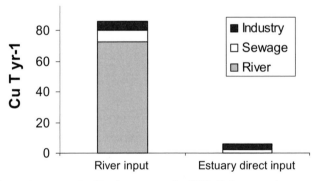

Fig. 7. Copper inputs to the Humber estuary in 2000. Industry and sewage data are from Environment Agency consent data for 2000. Total river data is from River Input Discharge (RID) data for 2000 reported to OSPAR. River input includes inputs to the tidal reaches

From scenarios to policy analysis

Cost-effectiveness is an important issue and the Water Framework Directive requires member states to carry out this analysis in all catchments. But there are also wider policy issues. Each of the options available to decision-makers to improve water quality has wider impacts in a range of areas: e.g. on biodiversity, on regional economic growth, on unemployment rates. Policy analysis is a tool that researchers use to analyse the wider impact of policy options, and how they are perceived by a variety of "stakeholders" or interested parties. Its aim is to compare policy options by reference to how different stakeholders are affected or perceive to be affected. In the context of EUROCAT, stakeholders provide input by identifying possible policy packages, as well as specific criteria within three broad categories (economic, environmental, and social). They can also, in principle, be asked to give weights to the evaluation criteria to determine their relative importance. The impact of each of the policy packages on the set of criteria can be assessed through scores, determined by modelling and expert opinion. Policy options can be ranked according to their impacts on the criteria, taking into account the stakeholder weights attributed to each criterion. But the dominant criterion, especially in the context of the Water Framework Directive, will be economic cost and cost-effectiveness.

Table 5. Policy packages

Sectors	Policy package 1	Policy package 2	Policy package 3
Agriculture	Small amounts of subsidies for extensive agriculture	Subsidies for good agricultural practices	Possible large scale support for green agriculture. Higher food prices
Industry	Non-restrictive emission standards	More restrictive emission standards, preceded by incentives	Eco-labelling and fall back emission standards
Urbanisation	Planning measures	Some SUD*	SUD wherever feasible
Sewage treatment	Minimal sewage treatment	Increased sewage treatment and water charges	Tertiary sewage treatment with high water charges
Contaminated sediments	No policy	No policy	Possible clean-up if new technologies allow it
Wetlands	Support for limited, opportunistic wetland creation	Support for medium scale wetland creation but no overall strategy.	Support for strategic large scale wetland creation; land banking

* SUD= Sustainable Urban Drainage

This section reports on the stakeholder consultation, which took place in the context of the EUROCAT Humber case study. The main objectives were to identify the policy packages likely to be implemented and the main criteria use to evaluate them.

A workshop was organised to bring together key stakeholders from the Humber estuary and catchment, and get their views on options to improve water quality in the Humber estuary. The day was structured in plenary brainstorming sessions and break-out in-depth discussion groups, with two main themes: policy options to improve water quality in the Humber, and criteria to evaluate these policy options.

In the first part of the workshop, participants were asked to provide three examples of policy options that they considered would contribute to improving water quality in the estuary, and to organise the list of options into three coherent policy packages. To avoid being constrained by present conditions, and to make sure policy packages were as different as possible to make the analysis worthwhile, they were asked to think about three world views, or mind sets, which would result in different approaches to dealing with water quality. Participants were therefore presented with the context of scenarios, although the scenario methodology was not introduced. The story lines of scenarios were presented only to frame the identification of policy packages: participants were asked to discuss the combination of instruments they would use in each different context using the list of options resulting from the brainstorming session as a starting point, and including others if necessary. Rationalising these preliminary discussions, and making packages consistent amongst each other could lead to three policy packages as described in Table 5, using the same instruments to varying degrees.

Towards integrated catchment management

The policy packages reported in Table 5 are fairly broad brush and are not spatially specific. Actual catchment management will need to address data and other requirements down to the sub-catchment spatial scale. This level of detail is a future priority for regulating agencies in the UK but so far only limited progress has been made. In the analysis below we sketch out some preliminary analysis which can lay a role in future catchment management. The analysis takes as a core starting assumption that coastal managed realignment policy will be an important component of any future planning for the Humber estuary and catchment. But the key insight highlighted at various points in this chapter is that managed realignment (and its impact in terms of increased intertidal habitat) carries with it a number of positive externality effects. It creates more habitat with potential biodiversity, amenity and recreational values; a more extensive nutrient and contaminants storage capacity; and a carbon sequestration function. All these potential economic benefits are in addition to its sea defence/coastal protection benefits in terms of increased flexibility in response to sea level rise and climate change and therefore reduced maintenance costs.

Form an economic efficiency perspective, the first test of managed realignment requires the scheme or programme to demonstrate net economic benefit i.e. that compared to the traditional hard engineering sea defence strategy "hold the line", managed realignment yields an efficiency gain in terms of net benefit or lower overall costs. Our analysis demonstrates that for a range of managed realignment schemes around the estuary and the tidal rivers, there is a net economic benefit i.e. that the costs of realignment over a 25 year time period or more are outweighed by the benefits created in terms of savings in maintenance costs and the positive environmental externality effects (Coombes and Turner 2004).

Coastal realignment can therefore form a key component of, for example, any wider catchment or estuary water quality policy. Cost-effectiveness analysis can show which combination of policy measures combined with realignment are most economic. Thus achieving higher water quality standards in the Humber estuary in terms of nutrient reduction will require managed realignment plus a set of measures directed at point sources e.g. sewage treatment plants and diffuse pollution e.g. agricultural run-off. Since managed realignment policy will be implemented anyway, and our analysis shows that it is an economically efficient strategy, the pragmatic response to water quality issues should be to factor in this policy context. Therefore the policy response to the problem of meeting future higher water quality standards should be to find the most cost-effective set of measures around the estuary and in the catchment that can provide additional nutrient reduction effects, once the baseline effect of increased intertidal habitat (via realignment) has been quantified. Our analysis has investigated improved sewage treatment and nitrate zoning and related measures in agricultural areas as elements of the overall pollution reduction programme, given a range of water quality targets in the estuary (Brouwer et al. 2004).

Conclusions

This chapter has presented a practical example of applying integrated assessment at the catchment scale, in the context of coastal water quality improvement. The existence of extensive data series from previous research programmes has been essential in scoping management issues in the Humber, and tracing the main drivers and pressures of environmental change in estuarine water quality, and their evolution over time. The evidence gathered so far indicates that although point source inputs have shown significant reductions over the last decade, atmospheric inputs to, and contaminated sediments from, the river catchment are likely to continue to have an impact on the water quality of the Humber estuary over the next few decades. The planned implementation of extended nitrate vulnerable zones (NVZs) throughout the catchment, combined with continuing improvements in sewage treatment, is likely to reduce nutrient and contaminant inputs to the estuary. Some of this reduction, however, may be buffered by the atmospheric and groundwater inputs. The continued availability of a supply of contaminated riverine sediments, together with expected increases in flooding due to climate change, is likely to limit improvements in water quality with respect to trace metals. Most of the sediments on the bed of the Humber are contaminated by trace metals resulting from long-term anthropogenic inputs. As the water quality in the estuary improves due to expected reductions in anthropogenic inputs, leading to a lowering in the concentrations of some elements in solution, there is potential for trace elements presently bound to sediments to go into solution, and for interstitial waters to try to equilibrate with overlying waters, leading to a flux of trace elements out of the sediments. Like the atmospheric and groundwater inputs, the bed-sediments may therefore act to buffer any improvement in water quality for many decades to come. However, setback of sea defences around the estuary may provide an additional sink for sediment-bound nutrients and contaminants.

Scenarios are an essential part of this research. The main aim is to aid decision-making in setting up river basin management plans in the context of uncertainty. Uncertainty stems from both socio-economic and political future conditions, and climate change. The Water Framework Directive requires the use of baseline scenarios to assess future impacts. Two of the scenarios used within EUROCAT can be related to this approach: the Business as Usual scenario can be interpreted as a pessimistic baseline, while the Policy targets scenario could be an optimistic one. The Deep Green scenario goes further and investigates the issue of how close we can get to "pristine" conditions, and what the socio-economic implications would be. This is an important issue in the current debate around the notion of what "good ecological status" means.

The first steps of the integrated assessment feed into the final stage of policy analysis. Now the physical relationships have been identified, the preferred management strategy can be further investigated. Cost-effectiveness analysis is an essential component of any river basin management plan and an explicit requirement of the Water Framework directive. As shown by the copper abatement cost curve methodology, it is a valuable tool for ensuring the efficient reduction of pollution, however, applying this analysis at a catchment scale is a real challenge.

Acknowledgements

This work is funded by the European Commission under the Fifth Framework Programme (project No EVK1-CT-2000-00044). The authors are grateful to the UK Environment Agency for facilitating communication with the Humber Shoreline Management Plan steering group, and for making available biogeochemical datasets for the Humber.

References

AEA Technology (1999) Guidelines for defining and documenting data on costs of possible environmental protection measures. European Environment Agency, Copenhagen

Beaumont NJ (2000) The Assimilative Capacity of the Humber Estuary for Copper: An Interdisciplinary Study. Ph.D. thesis, University of East Anglia, Norwich

Beaumont NJ, Tinch R (2003) Cost-effective reduction of copper pollution in the Humber estuary. CSERGE working paper ECM 03-04, University of East Anglia, Norwich

De Boer B, Bosch P (1995) The greenhouse effect: an example of the prevention cost approach. Second meeting of the London group on national accounts and the environment, Washington

Brouwer R, Cave R, Coombes E, Turner RK, Hadley D, Lorenzoni I (2004) Cost effectiveness analysis of alternative policy options to improve water quality in the Humber catchment in the UK. Forthcoming; CSERGE, University of East Anglia, Norwich

Burbridge P (2001) Sustainability and Human Use of Coastal Systems. LOICZ newsletter 21:3-5

Cave R, Ledoux L, Turner RK, Jickells T, Andrews J (2002) The Humber Catchment and its Coastal Area: from UK to European Perspectives. Sci Total Environ 314-316:31-52

Coombes E, Turner RK (2004) A cost-benefit analysis of managed realignment policy in England. CSERGE working paper, forthcoming; University of East Anglia, Norwich

Crooks S, Ledoux L (1999) Mitigation banking as a toll for strategic coastal zone management: a UK perspective. CSERGE working paper GEC 99-02

DETR (2000a) Modern Ports: A UK Policy. Department of the Environment, Transport and the Regions, London

DETR (2000b) Transport Statistics, Maritime Statistics: United Kingdom 2000. Department for Environment, Transport and the Regions, London

Edwards AMC, Freestone RJ, Crockett CP (1997) River management in the Humber catchment. Sci Total Environ 194:235-246

Environment Agency (1999) The Humber Estuary, State of the Environment 1998, London

European Commission (1999) Towards a European Integrated Coastal Zone Management (IZM) Strategy, General principles and Policy Options: A reflection Paper, Brussels

European Commission (2000) Communication from the Commission to the Council and the European Parliament on "Integrated Coastal Zone Management: a Strategy for Europe" COM/00/547, Brussels

European Environment Agency (2000) Cloudy crystal balls: an assessment of recent European and global scenario studies and models. Environmental issue report no. 17, European Environment Agency, Copenhagen

Gallopin GC, Raskin P (1998) Windows on the future: global scenarios and sustainability. Environment 40:7-11, 26-31

Greer L, Van Loben Sels C (1997) When pollution prevention meets the bottom line. Environ Sci Technol 31:418-422

Hammond A (998) Which World? Scenarios for the 21st century. Global Destinies, Regional Choices. Earthscan Publications Ltd., London

HEMS (1997) Humber Estuary Management Strategy. Environment Agency, Hull

Jarvie HP, Neal C, Robson AJ (1997) The geography of the Humber catchment. Sci Total Environ 194:87-99

Jickells T, Andrews J, Samways G, Sanders R, Malcolm S, Sivyer D, Parker R, Nedwell D, Trimmer M, Ridgway J (2000) Nutrient fluxes through the Humber estuary - Past, present and future. Ambio 29:130-135

Jung C, Krutilla K, Boyd R (1996) Incentives for advanced pollution abatement technology at the industry level: an evaluation of policy alternatives. J Environ Econ Man 30:95-111

Kaika M, Page B (2002) The making of the EU Water Framework Directive: shifting choreographies of governance and the effectiveness of environmental lobbying. Economic Geography Research Group, working paper WPG 02-12. School of Geography and the Environment, University of Oxford, Oxford

Kallis G, Butler D (2001) The EU Water Framework Directive: measures and implications. Water Policy 3:125-142

Kassler P (1995) Scenarios for world energy: barricades or new frontiers? Long Range Plann 28(b):38-47

Kram T, Schol E, Stoffer A, Rothengatter W, Guhnemann A, Sorensen MM, Stouge A, Suter S, Walter F (1995) External costs of transport and internalisation, Synthesis report on topic A: External costs of transport. Report to the European Commission, ECN-C-95-080. Energy Research Centre of the Netherlands (ECN), Amsterdam

Ledoux L, Cave R, Turner RK (2002) The use of scenarios in integrated environmental assessment of catchment-coastal zones: the Humber estuary, UK. LOICZ newsletter 23:1-4.

Ledoux L, Cornell S, O'Riordan T, Harvey R, Banyard L (2004) Managed realignment: towards sustainable flood and coastal management. CSERGE working paper, forthcoming, University of East Anglia, Norwich

Ledoux L, Crooks S, Jordan A, Turner RK (2000) Implementing EU Biodiversity Policy: UK experiences. Land Use Policy 17: 257-268

Longfield SA, Macklin MG (1999) The influence of recent environmental change on flooding and sediment fluxes in the Yorkshire Ouse basin. Hydrol Process 13:1051-1066

Lorenzoni I, Jordan A, Hulme M, Turner RK, O'Riordan T (2000a) A co-evolutionary approach to climate change impact assessment – Part I: Integrating socio-economic and climate change scenarios. Global Environ Change 10:57-68

Lorenzoni I, Jordan A, O'Riordan T, Turner RK, Hulme M (2000b) A co-evolutionary approach to climate change impact assessment – Part II: A scenario-based case study in East Anglia. Global Environ Change 10:145-155

Mance G, Brown VM, Yates J (1984) Proposed environmental quality standards for list II substances in water: copper. Water Research Council, technical Report 210, WRc, Swindon

Maya RS, Fenhann J (1994) Methodological lessons and results from UNEP GHG abatement costing studies. Environ Policy 22:955-963

Miles I (1981) Scenario analysis: identifying ideologies and issues. In: Methods for Development Planning: Scenarios, Models and Micro-studies. UNESCO Press, Paris

Millward GE, Glegg GA (1997) Fluxes and retention of trace metals in the Humber Estuary. Estuar Coast Shelf Sci 44:97-105

Murby P (2001) The Causes, Extent and Implications of Intertidal Change - A regional view from eastern England. Ph.D. thesis, University of Hull, Hull

Neal C, Davies H, (2003) Water quality fluxes for eastern UK rivers entering the North Sea: a summary of information from the Land-Ocean Interaction Study (LOIS). Sci Tot Environ 314-316:31-52

NRA (1993) The Quality of the Humber Estuary 1980-1990. National Rivers Authority, London

OECD (1993) OECD Core Set of Indicators for Environmental Performance Reviews. A Synthesis Report by the Group on the State of the Environment. OECD, Paris

Office for National Statistics (2002) Census 2001: First results on population for England and Wales. HMSO, London

Office of Science and Technology (1999) Environmental futures. Report for the UK's National Technology Foresight Programme, Department of Trade and Industry, DTI/Pub 4015/lk 3 99 NP, URN 99 647, London

Parry ML (ed) (2000) Assessment of Potential Effects and Adaptations for Climate Change in Europe: The Europe Acacia Project. Jackson Environment Institute, University of East Anglia, Norwich

Parry M, Carter T (1998) Climate Impact and Adaptation Assessment. Earthscan Publications Ltd., London

Pilcher R, Burston P, Davis R (2002) Seas of Change. The potential area for inter-tidal habitat creation around the coast of mainland Britain. Royal Society for the Protection of Birds, Sandy, Bedfordshire

Riege-Wcislo W, Heinze A (1996) The construction of abatement cost curves. Methodological steps and empirical experiences. German Environmental Economic Accounting (GEEA), Federal Statistical Office, Bonn

Turner, RK (2004) Integrated environmental assessment and coastal futures. In: Vermaat JE, Bouwer LM, Salomons W, Turner RK (eds) Managing European coasts: past, present and future. Springer, Berlin, pp 255-270

Turner RK, Adger WN, Lorenzoni I (1998a) Towards integrated modelling and analysis in coastal zones: principles and practices. LOICZ Reports and Studies No 11, NIOZ, Texel

Turner RK, Bateman IJ, Adger WN (eds) (2001) Economics of Coastal and Water Resources: Valuing Environmental Functions. Kluwer, Dodrecht

Turner RK, Lorenzoni I, Beaumont N, Bateman IJ, Langford IH, McDonald AL (1998b) Coastal management for sustainable development: analysing environmental and socio-economic changes in the UK coast. Geogr J 164:269-281

Wenborn M, King K, Boyd R, Johnson C, Marlowe I (1997) Towards guidelines for data collection on costs of possible environmental protection measures – technical report, Report number AEAT-1836/RHOD/20325001/ISSUE1. European Environment Agency, Copenhagen and National Environmental Technology Centre, London

Wickborn G (1996) Avoidance Cost Curves for NOx. Third London group meeting on environmental accounting, Stockholm

19. The impact of subsidence and sea level rise in the Wadden Sea: Prediction and field verification

Joop M. Marquenie[1] and Jaap de Vlas

Abstract

Gas exploration is expected to cause subsidence in the Wadden Sea. This in turn may have impacts on hydrodynamics, tidal geomorphology, benthic fauna, wading birds and salt marshes. This paper reports on two separate multidisciplinary studies: one predicting the longer-term effects of additional gas exploitation sites relative to sea level rise, and the other a longer-term verification monitoring at an existing site on the Wadden Sea island of Ameland. Both studies conclude that subsidence will cause local impacts on tidal flats, saltmarsh and dune geomorphology, on groundwater levels and inundation frequency of coastal marshes. The induced increased sediment needs, however, are small compared to those generated by sea level rise, and can be covered by present annual sediment fluxes unless sea level rise follows the most extreme IPCC scenario of a 100 cm rise in the coming century. The effects on the biota are insignificant compared to natural temporal variation.

Introduction

The Wadden Sea is a major European intertidal wetland comprising a network of tidal gullies, sand flats and coastal marshes protected by a chain of barrier islands. It extends from Den Helder in The Netherlands to Esbjerg in Denmark covering about 8000 km^2. Its importance as a stop-over for migrant birds and nursery for offshore fish is high (e.g. Wolff 1983) but other functions are also considered to be important, as indicated by the high total economic value estimated by De Groot (1992). Recreation, for example, is the major source of income for inhabitants of

---

[1] Correspondence to Joop Marquenie: joop.marquenie@shell.com

J.E. Vermaat et al. (Eds.): Managing European Coasts: Past, Present, and Future, pp. 355–363, 2005.

the barrier islands, and this tourism is tightly linked to the natural character of the area.

When natural gas was found under the island of Ameland in 1962, over 50 exploratory wells were drilled in the Wadden area. Application for exploitation permits led to severe protests in society, with several NGOs effectively stressing the uniqueness of the Wadden Sea as an area of outstanding natural beauty. Three concessions were granted to produce gas from the Wadden Sea: Groningen, Noord Friesland and Zuidwal. Production on Ameland (Noord Friesland) started in 1986, whilst a 10-year moratorium on further drilling was agreed upon in 1984. As of 1987, subsidence due to gas extraction and its possible effects on salt marshes, dunes and tidal flats has been monitored on this Ameland site. Within 50-100 km, the much larger Slochteren gas field on the mainland also causes locally variable subsidence in the Wadden Sea.

When the moratorium had passed, mining companies indicated their preparedness to commence exploitation of the Wadden Sea gas fields. Several exploitation permits have been granted under the condition of an extensive cumulative assessment of subsidence from present and new exploitation fields, including foreseen sea level rise. This assessment, or 'predictive' study was organised alongside the regular monitoring or field 'verification' programme. The comprehensive predictive study was carried out by a multidisciplinary team and is reported in Beukema et al. (1998). The long-term monitoring at the Ameland site has been reported in Eysink et al. (2000). The present paper summarises the main findings of these two multidisciplinary reports and explores the most probable effects of exploitation-related subsidence on coastal habitats and bird communities of the Wadden Sea. The work is placed in the perspective of a changing public perception, acknowledging that public and political support for gas exploration has dwindled.

Observed subsidence at Ameland-Oost after 13 years of exploitation

A first tentative estimate of the probable total subsidence that would occur at the Ameland site was made in 1985 using the NAM-model that was developed and calibrated earlier for the larger Groningen gas field. The predicted maximum surface depression above the centre of the field was then predicted to be 26±5 cm. This would produce a shallow crater of dish-like appearance with a volume of approximately 28 million m^3.

This estimate prompted the proprietor of the nature reserve, the NGO 'It Fryske Gea', to request a careful monitoring of the actual subsidence that would occur from this gas extraction on Ameland and the changes this would lead to in its nature reserve. A research team supervised by an independent scientific committee commenced a programme monitoring in 1987, which set out to provide:

- Accurate level measurements in order to determine the actual subsidence;
- Vegetation composition of the salt marshes and dunes;
- Detailed topography and water depths;

- Changes in the bird population;
- Data that can be used to establish economic damage to adjacent grazing areas, the Nieuwlandsrijd salt marsh and the Buurdergrie polder, or to the extraction concession for drinking water in the adjacent Buurderduinen.

Subsidence

The initial forecast of 1985 was adjusted in 1991 on the basis of more recent information about the reservoir characteristics. The ultimate drop in the centre was estimated to be approximately 18 cm. However, measurements showed that the rate at which subsidence was taking place in the near vicinity of the NAM location was higher than forecasted in 1991. In fact the subsidence rate corresponded almost exactly with the 1985 forecast. This led to a thorough revision of the applied models and the improved model reproduced the observed subsidence more accurately. The following predictions resulted:

- The ultimate subsidence in the centre of the dish would be 28 cm, and
- The subsidence bowl will be slightly deeper in the middle and slightly shallower at the edges than originally predicted in 1985. The total subsidence volume will ultimately be significantly less than predicted in 1985 (Table 1).

Table 1. Evolution in the subsidence depth estimates for the Ameland-Oost gas field

Date	Maximum subsidence (cm, range)	Total subsidence volume (10^6 m^3)
1985	26 (21-31)	28
1991	18 (14-22)	18
1998	28 (22-34)	14-18

The actual subsidence observed in February 1999 at the deepest point was approximately 22 cm, i.e. 70% to 75% of the ultimate drop now expected. At that time the volume of the dish was 9×10^6 m³, which is substantially less than predicted. The subsidence closer to the edges still remains less than the 1985 prediction, demonstrating that the exact shape of the dish is difficult to predict due to geological complexity.

Sea level and groundwater levels

Mean sea level was estimated from tidal gauges at Den Helder, in the Western part of the Wadden Sea, and at Nes, Ameland, in the vicinity of the exploitation site. Over the course of the observation period, i.e. since 1987, mean sea level has increased by about 2 cm, but there is no indication of any systematic difference between the two tidal gauging stations.

Variability in rainfall and storm patterns was substantial over the observation period: 1992-1995 formed a succession of wet years with a high frequency of storm tides. This affected inter-annual variation in water level in the piezometers around the exploitation site. Within-year variation in ground water level varied between 0.5 and 1.0 m. This annual pattern is recognizable across the island and spatial patterns are simple: gentle gradients exist from higher to lower-lying areas. In areas of subsidence, groundwater level increased relative to local surface. This effect is permanent in areas where accretive compensation is absent, such as polders that are not inundated by the tides. One of the polders, the Buurdergrie, had subsided between 1 and 4 cm in 1999, the range depending on the distance from the exploration site.

Table 2. Net and gross subsidence on salt marsh two transects on the eastern tip of Ameland. Net subsidence is after compensation by accretion. Subsidence is presented in cm between 1986 and 1999 for the western and eastern ends of the transect: W-E

Subsidence:	gross	net
Transect:		
Nieuwlandsrijd	13 – 3	5 – 0
De Hon	22 – 9	10 - 0

Geomorphology

Since 1990, beach nourishment is carried out all along the Dutch coast including Ameland to maintain the coastline, using sand dredged from – 20 m offshore. This practice is not executed east of the NAM location, since it is policy to have natural coastal dynamics prevailing here. Both hydraulic modelling and observational data show substantial natural change in coastal morphology along the eastern side of the island. Channel migration as well as de novo genesis with associated erosion and sedimentation occur at a large scale and make it difficult to identify any, comparatively small, impact of subsidence. Clearly, locally increased subsidence on the flats could easily be compensated for by the transport of large volumes of sediment across the system.

Salt marsh level measurements did allow a detailed analysis of sedimentation and subsidence, since geomorphological change is less pronounced. Sedimentation gradients were apparent across the transects (Table 2). Accretion in the salt marshes depends on the salt marsh level and, probably, the distance to the Wadden Sea or a salt marsh creek. As a rule the subsidence in the low-lying salt marshes (lower than NAP +1.25 m) was more than compensated for by accretion. On the higher marsh, this compensation gradually decreased to zero at a level of 2 m above NAP relatively close to the coast or 1.7 m above NAP at greater distances from the sea. In other words, the salt marsh had become somewhat flatter.

Further inland, the complex of low-lying dunes and slacks have a stable geomorphology with very little sand movement. Here, subsidence was permanent and increased groundwater tables resulted that were locally considerable.

Salt marsh and dune vegetation

Locally, negative accretion balances were observed on the higher salt marshes. This, however had virtually no effect on the vegetation as only one of the 50 plots displayed 'regression' to predominance of lower marsh species. Local drainage patterns were found to have important modulating effects on the predictions based on elevation alone. In the higher dunes, vegetation change had been minor and could not be attributed to subsidence. Outside the permanent plots mass mortality of three species of shrubs was observed in 1995 (i.e. *Crataegus monogyna, Sambucus nigra and Hippophae rhamnoides*). Since similar mortality was observed on other adjacent islands without gas exploration, subsidence cannot have been the sole cause. Mortality was found to have occurred after extreme flooding events with seawater in springtime 1986 and 1991, and by inundation with groundwater in extremely wet years of 1993/1994.

Birds

Migratory bird counts were compared for the Hon on Ameland and the adjacent and similar Boschplaat area on Terschelling using available time series data from 1984-1999. Trends were practically identical. Also summer nesting bird counts on the Hon salt marshes showed no change over the observation period, which is in accordance with the absence of any appreciable change in vegetation.

Grazing and drinking water: Economic losses

Enclosed polders subject to subsidence did not face an increase in inundation risk. Also, the limited subsidence (4 cm at most) would not have affected the grass yield. Only salt marsh areas open to inundation incurred an increased inundation frequency. This was estimated to reduce the time that these marshes are accessible for cattle grazing, but the loss was limited. Estimates were 2% in 1994 and 3% in 1998.

An earlier interim report (cf. Eysink et al. 2000) demonstrated that coastline movements in the order of 50 m would not affect the chloride content of the extracted drinking water through changes in the interface between the island's freshwater lens and deeper saline groundwater. Whilst the government maintains an active policy of coastline fixation through beach nourishment, no effects on the volume or quality of the extractable drinking water are foreseen.

Predicting environmental impacts of future gas exploitation

The 'predictive' study (Beukema 1998) involved an environmental impact assessment of future, new gas exploration fields and had to include possible effects of other reserves and field s that were already in exploitation. Uncertainty over predictions on sea level change were coped with using three scenarios: the present sea level rise at 20 cm per century, a most realistic estimate of 60 cm and an extreme scenario of 100 cm (cf. Nicholls and Klein this volume). Other natural developments and human interference that may affect the Wadden Sea sand balance and the persistence of saltmarshes were also included where possible. Status quo assumptions were made for the present coastline, area of salt marsh and tidal channel depths. Time frame for the study was 1999-2050, i.e. till the end of the gas extraction from the Ameland field.

Uncertainty in predicting the form and volume of a subsidence bowl remains substantial, since it depends on the form of the gasfield and the structure of the gas-containing deposits, but also on the variable capacity of covering strata to cope with change in pressure. All these are imperfectly known as well as hard and expensive to quantify. It is difficult, therefore, to attach confidence intervals to our subsidence estimates over the larger Wadden Sea basin. This is also apparent from the subsidence depth updates reported above (Table 1).

Table 3. Balance of Wadden Sea sediment needs ($x10^6$ m^3) apportioned to different sinks for the period before 2000 and that of 2000-2050

	Per year	1960-2000	2000 – 2050
Sea level rise (20 cm per century)	4	160	200
Sea level rise (IPCC: 60 cm per century)	12	480	600
Closure of the Lauwerszee (1970-1990)	2	36	
Shell extraction (>2000)	0.25	5.6	12.5
Sand extraction (<1997; has stopped)	0.6	24	
Overfishing mussel beds (1960-1980)	0.5-1	10-20	
Gas extraction, existing (1960-2000)	0.7	27	
Gas extraction, existing (2000-2050)	0.6		32
Gas extraction, new, probable scenario (maximum scenario)*	0.2		12 (25)

* Future probable and maximum gas extraction scenarios differ in the proportion of drilled wells that actually strike exploitable gas. The maximum scenario assumes that all wells will become productive. The present rate of success to strike gas is < 50%.

Estimating subsidence and sediment requirements

All possible causes of subsidence and other volume changes to the Wadden Sea were quantified that were considered to have consequences for the sediment balance (Table 3). Estimates were expressed per year as well as over the total period. Two periods were compared: the decades before 2000, and 2000-2050. Present extraction of gas is already causing subsidence in parts of the Wadden Sea. This source was included. For the comparative balance, sea level rise was assumed to proceed as present.

Overall, subsidence due to gas extraction was estimated to cause a sediment requirement in the order of 40 to 60 million m^3. This is small compared to the volume needed to compensate sea level rise, even when the latter is assumed not to accelerate over the coming decades. Hence, it is even smaller when compared with the presently most realistic estimate of 60 cm over the next century. Furthermore, subsidence will be restricted to the period of extraction and have a more localised impact than sea level rise: particularly the tidal basin draining through the channel east of Ameland will be affected.

Because subsidence causes extra sediment to be brought from the coastal zone into the Wadden Sea, the estimated lowering of the mudflats amounts to 6-8 cm in the area with the greatest subsidence, i.e. the tidal basin between Ameland, Schiermonnikoog and the mainland). The area of mudflats would be shrinking temporarily by a few square kilometres. The lowest point will be reached 10 -15 years after the start of gas extraction and this will gradually be recovered again because sedimentation will gradually compensate subsidence. The additional sediment brought from the North Sea amounts to a mere 50 million m^3 for the period 2000-2050, or about one million m^3 y^{-1}. As a comparison, the sediment required by the entire Dutch Wadden Sea as a result of the current relative rise in sea level of 20 cm/century is in the order of 4 million m^3 y^{-1}. More rapid rates of sea level rise would delay the filling-up of the subsidence depression, or phrased otherwise, subsidence would locally enhance the effects of such high rates of sea level rise.

Salt marshes

It appears that the rate of sedimentation in the salt marsh zone is probably almost always sufficient to absorb the maximum subsidence, even in the event of rapid sea-level rise (Dijkema 1997). In the transitional zone between salt marsh and mudflat (the pioneer zone), the rate of sedimentation is often slower. This may create potential problems on the Groningen salt marshes, but these can be tackled by measures to promote sedimentation or by local restrictions on the rate at which gas is extracted.

Benthic fauna of the tidal flats

Macrobenthos biomass and species richness in the Wadden Sea displays an optimum curve with respect to tidal depth. The maximum for both are somewhere midway mean sea level and mean low tide, which presently is at –0.5 m NAP (Dutch Ordinance Level). Model predictions suggested that biomass decreases would be in the order of 1-2 % in the tidal areas with maximal subsidence. This is negligible compared to natural inter-annual variation observed in long-term time series data from the Wadden Sea and elsewhere (e.g. Frid et al. 2000; Beukema et al. 2002).

Effects on wader numbers in the Wadden Sea

Sea level rise as well as subsidence would temporarily lead to submergence of a limited proportion of the tidal flats. Biomass of benthic fauna, the main food for waders, appears not to be affected significantly (see above), hence mud flat area would serve as a valid indicator of food availability and hence the capacity of the mudflats to host wading birds. Estimates were made for the Pinkegat tidal basin, the area most affected by subsidence, assuming no sedimentary compensation (worst case). The different bowls resulted in an average planar lowering of this basin of 8 cm. The theoretical impact was calculated using a feeding wader habitat model (Ens and Brinkman, pers. comm.) and an enforced sea level rise of 8 cm on the existing tidal regime. Total wader numbers were estimated to decline with some 10%, and individual species declined (Grey or Black-bellied plover, *Pluvialis squatarola* -23%) or even increased (Greenshank, *Tringa nebularia* + 9%) in numbers. This decrease must be considered small compared to inter-annual variation, which is in the order of 30% (Smit and Zegers 1994; Ens and Brinkman unpubl.).

Options for mitigation and compensation

As (underwater) beach nourishment is presently carried out in the troughs between the sandbanks to maintain the coastline of the islands and considered a satisfactory compensation tool, this might also be considered as an option to fulfil the extra sediment requirements in the Wadden Sea due to subsidence, provided these are not met in a natural way. Suppletion in the flood channels may be a more attractive option to maximise the natural sediment redistribution capacity of the tidal basin system and not to impact the trough shell community. Where subsidence may have permanent effects on e.g. infrastructure such as sea defence works and pipelines, or on economic exploitation of the sea or its coastal marshes, mitigation options have been specified in Beukema (1998) and the major conclusion is that these measures need to be taken timely so that no measurable effects of subsidence will be observed.

Conclusions

Both studies conclude that subsidence will lead to local impacts on tidal flat, salt marsh and dune geomorphology, on groundwater levels and inundation frequency of coastal marshes. The observed subsidence was bowl-shaped as predicted, and depth as well as volume of the bowl was estimated comparatively well, whereas the exact form was more difficult to predict. Subsidence will cause locally increased sediment needs which are small compared to those generated by sea level rise, and can be covered by present annual sediment fluxes unless sea level rise will follow the most extreme scenario of a 100 cm rise in the coming century. The effects on the biota of tidal flats and salt marshes are insignificant compared to natural temporal variation. Mitigation or compensation of locally emerging subsidence effects is driven by monitoring programs.

In addition, the results of this multidisciplinary effort were reviewed by an external panel of experts, published and presented in public. Still, however, public opinion and the position of policy makers are not strongly altered after the rationalisations of the presented studies. Communication and public opinion assessment must be seen as important aspects when stakeholder benefits appear unbalanced.

Acknowledgements

We like to thank Brian Ward for his valuable comments and constructive remarks concerning the verification study. Jan Vermaat is most greatly thanked for his stimulation to complete this chapter and rigorous editorial support.

References

Beukema JJ, Cadee GC, Dekker R (2002) Zoobenthic biomass limited by phytoplankton abundance: evidence from parallel changes in two long-term data series in the Wadden Sea. J Sea Res 48:111-125

Beukema JJ, Brinkman AG, Buijsman M, Dankers N, De Vlas J, Dijkema KS, Van Dobben HF, Van Duin WE, Ens BJ, Eysink WD, Fokkink RJ, Gussinklo HJ, Kers AS, Kruse GAM, Marquenie JM, Onassis J, Oost AP, Smit CJ, Steyaret F, Stive MJF, Van der Wijk J, Van Wijnen HJ, Verboom BMJ, Verburgh JJ, Walrecht G, Wang ZB, Zegers J (1998) Integrale bodemdalingstudie Waddenzee. NAM. Assen (in Dutch: Integral subsidence study in the Wadden Sea)

De Groot RS (1992) Functions of nature – evaluation of nature in environmental planning, management and decision making. Wolters-Noordhoff, Groningen

Dijkema (1997) Impact prognosis for salt marshes from subsidence by gas extraction in the Wadden Sea. J Coast Res 13:1294-1304

Eysink WD, De Vlas J, Dijkema KS, , Slim PA, Smit CJ, Sanders ME, Schouwenberg EPAG, Wiertz J, Schouten D, Van Dobben HF (2000). Monitoring effecten van bodemdaling op Ameland-Oost. – Evaluatie na 13 jaar gaswinning.NAM Assen (in Dutch: Monitoring the impact of subsidence on the island of Ameland – Evaluation after 13 years of gas exploitation)

Frid CLJ, Harwood KG, Hall SJ, Hall JA (2000) Long-term changes in the benthic communities on North Sea fishing grounds. ICES J Mar Sci 57:1303-1309

Nicholls RJ, Klein RJT (2005) Climate change and coastal management on Europe's coast. In: Vermaat JE, Bouwer LM, Salomons W, Turner RK (eds) Managing European coasts: past, present and future. Springer, Berlin, pp 199-225

Smit CJ, Zegers PM (1994) Shorebird counts in the Dutch Wadden Sea, 1980-91 - a comparison with the 1965-77 period. Ophelia Supp 6:163-170

Wolff WJ (1983) Ecology of the Wadden Sea, Balkema, Rotterdam

20. The need for integrated assessment of large-scale offshore wind farm development

Andreas Kannen[1]

Abstract

This paper describes ongoing developments and future plans regarding the implementation of large offshore wind farms in the German Exclusive Economic Zone (EEZ) in the North Sea. Based on these developments, which form an important part of the German energy and emissions reduction policy at the Federal level, potential ecological as well as socio-economic impacts of these new permanent structures are described and the need for an Integrated Assessment as part of Integrated Coastal Zone Management is outlined. As a regional example for impacts and especially socio-economic opportunities and risks, the North Sea coast of Schleswig-Holstein will be used. It is argued that Integrated Assessment should be a prerequisite and Integrated Coastal and Marine Management an accompanying activity for an ecological, economical and socio-cultural sustainable development of this type of generating renewable energy in the North Sea.

Offshore wind farm development in the German North Sea

Around Europe more than 70 wind farm projects are under development in European coastal waters or marine areas, 31 of these in front of the German coast (Paul and Lehmann 2003). 11 wind farms were connected to the electricity grid by the end of 2002, including Horns Rev, sited 14-20 km into the Danish North Sea, west of Blåvands Huk. It is the largest farm with a capacity of 160 Mega Watt (MW).

In Germany, the Federal government has set the goal to increase the share of renewable energies to 12.5% by 2010 and to 50% by 2050 (BMU 2002a). According to the Environmental Ministry this requires the installation of a minimum ca-

[1] Correspondence to Andreas Kannen: kannen@ftz-west.uni-kiel.de

J.E. Vermaat et al. (Eds.): Managing European Coasts: Past, Present, and Future, pp. 365–378, 2004.
© Springer-Verlag Berlin Heidelberg 2005.

pacity of 20.000 MW in offshore wind farms by 2030, covering an estimated area of 2000-2500 km² (BMU 2002b). The different phases of this National Strategy for the Use of Wind Energy in the Sea are shown in table 1.

Table 1. Phases of the National Strategy for the Use of Wind Energy in the Sea and related capacity (BMU 2002a)

Phases	Time	Capacity
1. Preparation Phase	2001-2003	-
2. Starting Phase	2003/4-2006	minimum 500 MW
3. First Extension Phase	2007-2010	2000-3000 MW
4. Additional Extension Phases	2011-2030	20,000-25,000 MW

Because available space in the Baltic Sea is limited and conflicts with other human activities exist, this planning is largely based on capacities in the Exclusive Economic Zone (EEZ) of the German North Sea.

Economically renewable energies and especially the wind power sector have experienced dramatic growth in recent years, which is shown by a strong increase of installed wind power capacities in Germany (Fig. 1). Based on expected investments of 9 billion EUR until 2020 for developing offshore locations it is estimated by the Federal Government up to that about 11,000 new jobs will be created in Germany including a substantial share for the coast.

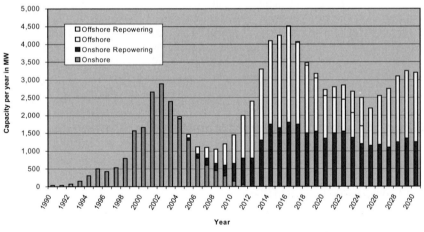

Fig. 1. Assessment of future development of wind power capacities; Repowering means the exchange of old wind turbines against new ones (usually less turbines but more powerful ones) (Source: Volmari 2002, based on material of Bundesverband Windenergie e.V. (BWE)

Planning context and legislative background

Compared to other countries like Denmark, one constraint on planning offshore wind farms in the German North Sea is formed by the designation of National Parks in the Wadden Sea, in which the construction of such installations is prohibited. As a result, near-shore areas with lower water depths cannot be considered for wind farm construction, so German wind farm developers focus on areas further offshore in deeper water, which raises the costs and enhances technical problems. This implies that experiences from e.g. the existing wind farms in Denmark are not directly transferable and of limited use for German investors, resulting in time consuming planning and development, high economic risks and slow development of pilot projects.

Planning for several major wind farm projects in the German Exclusive Economic Zone (EEZ) has been under way since 1997. At the moment 24 project applications in the North Sea (and 6 in the Baltic Sea), some of them comprising several hundred wind turbines, are handled by the Federal Maritime Agency (BSH, http://www.bsh.de). Figure 2 illustrates the demand for spatial areas in coastal and marine waters, which are currently claimed for wind farm projects.

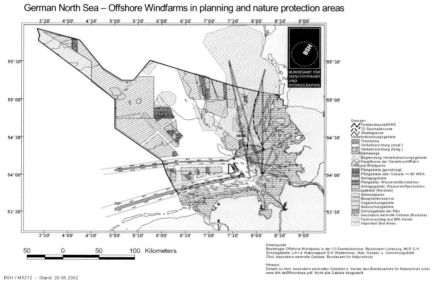

Fig. 2. Offshore wind farms under approval and nature protection areas (Source: BSH, CONTIS information system)

In November 2001 the first approval for the installation of 12 wind turbines in a pilot phase had been granted. These wind turbines will be located 45 kilometres northwest of the island of Borkum at a water depth of about 30 m. On 18 December 2001 the project "Butendiek", located 34 kilometres west of the island of Sylt and consisting of 80 turbines, received approval.

While in the EEZ the Federal Maritime Agency (BSH) is responsible for the approval procedures, the German Länder are responsible for the approval in the area of the territorial sea up to the 12 nautical miles limit. Permits for the cables and grid connection are handled separately in an additional planning procedure in which the national and the federal level both need to cooperate closely with each other and where also the municipal level (on the North Frisian islands if cables touch their ground or at the location where the cable reaches the land) is affected. The approval procedure is based on the Seeanlagenverordnung (SeeAnlV – Marine Facilities Ordinance), until now the only legislative instrument for planning in the EEZ. According to the SeeAnlV vessel traffic routing and risk of ship accidents on one hand and impacts on the marine environment are the subjects on which the decision is grounded.

Pollution in the sense of the United Nations Law of the Sea and risks for bird flyways form particular environmental reasons for not issuing a permit (Tiedemann 2003). Generally this legislative background is vague, with the exception of the relevant EU Directives like the Flora and Fauna Habitat and the Birds Directive (Koch and Wiesenthal 2003). On the other hand, the assessment and valuation of potential impacts on areas protected by these directives can create tremendous problems due to uncertain and missing knowledge and due to different interpretation by multiple stakeholders. The currently existing legislative background does not allow sufficient freedom for long term goals and future oriented planning of marine area use (Koch and Wiesenthal 2003, Czybulka 2003). Therefore a broad range of actors and stakeholders as well as scientists argue for instruments and legislation that allow spatial planning similar to terrestrial areas for the marine areas including the EEZ in Germany (Buchholz 2002, Koch and Wiesenthal 2003, Rat von Sachverständigen für Umweltfragen 2003).

In order to assist planning and permitting procedures as well as spatial planning in offshore areas there is a need for accompanying tools that integrate information, are flexible in terms of incorporating newly gained knowledge and also in terms of new uses that might come up in the future (Kannen 2000, Gee et al. forthcoming). From the view of the author this is the only way to handle knowledge gaps, uncertainty and active involvement of local stakeholders in form of a "management of change" process, as outlined in Kannen et al. 2000. This includes incorporation of wind farm development and other human uses in offshore areas into national ICZM strategies (Gee et al. forthcoming). These national strategies are currently under development in several EU Member States following the EU recommendations for ICZM (European Commission 1999, 2000, European Parliament and Council 2002).

Conflicts between wind farms, other coastal users and the environment

Strategically it must be recognised that each offshore wind energy project brings a global environmental benefit by reducing fossil fuel usage and consequently carbon dioxide emissions, thereby reducing the risk of catastrophic hazards due to climate change. In addition the impacts from offshore wind energy regeneration have to be seen in the light of impacts of other forms of energy supply like oil and gas exploration.

Impacts of offshore wind farms on the ecological systems need to be divided into impacts during the construction, due to the connection to the electricity grid and during normal wind farm operation. One of the big unsolved problems is the estimation of cumulative impacts from a series of large wind farms as well as missing information and uncertainty in knowledge regarding these effects. Thus, one of the core problems for an integrative analysis is the question "How is it possible to deal with complexity, uncertainty and missing knowledge?"

Impacts during the installation include effects on soil, seabed and water, on benthos, fishes and sea mammals, birds and marine archaeological sites. Impacts for the grid connection are mainly related to cable construction and in the long-term to pipelines for hydrogen transport. Impacts during wind farm operation include effects on water and seabed, change of material fluxes, electric fields and heat generation from cable links as well as impacts on habitats and species like sea mammals and birds. In summary, the following ecological effects are part of on-going investigations (Garthe, personal communication; see also several articles in Merck and von Nordheim 2000):

1. Birds:
- Mortality due to collisions of birds with rotors;
- Disturbance by wind farms, leading to barrier effects for migrating birds (both landbirds and seabirds) during migration and to short- and long-term habitat loss for seabirds.

2. Marine mammals:
- Habitat loss due to disturbance;
- Noise emissions leading to behaviour changes, stress and eventually mortality.

3. Fish:
- Disturbance by electromagnetic fields near cables;
- Disturbance and behavioural changes during construction;
- Changes in fishing activity, leading to alterations in fish communities.

4. Zoobenthos:
- Habitat loss during construction of wind mills and cables;
- Changes in sedimentation and increasing turbidity during construction, leading to alterations in zoobenthos communities.

Other effects include the pollution by oil and other substances due to increased traffic and increased risk of ship accidents. These risks are some of the most important fears of local people, especially on the Wadden Sea islands, which depend on tourism as most important source of local income.

Visual impacts that affect the local population as well as tourists are another issue of high importance at the local level. These fears are connected to the question whether the locals also receive positive economic benefits or whether they are left with risks and burdens while the benefits occur in other, not even coastal, regions. It should be noted however that cultural and political background conflicts might play a role in the argumentation of different stakeholders as well as a traditional perception of the sea as an open space without permanent installations.

In addition, offshore wind farms impose restrictions for fishermen because – due to the risk of ship accidents – fishing is going to be prohibited in wind farm areas. A positive alternative on the other hand could be the use of offshore wind farm areas for mariculture activities in form of an extensive Open Ocean Aquaculture (OOA), which could be an opportunity to create win-win situations between locally based traditional uses of coastal resources and use of wind as a source of renewable energy (Buck 2002). A general system of the most important interactions between wind farms and the coastal system is shown in simplified form in Figure 3.

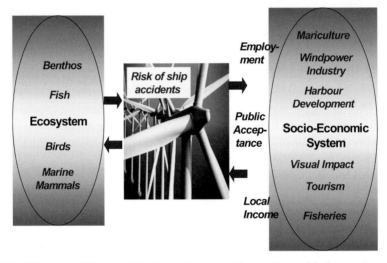

Fig. 3. Offshore wind farms and their most important interactions with the ecosystem and the socio-economic system (adapted from Kannen et al. 2003)

Managing these system interactions calls for integrative concepts that can handle socio-economic issues as well as ecological issues taking a range of future societal developments into account. These concepts should allow assessment of management options and alternatives according to their balance between ecosystem resilience or ecosystem integrity on one hand and the economic and socio-cultural system resilience on the other hand. In addition, not all information necessary to assess and evaluate policy and management options can be expected to be quantifiable, which requires tools that allow the combination of qualitative and quantitative data for holistic assessments.

Wind industry as a chance for economic development:
The North Sea coast of Schleswig-Holstein as case study

Aside from the critical impacts on the ecosystem and the shipping, fishing and tourism sectors (which are to a large degree not yet fully evaluated in terms of the strength and sometimes even the direction of the impact), wind energy has been a positive stimulus for regional economy in some locations along the North Sea coast in Germany. As already outlined above it is expected that the wind energy sector will create ongoing economic growth in the future if the wind farm developers are successful in their implementation of large scale offshore wind farms.

In the northwestern part of Schleswig-Holstein wind energy has grown to represent a significant economic cluster of manufacturers, operating consortia, planning companies and financial service providers over the last 10 years. The regional chamber of commerce estimates the gross surplus share of wind power in the North Sea region to be between 12 and 15% (Volmari 2002). Together, this cluster provides more than 1200 jobs in Husum alone, a town of 25.000 inhabitants and host of the world's largest trade fair for wind power. In 2002 the three turbine manufacturers located in and near Husum had a share in the German market of 33% (Volmari 2002). According to these figures the wind industry cluster forms behind tourism the second most important economic activity along the North Sea coast in Schleswig-Holstein with a regional focus in the northern part. In June 2003, a comprehensive questionnaire survey of local companies and enterprises as part of a regional economic study was carried out. By the year 2030, the study concludes, the region is likely to reach a net value in the range of billions of Euros and create over 3000 jobs in the northwestern part of Schleswig-Holstein (Hohmeyer 2003).

The realisation of this potential however requires prior investments in infrastructure, for instance the expansion of Husum Port, as well as minimisation of intraregional competition in order to win through against competing locations in other German States and Denmark. Alongside port development, infrastructure to stimulate offshore wind industry encompasses improved road infrastructure (port access) and provision of storage areas in ports, areas for test fields for offshore turbines and areas near harbour facilities for final steps of windmill production.

Used successfully, these potentials could yield significant economic multipliers particularly for this structurally weak coastal region. On the other hand the development of the wind power industry can be expected to imply changes in the regional economic, social and institutional setting including a shift of power from traditionally strong societal groups to new actors. Therefore, at the same time, a balance is required with other forms of use, specifically tourism and nature conservation. To achieve this, a transparent and participatory development process would be definitely helpful (see O'Riordan in this volume).

Tourism is nowadays the dominating economic force in this region, especially on the North Frisian Islands. An equivalent of 9,000 full time jobs is estimated to be linked with the tourism sector (Gätje 2003). Since 1985 the Wadden Sea is protected as a National Park and forms part of the Trilateral Wadden Sea Cooperation between Germany, Denmark and the Netherlands. Following heavy conflicts between nature conservationists and local people until the end of the 1990s today the National Park is now beginning to become acknowledged as a major tourist attraction of the region.

Traditionally agriculture forms the major land use in the Wadden Sea areas of Schleswig-Holstein. Even though its economic importance has experienced a steady decline over the last 30 years, agriculture still is a politically important actor. Within this regional setting the use of marine areas for large scale wind farms is perceived by a considerable number of people as an industrial use that endangers existing forms of economy, especially tourism and increases the risk of accidental oil spill from shipping.

Therefore, regional stakeholders generally discuss this issue in a polarised fashion. Especially on the islands considerable resistance can be observed. It can be summarised that development of offshore wind farms is a controversial issue which requires a valuation of potential economic gains against economic risks (for other sectors), an assessment of impacts across different scales as well as different spatial units (e.g. islands vs. mainland) and an assessment of economic vs. intrinsic cultural values (e.g. perception of the sea as energy resource vs. perception as open sea).

To maintain ecological, economical and social sustainability of coastal and marine systems, planning instruments are required to be able to weigh potential developments with the existing and likely pressures on space in the sense of a holistic systems-based approach. Therefore it is an elementary objective of the research project Zukunft Küste – Coastal Futures (Kannen et al. 2003), which is planned to start in spring 2004, to develop and document the required methods and instruments to enable a holistic systems analysis.

Other measures discussed in the area to strengthen the regional economic share, but also to achieve local acceptance for offshore wind farms, are incentives to increase the regional shares in wind farm consortia. An example is the "Bürger-Windpark" Butendiek, which is owned by local shareholders (10.000 shares distributed widely within the region). Another potential measure involves multi-use concepts for the wind farm areas. As an example, assessments are under way which try to evaluate the feasibility for mariculture within wind farms by using the permanent structures of the windmills for seed mussel cultures. This can be extended towards other candidate species for the so-called Open Ocean Aquaculture (Buck 2002). Such win-win situations between wind farms and traditional coastal activities could form a way to achieve a consensus-based development of offshore wind farms.

Addressing offshore wind farms within ICZM

Sustainable use of offshore wind energy implies the need to combine goals for climate policy with policy targets for nature protection, but also including the needs of traditional and established activities in coastal and marine waters. In addition the local or regional strength of wind power industry can form an additional factor for the assessment. Taking sustainable development into account as a policy target, the development and assessment of offshore wind farms needs to be evaluated according to regional development needs, impacts on the ecosystem at the regional level and socio-cultural changes in regional structures.

Politically, local acceptance is based on the assessment of ecological and economical risks, e.g. conditioned by the perceived risk of shipping accidents and chances for local economy and society as outlined above. Therefore, local institutions ask for active involvement of local people and assistance to local development needs in order to avoid mistrust, create a local sense of ownership and to minimise conflicts. Thus, the spatial distribution of positive impacts versus negative burdens forms an important factor for local positions, even for generally positively perceived policies like mitigation of climate change.

Following the results of the EU demonstration programme for Integrated Coastal Zone Management (European Commission 1999, 2000) and the recommendations of the EU Parliament and the EU Council (European Parliament and Council 2002), Integrated Coastal Zone Management (ICZM) is expected to handle this type of conflicts (Kannen forthcoming). The general role of ICZM for coastal development is outlined in Figure 4.

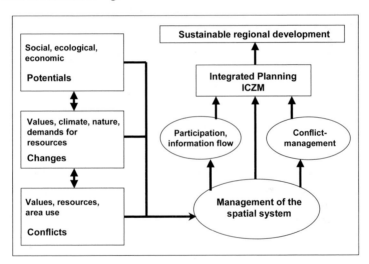

Fig. 4. ICZM in the context of spatial planning (from Kannen 2001a, adapted from Kannen et al. 2000)

ICZM literature has described ICZM within the context of several human activities in coastal areas. Generation of renewable energies in marine areas has not been analysed in this context up to now. When taking the size and the expected future speed of the offshore wind farm development in the German EEZ into account, it becomes obvious that is a major issue that needs to be addressed within a concept for marine area spatial planning as well as within the context of ICZM strategies at the level of the Federal government, the Länder level and the local level in those islands and municipalities that are affected in one or another way. For the national ICZM strategy in Germany this demand has been formulated by

Gee et al. (forthcoming) as a result of the national stocktaking exercise on behalf of the Federal Office for Building and Regional Planning (BBR).

According to the problems with offshore wind farm development described in this article, an ICZM, which is adapted to offshore wind farm issues should take the following elements into account:

- Developing concepts and tools for a comprehensive Integrated Assessment including, but not limited to, Strategic Environmental Impact Assessment (SEA);
- Developing mechanisms that allow to handle the diverging perceptions and information needs from planning at different scales from local to European and global in an integrative way.
- Developing policy dialogues and adaptive spatial planning structures for marine areas, preferably at the trans-national level;
- Applying participatory dialogue techniques and information approaches at the local level in order to include local fears as well as local development issues;
- Developing win-win solutions with other resource uses like mariculture, tourism and nature protection, e.g. framed by multiple use concepts for wind farm areas;
- Designing monitoring schemes which take cumulative effects into account including developing necessary models or other tools.

Design of integrated assessment for offshore wind farms

Integrated Assessment should not be misunderstood as another word for formal planning instruments, but more as a long-term policy assessment taking potential future policy changes into account. Following this logic, the Integrated Assessment needs to consider future developments and policies, e.g. given the impact of changes in governance regimes. Regarding the issue of offshore wind farms, policy changes in Germany might be important as well. The degree to which a new government consisting of the current opposition parties would change the current policy, which is in favour of developing offshore wind farms, needs to be taken into account.

When screening different plausible future developments in society by scenario techniques, the result will be different potential policy targets, usually related to societal risk perception and risk aversion, and consistent with the scenario storyline. Given the uncertainties about the future needs of humanity, the future development of ecosystems in the light of global change and the current limited insight into the complexities of the ecological systems as well as of the socio-economic system, it is not feasible to identify exact values for ecosystem functions, which are critical in the sense, that risks for the ecosystem services may be avoided (Colijn et al. 2002).

On the other hand it seems possible and is in line with the Sustainable Development paradigm to follow the precautionary principle (Turner et al. 2001), and to develop management strategies that allow a maximum use of ecological services while keeping the ecosystem integrity at least at the present level, thereby reducing the risk of hazardous natural developments. Thus the target of the integrated approach is to inform the society about possible future trends and the connected risks, allowing a balance between "ecosystem use" and "ecosystem squeeze" (Colijn et al. 2002).

To conclude, it needs to be noted that the use of wind energy in offshore areas and development of the related economic sector occurs in a complex system of human and societal interactions but also within a complex ecosystem. To perform an Integrated Assessment which has relevance on a regional and local scale therefore requires a combination of scientific tools from natural as well as social and economic sciences.

The Vision: A Toolbox for Integrated Assessment in ICZM

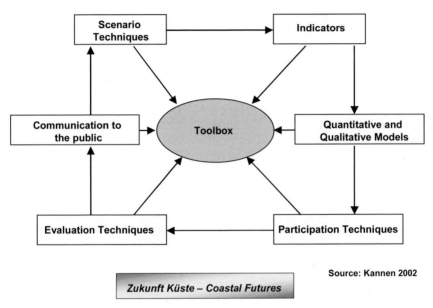

Fig. 5. The concept of an integrative toolbox for tools and methods within an Integrated Assessment (adapted from Kannen, forthcoming)

Instead of trying to develop an ideal integrated model, from the view of management, which asks for decisions now, the vision of an integrated toolbox (Kannen, forthcoming; Figure 5) seems to be a suitable approach for integrated assessments within an ICZM framework. This toolbox should include tools for:

- Future assessments (e.g. foresight approaches and scenario techniques);
- Indicators, suitable to describe effects of interactions on different scales;
- Qualitative and quantitative models for impact assessment;
- Expert Systems like SimCoast[TM] as integrating tools, combining interaction modelling and participatory dialogues;
- Stakeholder dialogue approaches.

This approach will be further developed and tested within the BMBF funded research project Zukunft Küste – Coastal Futures, expected to start in April 2004. While the toolbox is useful in order to integrate methods and assessment tools from different scientific disciplines, the DPSIR (Driver-Pressure-State-Impact-Response) concept as developed by OECD (as PSR) and EEA (as DPSIR) has proved to be applicable as an integrating methodological structure. It allows the analyst to structure information from different scientific disciplines in an integrative way. The DPSIR approach has been successfully used within LOICZ basins assessments (see several reports, http://www.nioz.nl/loicz) and within the EU funded LOICZ project EUROCAT.

Outlook

This paper draws a line from an ongoing issue for the North Sea coast of Schleswig-Holstein, which might reach far into the future. The development of wind farms in offshore areas includes plans for large scale investments until 2030 and is connected to long term policies for mitigation of climate change effects and an energy policy that is based on assisting a transfer from non-renewable to of renewable sources of energy. From the analysis of current developments regarding the installation of large scale wind farms in offshore areas, the paper derives the strong need for offshore wind farm development to be included in ICZM concepts and strategies at the local, regional and national level. To assist societal decisions at all levels regarding a balanced assessment of interactions between offshore wind farm development with other, partly conflicting, coastal activities as well as with the impacts on the ecosystem it is argued, that Integrated Assessment should be used as a tool to stimulate and assist discussions at the local level as well as at the policy level (regional, national, transnational).

Acknowledgements

This work was funded by the EUROCAT project of the European Union (EVK1/2000/00510) and assisted by the project "Integrated Coastal Zone Management: Strategies for spatial planning in the coastal and marine environment" funded by the Federal Office for Building and Regional Planning (Az: Z6 – 4.4 – 02.119).

References

BMU (2002a) Strategie der Bundesregierung zur Windenergienutzung auf See, Berlin
BMU (2002b) Weiterer Ausbau der Windenergienutzung im Hinblick auf den Klimaschutz, Teil 2. Erarbeitet im Rahmen des F&E Vorhabens 99946101 von Deutsches Windenergie-Institut GmbH, Wilhelmshaven

Buchholz, H (2002) Strategien und Szenarien zur Raumnutzung in den deutschen Ausschließlichen Wirtschaftszonen in Nordsee und Ostsee. Auftrag des Bundesamtes für Bauwesen und Raumordnung, K & M Consult Büro für räumliche Planung und Entwicklung in Küstenzonen und Meeren

Buck, BH (2002) Open Ocean Aquaculture und Offshore-Windparks: Eine Machbarkeitsstudie über die multifunktionale Nutzung von Offshore-Windparks und Offshore-Marikultur im Raum Nordsee, Reports on Polar Research, Alfred Wegener Institut for Polar and Marine Research 412

Colijn F, Kannen A, Windhorst W (2002) The use of indicators and critical loads, EUROCAT Deliverable 2.1, www.iia-cnr.unical.it/EUROCAT/project.htm

Czybulka, D (2003) Meeresschutzgebiete in der Ausschließlichen Wirtschaftszone (AWZ). Z Umweltrecht, 5: 329-337

European Commission (1999) A European strategy for integrated Coastal Zone Management (ICZM): General principles and political options, Brussels

European Commission (2000) Communication from the Commission to the council and the European Parliament on Integrated Coastal Zone Management: A strategy for Europe, COM (2000) 547 final

European Parliament and Council (2002) Recommendation of the European Parliament and of the Council of 30 May 2002 concerning the implementation of Integrated Coastal Zone Management in Europe. (2002/413/EC). Off J Eur Communities L 148/24 EN

Gätje, Ch (2003) Tourismus und Erholung im Wattenmeer. In: Lozan, JL, Rachor, E, Reise, K, Sündermann, J, v. Westernhagen, H (Ed.) Warnsignale aus Nordsee und Wattenmeer, pp. 117-121

Gee, K, Kannen, A, Glaeser, B, Sterr, H (forthcoming) IKZM-relevante Strukturen, Instrumente und Institutionen in Deutschland (Bericht zur nationalen Bestandsaufnahme), Berlin

Hohmeyer, O (2003) Gutachten über Regionalökonomische Auswirkungen des Ausbaus einer Offshore Struktur des Husumer Hafens, Zusammenfassung

Kannen, A (2000) Analyse ausgewählter Ansätze und Instrumente zu Integriertem Küstenzonenmanagement und deren Bewertung. FTZ-Berichte 23, Büsum

Kannen, A (forthcoming) Schlussfolgerungen aus dem Europäischen Demonstrationsprogramm zum Integrierten Küstenzonenmanagement für eine nachhaltige Entwicklung der deutschen Küsten. In: Glaeser, B (Hrsg.): Küste, Ökologie und Mensch: Integriertes Küstenmanagement als Instrument nachhaltiger Entwicklung. Schriftenreihe der Deutschen Gesellschaft für Humanökologie Band 2

Kannen, A, Gee K, Ulich E, Schneider E (2000) "Management of Change" und nachhaltige Regionalentwicklung in Küstenzonen am Beispiel der Nordseeküste Schleswig-Holsteins. In: Blotevogel, Oßenbrügge, Wood (eds): Lokal verankert – weltweit vernetzt. Verhandlungsband des 52. Deutschen Geographentages 1999, pp 130-135

Kannen, A, Windhorst, W, Glaeser, B, Ahrendt, K, Colijn F (2003) Zukunft Küste – Coastal Futures Project Proposal, Unpublished, FTZ, Büsum

Koch, H-J, Wiesenthal, T (2003) Windenergienutzung in der AWZ. Z Umweltrecht 5: 350-356

Merck, T, von Nordheim, H (ed) (2000) Technische Eingriffe in marine Lebensräume. Bundesamt für Naturschutz

O'Riordan T (2004) Inclusive and community participation in the coastal zone: opportunities and dangers. In: Vermaat JE, Bouwer LM, Salomons W, Turner RK (eds) Managing European coasts: past, present and future. Springer, Berlin

Paul, N, Lehmann, KP (2003) Big Plans Off the Shores Of Europe. Sun and Wind energy, Spec Int Issue 2003: 111-116

Rat von Sachverständigen für Umweltfragen (2003) Windenergienutzung auf See, Aktuelle Stellungnahme, Berlin

Tiedemann, A (2003) Windenergieparke im Meer. In: Lozan, JL, Rachor, E, Reise, K, Sündermann, J, v. Westernhagen, H (ed.) Warnsignale aus Nordsee und Wattenmeer, Geo, Hamburg, 142-148

Turner RK; Ledoux, L, Cave, R. (2001) The use of Scenarios in Integrated Environmental Assessment of Coastal-Catchment Zones: the case of the Humber Estuary. Unpublished, CSERGE, Norwich

Volmari, M. (2002) Positionspapier zur Windkraftbranche in Nordfriesland, Husum

21. Group report: Reflections on the application of integrated assessment

Laurens M. Bouwer[1], Saara Bäck, Guiseppe Bendoricchio, Stavros Georgiou, Andreas Kannen, Areti D. Kontogianni, Joop M. Marquenie, Laurence D. Mee, Corinna Nunneri, Tim O'Riordan, Wim Salomons, Raphael Sardá, Mihalis S. Skourtos, Paul Tett, Jos Timmerman, R. Kerry Turner, Tiedo Vellinga, Jan E. Vermaat, Maren Voss, and Wilhelm Windhorst

Abstract

This group report gives an overview of applications of and issues in integrated assessment (IA) applied in the coastal zone area in Europe. We conclude that there are various reasons why IA is sometimes not successful. For instance, integrated assessment tends to be highly specific, dialogues are seldom an integral part and environmental thresholds are uncertain. A way forward would be to have an alternative framework that could fulfil some of these needs, which is proposed at the end of the chapter.

Introduction

Previous chapters (16-20) have highlighted the views on and use of integrated assessment (IA) by a number of important sectors and stakeholders along the European coasts. Most are local cases and have a sectoral perspective. The need was felt to place these in a wider, reflective, framework. Justification of our choice for real-world cases of IA lies in the practical needs of coastal practitioners. What can be learned? What can be improved? What will be necessary in the light of probable future changes? (e.g. Rotmans 1998).

In previous chapters examples of IA were given from different sectors and regions. Here we present some lessons learned and a list of basic requirements for a specific model for inclusive participation within IA. Our discussion attempts to aggregate sector-wise cases from the previous chapters into a comparative analysis

---

[1] Correspondence to Laurens Bouwer: laurens.bouwer@ivm.falw.vu.nl

J.E. Vermaat et al. (Eds.): Managing European Coasts: Past, Present, and Future, pp. 379–387, 2004.

of variation in issues and impact mechanisms as well as common gaps and successes. In contrast to previous group reports (Moschella et al. this volume, Lise et al. this volume, Rochelle-Newall et al. this volume, Nunneri et al. this volume) the discussions on IA were only structured in a general sense in order to facilitate the brainstorm character of the group session. With hindsight, we notice that our considerations remain close to the cases presented in previous chapters (16-20) and the compromise between overarching generalisations and realistic detail was difficult to forge. The present chapter gives more general conclusions that were drawn from the accumulated experience of the group rather than conclusions based on published literature, which makes this chapter somewhat different from previous ones.

In this concluding chapter we chose to use the framework for analysis and drivers that were identified by Turner (this volume). We have aggregated tourism with the expansion of the built environment, and have not included the issue of climate change (Table 1). We will use these major drivers to first assess the major sectoral problems. Second, we discuss the success of IA. Finally, we present some lessons learned and a possible way forward.

Table 1. Seven of the major drivers for changes in the coastal area (adopted from Turner this volume)

Primary	Secondary
Expansion of the built environment	Climate change
Tourism	Agricultural change
Industrial development and trade	Provision and supply of energy
Fisheries and aquaculture	

Experiences in different sectors

Tourism and the expansion of the built environment

Expanding tourism over the last centuries has emphasised and exacerbated weaknesses in national level spatial planning and regulation, for instance particularly apparent in Mediterranean countries (see e.g. Sarda et al. this volume). Along the Atlantic shores the tourist centres are conditioned by accessibility from the urban centres and traditional tourist seasonality. The pressures of urbanisation and expansion of the built-up area as a result of tourism are localised, since authorities at the local level act as regulators. People from northwestern Europe tend to buy secondary houses along the Mediterranean coast, and these are also increasingly often used for more permanent residence. This trend of an increasing number of secondary residences has also been encouraged by the attempts of local politicians to increase the number of tourists in the low season (winter). At the same time the peak in the number of visitors in summer remains constantly high.

The tourism sector is flexible and can adjust to changes in for instance demands from tourists. In Italy the sector swiftly responded to changing demands as people from Eastern Europe started to visit the country more frequently. Tourism is diversifying, e.g. scientific tourism that is being attracted by the building of conference halls.

Individuals as well as companies are buying out large parts of villages. This can be observed along the Mediterranean, but also along the Black Sea, where people buy secondary houses. In Greece, a new law on spatial planning appears to be quite promising and also includes a chapter on coastal planning. While in Finland the problems with the development of secondary (vacation) houses occurred about 50 years ago. New regulations determine the standards for turning secondary houses into permanent houses.

Because of high prices along the coast tourism urbanisation expands into the interior. The apparent deadlock situation in spatial planning leads to unsustainable inland expansion.

Industrial development and trade

Both industry and trade concentrate along the facilities of major ports. Sea-ferry transport is concentrating towards large ports because of the possibilities for scale expansion in logistic facilities. Ports and harbours around Europe are expanding because of increasing volume of goods transported worldwide and because of increasing competition among ports. Safety and waste disposal issues that arise because of this expansion can be solved, but regulation and legislation are needed to force harbour management and industry to take action (see also Vellinga and Eisma this volume), a legal basis is thus conditional. It should also be noted here that industry and port authorities do not necessarily share interests and responsibilities equally.

A case of particular interest in the UK is the locating of a new harbour in Southampton, where environmental issues are to be taken into account. Germany utilises a comparable legal compensation mechanism to that of harbours in The Netherlands (Rotterdam) and Belgium (Antwerp), but expansions of harbours is not so pressing an issue in these countries. For most German harbours, such as Hamburg, Bremerhaven and Wilhelmshaven, the maintenance and increase of access depth and consequent dredging is the largest environmental problem.

When looking at regulation, the infrastructure to control the development and expansion of ports and harbours in Europe appears to be in place, also because there is usually a (single) authority to address. This is in contrast with tourism where a wide range of sub-sectors and enterprises are confronted with considerable local and temporal variation in regulations and management institutions.

Fisheries and aquaculture

Near-shore and coastal fisheries often have remained small-scale across Europe with few notable exceptions (e.g. cockle trawling in The Netherlands). The larger high sea fisheries of Europe have expanded their reach across the Atlantic Ocean, whilst being subject to sectoral adjustment as well as strong attempts to regulate overexploitation (e.g. EC 2002). Employment losses due to restricted fishery quota underline the need for alternatives. Such alternatives can be for example tourism, resulting in a new set of pressures (which are not the same as the pressures of fisheries). This shift from fisheries activities to tourism and the consequent urbanisation pressure has been occurring along the Mediterranean coast since the 1950s.

Fisheries are turning to aquaculture across the Mediterranean. Aquaculture is a competitive sector, which is sensitive to subtle market changes such as enterprise upscaling: small groups have to work together to make it profitable. Both as an economic activity and its resulting pressures on natural ecosystem, aquaculture has little more in common with fisheries than its basic resource.

Fisheries in Europe have moved from small-scale community-based to corporate organisations, leading to short-termism which rapidly turns natural capita into potential short-term financial profits. In macroeconomic terms, the European fishery sector is not economically very significant, but at the same time it is a powerful political force as witnessed from the quota negotiations. Disappearance of the community-based fisheries sector allows the big companies to enter the market.

At the World Summit on Sustainable Development in Johannesburg it was decided that a new marine stewardship should be put in place by the year 2005. This could give a new opportunity for European researchers to liaise with stakeholders and major players. An example is the support that could be given to companies, for example large food enterprises to review future environmental economics and resource acquisition, also in relation to agriculture (Gerbens-Leenes and Nonhebel 2002). Sustainability aspects, such as the linkage between the driving forces of climate change and energy production and provision should be looked at closer.

Agricultural change

Major developments in the agricultural sector in Europe will include a continued movement towards "green" farming with a reduced nutrient and pesticide consumption, and the expansion of the EU in 2006 that will lead to a reduction in subsidies. There is a trend within industrialised urban regions of the EU to assign the farmer the task of steward of the countryside landscape and to cater for recreational needs in urban centres.

How declining EU subsidies will impact on the coastal area is uncertain, but much depends on the location and the precise configuration of the changes. One scenario is that the accession countries will suffer an intensification of agriculture over large areas, leading to increasing nutrient input from Central and Eastern Europe to the Baltic Sea and Black Sea.

Provision and supply of energy

In the context of energy provision windmill development in particular is relevant for the coastal area. Cross-boundary issues and communication, e.g. between Denmark and Germany complicate the formulation and implementation of the required integrated assessment, (see also Kannen this volume). The development of windmill parks can lead to an uneven distribution of risks and benefits at different scales. Benefits are experienced at the national level, or international level e.g. greenhouse gas emission reductions, while the risks, damages and nuisances are experienced at the local scale.

Success of integrated assessment

When looking at integrated assessment, some important factors appear to determine its success:

- Integrated assessments tend to be less successful when they are highly specific and are too narrowly framed in time and scale;
- There appears to be a lack of dialogue with and identification of opinion formers and politicians. Perhaps integrated assessments should be taken on in another way; by going to stakeholders, and assess what they exactly need;
- The understanding and communication of the ecosystem and its thresholds is an important part of the process of an integrated assessment, in order to be able to communicate outcomes of different strategies and scenarios with the stakeholders.

The message should be logical and simple; overly complex situations are difficult to understand, more communication and more simple regulation is needed. For example, NAM/Shell failed in the discussion about gas exploration in the Waddensea area, because although the technical details were perfectly in order and dialogues had been set up, public opinion changed because of failing political support, first on the regional and then on the national level. Stakeholder mapping should be maintained throughout the exercise, this was not done in the NAM/Shell case.

The science is often pushed aside in discussions where large economic interests are at stake; e.g. in the case of the port of Venice. In the end the stakeholders have to execute the plans; good "scientific" arguments do not necessarily cover the interests of the manager.

The Black Sea integrated assessment was successful in the sense that it generated $0.5 billion of new resources. But it was not successful in other terms: local stakeholders have not been involved, and the institution that was created (the ministry of environment) had isolated itself (Mee 2001). IA could identify the root causes and create ownership, not only scientific excellence. For instance, from stakeholder mapping preceding the G8 Summit in Evian, France in 2003 it became clear that second subject of concern in almost all countries is the environment, but this was not addressed by the summit (G8 2003).

A code of practice is needed for integrated assessment in the coastal zone, or a framework for IA. We could be more sensitive to (changing) social and ecological interests. The stakeholder participation should not be a sequential part but it integral to the process. More room could be created for different options, for different styles of governance and different styles of delivery of products and outcomes. Integrated assessment could also be more "interactive", not merely represent a set of "ready-made" alternatives. The limits of tolerance of ecosystems should be a given, and then strategies to best adapt the most affected livelihoods should be explored (e.g. alternative use of boats in case they can't be used for fishing for some time).

A classic case is the North Sea fisheries; although the EU regulations have changed, negotiations have resulted in a continued over-catch. The lesson is that the stakeholder should be involved in the setting of the regulation, and only when dialogue has failed should regulation be imposed.

Another example is the habitat directive; activities have to be justified economically, and consequently impacts have to be compensated for and mitigated. This forces people to create win-win situations where they otherwise would remain dormant. An example is the compensation of every hectare of port reclamation.

If the stakeholder is convinced of the long-term interest of conservation and if alternatives from livelihood and income are sought, they are more willing to comply. The process should be looked at as a self-organising response mechanism; guidance and adaptive responses from local stakeholders should be encouraged

Conclusions, lessons learned, and a possible way forward

Past and current developments in particular locations in Europe can serve as examples for integrated assessments that can be or should be undertaken elsewhere. One example is the regulation in Finland for secondary houses, which aims at specific requirements for water and energy use. Such a regulation might serve as an example for other coastal regions where the increased water and energy use is not yet an issue, but might soon become one. Another example is the shift in Spain of employment from fisheries to tourism, which might relieve environmental pressure due to fisheries activities. Such a shift can set an example for regions where tourism is not (yet) well developed.

Across Europe a common framework could be developed, that takes into account the inclusion of stakeholders. This framework needs to be supported with a code that will ensure consideration of cultural and political variation across Europe. Such a protocol would help to legitimise the political process, and to build trust and accountability. There could also be a notion of mutual respect for different opinions, a notion of shared understanding of a vision of coastal future and a notion of fairness of treatment and a feeling for joint responsibility for outcome. Cultural and political variations across Europe are very important. A particular problem here is the fact that in the southern parts of Europe there is a strong mistrust in the government. Here the stakeholder participation should or could perhaps play a different role, namely the gaining of empowerment. The aim is to look for a general framework and then build in the sensitive local elements.

Such a protocol is an example of a more or less ideal and harmonic situation, and the actual situation often proves much more difficult. Alternatives include for example a conflict model. A number of basic requirements are listed below (Table 2).

Table 2. Suggested requirements for a code of practice for integrated assessment

1.	Initiation of the procedure
2.	Vision exercise
3.	Mutual trust
4.	Openness and transparency
5.	Accountability and adjustment
6.	Media friendly
7.	Creation of networks of community forums (spatial and functional)
8.	Independent audit and validation
9.	Set specific political and cultural characteristics
10.	Aim for an outcome that all parties (can) accept

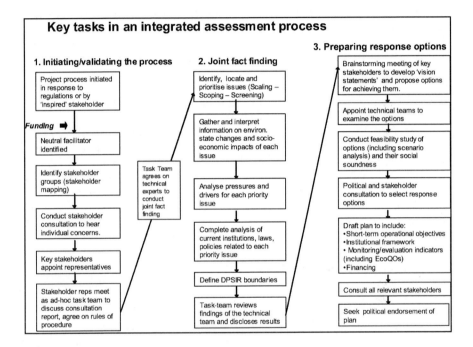

Fig. 1. Key tasks in integrated assessment (adapted from Mee 2004)

Consensus is not necessarily needed; an exploration of the differences that exist already can be very useful. Items that could be added include the necessity to identify the relevant stakeholders in the earliest stages, including national and international bodies and groups.

A stakeholder that has an interest usually initiates an IA, or there is a problem that forces policymakers to act. The initiation can be forced by an economic incentive or a statutory obligation. The process should then be taken over in at a very early stage by a neutral facilitator, in order to maintain equality.

The scale of the problems and issues is an important aspect for the code for integrated assessment, as models and processes have different boundaries and boundary conditions at different scales. Meaningful scales for the various parameters should carefully be chosen.

In additional to this list of requirements for a code of practice, different steps and tasks in the process of IA can be suggested. In each of the steps, ranging from agreeing on the rules of the procedure in the first step, setting up of joint-fact finding to the preparation of response options, intensive consultation between representatives of the different stakeholder groups should take place.

References

EC (2002) Council Regulation EC No 2371/2002 on the conservation and sustainable exploitation of fisheries resources under the Common Fisheries Policy

G8 (2003) Evian Summit 2003, http://www.g8.fr

Gerbens-Leenes PW, Nonhebel S (2002) Consumption patterns and their effects on land required for food, Ecol Econ 42: 185-199

Kannen A (2005) The need for integrated assessment of large-scale offshore wind farm development. In: Vermaat JE, Bouwer LM, Salomons W, Turner RK (eds) Managing European coasts: past, present and future. Springer, Berlin, pp 365-378

Lise W, Timmerman J, Vermaat JE, O'Riordan T, Edwards T, De Bruin EFLM, Kontogioanni AD, Barrett K, Bresser THM, Rochelle-Newall E (2005) Group report: Institutional and capacity requirements for implementation of the Water Framework Directory. In: Vermaat JE, Bouwer LM, Salomons W, Turner RK (eds) Managing European coasts: past, present and future. Springer, Berlin, pp 185-198

Marquenie JM, De Vlas J (2005) The impact of subsidence and sea level rise in the Wadden Sea: prediction and field verification. In: Vermaat JE, Bouwer LM, Salomons W, Turner RK (eds) Managing European coasts: past, present and future. Springer, Berlin, pp 355-363

Mee LD (2001) Eutrophication in the Black Sea and a basin-wide approach to its control, in: Von Bodungen B, Turner RK, Science and integrated coastal management, Dahlem University Press, Dahlem, pp 71-112

Mee LD (2004) The GEF IW TDA/SAP process: a proposed best practice approach, manuscript in press

Moschella PS, Laane RPWM, Back S, Behrendt H, Bendoricchio G, Georgiou S, Herman PMJ, Lindeboom H, Skourtous MS, Tett P, Voss M, Windhorst W (2005) Group report: methodologies to support implementation of the Water Framework Directive. In: Vermaat JE, Bouwer LM, Salomons W, Turner RK (eds) Managing European coasts: past, present and future. Springer, Berlin, pp 137-152

Nunneri C, Turner RK, Cieslak A, Kannen A, Klein RJT, Ledoux L, Marquenie JM, Mee LD, Moncheva S, Nicholls RJ, Salomons W, Sardá R, Stive MJF, Vellinga T (2005) Group report: integrated assessment and future scenarios for the coast. In: Vermaat JE, Bouwer LM, Salomons W, Turner RK (eds) Managing European coasts: past, present and future. Springer, Berlin, pp 271-290

Rochelle-Newall E, Klein RJT, Nicholls RJ, Barrett K, Behrendt H, Bresser THM, Cieslak A, De Bruin EFLM, Edwards T, Herman PMJ, Laane RPWM, Ledoux L, Lindeboom H, Lise W, Moncheva S, Moschella P, Stive MJF, Vermaat JE (2005) Group report: global change and the European coast – climate change and economic development. In: Vermaat JE, Bouwer LM, Salomons W, Turner RK (eds) Managing European coasts: past, present and future. Springer, Berlin, pp 239-254

Rotmans J (1998) Methods for IA: the challenges and opportunities ahead, Environ Model Assess 3(3): 155-179

Sardá R, Mora J, Avila C (2005) Tourism development in the Costa Brava (Girona, Spain) – how integrated coastal zone management may rejuvenate its lifecycle. In: Vermaat JE, Bouwer LM, Salomons W, Turner RK (eds) Managing European coasts: past, present and future. Springer, Berlin, pp 291-314

Turner, RK (2005) Integrated environmental assessment and coastal futures. In: Vermaat JE, Bouwer LM, Salomons W, Turner RK (eds) Managing European coasts: past, present and future. Springer, Berlin, pp 255-270

Vellinga T, Eisma M (2005) Management of contaminated dredged material in the Port of Rotterdam. In: Vermaat JE, Bouwer LM, Salomons W, Turner RK (eds) Managing European coasts: past, present and future. Springer, Berlin, pp 315-322